Sounding Bodies

Sounding Bodies: Music and the Making of Biomedical Science

Peter Pesic

The MIT Press
Cambridge, Massachusetts
London, England

The MIT Press would like to thank the anonymous peer reviewers who provided comments on drafts of this book. The generous work of academic experts is essential for establishing the authority and quality of our publications. We acknowledge with gratitude the contributions of these otherwise uncredited readers.

This book was set in Times New Roman by New Best-set Typesetters Ltd. Printed and bound in the United States of America.

Library of Congress Cataloging-in-Publication Data

Names: Pesic, Peter, author.
Title: Sounding bodies : music and the making of biomedical science / Peter Pesic.
Description: Cambridge : The MIT Press, [2022] | Includes bibliographical references and index.
Identifiers: LCCN 2021046199 | ISBN 9780262046350 (paperback)
Subjects: LCSH: Science—History. | Music and science—History.
Classification: LCC Q172.5.M87 P473 2022 | DDC 509—dc23/eng/
LC record available at https://lccn.loc.gov/2021046199

10 9 8 7 6 5 4 3 2 1

For Gerald Holton,
teacher, mentor, friend,
who encouraged me
to spread my wings

Contents

Introduction

"Why don't biologists care as much about music as physicists do?" With this provocative question, an eminent physicist and historian of science set me on the quest that led to this book. I had just written *Music and the Making of Modern Science*, which argued that music has had a deep effect on the development of mathematics and physics, both ancient and modern. In my next book, *Polyphonic Minds: Music of the Hemispheres*, I turned to the history and significance of polyphony, the music of many voices, in particular its relation to the science of the mind and the human sciences in general. Finally addressing my friend's playful question, this book considers the relation between music and the development of the medical and life sciences. Though each of these books was intentionally written to stand on its own, not requiring that they be read in sequence, they are in fact siblings that together form a trilogy about the significance of music for the sciences.

Turning to my friend's question, it seems impossible to determine whether biologists and physicians care about music as much as (or in the same way that) physicists do. Many life scientists would dispute that they are less musical than physical scientists. For instance, the long tradition of musically gifted physicians could be set against the many physicists who were passionate about music.[1] Surely I cannot settle this intriguing question, which would require sociological study and probably has changed over time. Perhaps others might be able to address it fruitfully, but however many examples they could amass, in the end it is difficult to see how so many different individual cases could be weighed against each other.

Instead, I realized that my friend was playfully asking a deeper question: "Did music affect the development of biology and medicine in the same ways as it affected mathematics and physics?" In response, I will argue that the effects of music on the biomedical sciences are consequential and important but different in kind than for the physical sciences. The reasons for this go back to antiquity but reverberated for two millennia, continuing to the present day. Traditionally, medicine stood apart from the "quadrivium," the "four-fold way" of arithmetic, geometry, music, and astronomy, the "liberal arts" that began with the Pythagoreans. *Music and the Making of Modern Science* held that the physical and mathematical sciences were children of these ancient liberal arts, learning from music and her sisters how to connect number with physical reality.

This book argues that biology and medicine also came under the influence of Pythagorean mathematical and physical approaches. Indeed, from their earliest days, the Pythagoreans practiced a medicine that did not depend on divine influence, paralleling their application of number to the physical world. Thus, Western medicine was another "child" of the Pythagoreans, a child that gradually came to know its siblings, the physical and mathematical sciences. The earliest links developed between music, mathematics, and medicine as arithmetic was applied to the pulse and to the timing of "crises" in disease. Beginning about 1700, new ideas, tools, and techniques gradually transformed medicine and biology by seeking physical and mechanical explanations. Among these was an important "sonic turn" in these fields, by which I mean the use of sound to give new insights and interventions into living organisms. The ancient pursuit of Pythagorean ratios felt in the pulse ultimately led to new practices of listening to the body involving what I will call *sonic* and *rhythmic knowledge*

By comparison with the physical sciences, this sonic turn came later and had a different form and dynamic; it was not a single event but comprised successive waves. Sound offered a way to understand the hidden workings of living bodies that was not available even to the trained medical gaze or to observational biology based only on visible evidence. This sonic turn led to the development of such techniques as mediate auscultation and ultrasound, which became essential tools of the biomedical sciences. This turn toward sound also affected the superstructure of concepts these life sciences used to understand the bodies they were sounding. Most important, the ancient concept of living bodies as composed of humors gave way to a new concept in which those bodies were essentially formed by sensitive, resonant, and vibrating fibers, particularly understandable through their relation to sound. Also, in several instances, the structure of musical compositions (different between classical and Romantic works) provided different patterns for the structure of cure itself, patterns that fundamentally changed the nature of medical practice from outward observation to inward intervention by analogy with the ways music could affect and transform its listeners.

If music affected the physical sciences primarily through pitch or melody, it affected the biomedical sciences more through *rhythm*.[2] To be sure, the sonic turn involved all kinds of actual pitches and sounds, including a consequential move beyond human hearing to ultrasound. Yet even in ancient times, medicine had connected the musical ratios of rhythms to physiological phenomena, particularly heartbeats and the timing of crises in disease. The reconsideration of those rhythmic connections energized the sonic turn in the eighteenth century. Subsequently, the search to understand nerve signals extended to frequencies so low that they are perceived as rhythms rather than heard as pitches. Thus, the sonic turn led to a rhythmic turn as the analog perception of bodily sounds led to the discovery of binary, all-or-nothing neural rhythms.

The work presented here reflects a long development that involved the work of many before me. Daniel P. Walker, Penelope Gouk, Werner Friedrich Kümmel, and Jamie Kassler pioneered the study of music's relation to medicine, especially the use of music in healing.[3] Because they devoted so much attention to this topic, this book focuses on other aspects of the relation between music and sound and the biological sciences. Building on the work of these pioneers, James Kennaway, Ellen Lockhart, Mark Pottinger, Carmel Raz, Benjamin Steege, David Trippett, Axel Volmar, Janina Wellmann, and Rebecca Wolf settled in the new territory and noted new connections between music, medicine, and biology.[4] My work also builds on a growing wave of work by other historians of science and music such as Karin Bijsterveld, Robert Brain, Francesca Brittan, Stanley Finger, Heather Hadlock, Alexandra Hui, Myles Jackson, Julia Kursell, David Pantalony, Jacomien Prins, Tim Rice, Emily Thompson, and Viktoria Tkaczyk, who in different ways have illuminated interrelations of music, sound, mind, and body, particularly through the development of various technologies and devices.[5] The writings of Veit Erlmann and Jonathan Sterne helped me reflect on the connection of my work to sound studies, which concerns the neglected salience of sound for many sciences, technologies, and social institutions.[6] This book also contributes to disability studies by exploring the role of music and sound in the changing understanding of "melancholia," "mania," and "obsession" before they were viewed as medicalized disabilities or diseases.[7] I have also taken special care to draw attention to the contributions of women as well as to forgotten or little-known figures.

Because my topic required expertise not possible for any single person, I have relied on the specialized studies of many scholars: James Longrigg, Jacques Jouanna, and Vivian Nutton on ancient medicine, Nancy Siraisi on Renaissance medicine, Patrick Boner on Kepler, Tobias Cheung on fibers, Jacalyn Duffin on Laennec, Stanley Finger on the history of neuroscience, Axel Volmar on telephonic observation and sonification, Malcolm Nicolson and John Fleming on medical ultrasound, as well as many others.[8] I am glad to acknowledge my debt to them; I have tried to synthesize their contributions and add something further. The notes detail my sources so that readers may pursue these matters themselves.

In order to keep this book within manageable length, I have restricted its scope in various ways. Though I briefly consider various kinds of music therapy in historical perspective, my real concern is how music and sound have influenced the fundamental structure and practice of the biomedical sciences, a convenient modern term encompassing medicine and the various branches of biology. In order to focus attention on those fundamental aspects, I decided not to discuss most of the obviously sound-related aspects of animal behavior, such as hearing and otology or the songs of birds or whales, topics well discussed by others.[9] Likewise, I have not addressed the rhythms of chronobiology, which seem more dependent on light than sound, or rhythmic entrainment, or the rhythmic aspects of embryology extensively discussed by Janina Wellmann.[10] Nor have I discussed the biological theories of Jakob von Uexküll, who used music as a guiding metaphor to understand the *Umwelt*, the "surrounding world" experienced by animals.[11] Nevertheless, I have included bat echolocation as part of the sonic turn leading from animals to humans through ultrasound.

I have tried to keep in mind the adage that someone with a hammer sees nails everywhere.[12] Some of the most important chapters in the history of medicine and the life sciences (for instance, the development of the germ theory) do not involve sonic issues as far as I can tell. Though this book presents various—especially less familiar—cases of musical or sonic influence, I do not wish to make excessive claims about the ultimate place of music and sound in the biomedical sciences. I would like, though, to argue that sound and rhythm are significant threads in the intricate tapestry formed by the unfolding of these fields.

In that light, the term *music* will reach further than its ordinary connotations in common usage. For the ancient Greeks, *mousikē* included poetry and dance as well as what we consider "music" played by singers or instruments. Likewise, *rhythmos* began with the meters of music, poetry, and dance, then extended to the "rhythms" of sculpture and architecture, and eventually (as we shall see) to neural signals. Music along with arithmetic, geometry, and astronomy formed the quadrivium (as it later came to be called), meaning the "crossroads" or intersection of the four arts, which for the Pythagoreans and Plato constituted the central nexus of learning and the gateway to philosophy. As our story develops, "music" will also reach out to include all kinds of sounds, not just the restricted group of what once were considered "musical sounds," as well as ultrasounds and infrasounds audible to other organisms (or the appropriate man-made instruments) and the use of such vibrations in probing and even altering bodies.

The book is organized into four large parts. Part I considers the connections joining music, biology, and medicine as they developed from ancient Greece to the seventeenth century. Part II presents the sonic turn emerging from the growing understanding of the body in terms of fibers that were not only static structural elements but also enabled dynamic motion and even vibration. Growing since Greek and Roman medicine, these fiber theories led to a revival of musical approaches to cardiology during the eighteenth century that then grew into the new forms of sonic physical diagnosis, percussion and mediate auscultation. Part III turns to the influence of music and sound on reconsidering the typology of mental illness and on new therapies such as mesmerism. Part IV considers the varied array of contemporary biomedical sciences that rely on sonic information, including echolocation, ultrasound, sonic methods in neurophysiology, and other emerging approaches. In them, the sonic turn led to a rhythmic turn, in which the universal language of the nerves turned out to be a binary, all-or-nothing code whose changing rhythms—and frequencies—encode all bodily motions, sensations, and mental states.

Part I begins with the interrelations of music, medicine, and biology as they emerged somewhat less directly than those between music and the physical sciences, which reflected the intimate connections between music, arithmetic, geometry, and astronomy within the quadrivium. Chapter 1 outlines the relation between Greek medicine and natural philosophy, especially the role played by Pythagorean thought. Alongside their mathematical and musical studies, the Pythagoreans also emphasized medical practice. Though the healing arts initially stood apart from the quadrivium, Pythagorean thought influenced the rational medicine of the Greeks as carried forward by Alcmaeon, Philolaus, and the Hippocratics. Some of them described the body in terms of four liquid humors, a concept that only became central in late antiquity. For the Hippocratics, number regulated the timing of the crises of illnesses, the crucial moments in which the patient would either turn the corner toward renewed health or worsen. Important categories of disease were governed by fixed patterns of crisis, knowledge of which was crucial for physicians, whose understanding of those patterns they transmitted to the patients and their families, even if their scope of therapeutic action was limited. In light of its special quest for knowledge and healing, Plato held up medicine as an exemplar for the practice of philosophy, considered as the healing of souls from ignorance and delusion.

The controversies surrounding the Pythagorean program are the subject of chapter 2. Aristotle argued that *physis*, the realm of growth and change that included what we now call biology and physics, is not governed by numbers; instead, he advocated empirical but qualitative investigations. His arguments against the possibility of applying mathematics to physics tended to separate biology from music, as well as from the other mathematical sciences. Still, when Greek medicine moved to Alexandria, it both continued Aristotle's empirical work and also drew on Pythagorean ideas. In particular, Herophilus connected musical ratios with the functioning (or diseases) of the pulse, on the analogy with the various musical intervals those ratios represented when applied to vibrating strings, glasses, or pipes.

As chapter 3 relates, these connections between ratios and pulse patterns grew in importance during and after Roman times because Galen paid them considerable attention in his vast and influential compilations of medical knowledge. Then too, writers on music such as Aristides Quintilianus discussed Herophilus's ideas and described the fundamental processes by which music therapy operated. Authors as diverse as Martianus Capella and Augustine paid attention to the music of the pulse. As Rome was overtaken by invaders, Boethius described *musica humana* as the correlate in the human body of the cosmic harmonies. By taking up this influential concept as well as Herophilus's ideas, Muslim translators and commentators such as al-Kindī, Hunayn, al-Rāzī, and Ibn Sīnā played a crucial role in keeping these ideas alive and transmitting them to the West.

Chapter 4 presents the continued reverberation of Herophilus's teachings throughout the Middle Ages and Renaissance. Even before Galen's texts were generally available, many writers (such as Hildegard von Bingen) discussed the use of music to treat melancholia, a particularly important example of *musica humana* applied to medical practice.

Harking back to the ancient connection between astronomy and music, Robert Grosse-teste and many others connected these interventions with the influence of the planets; astrology was a common tool of medieval medicine. After the writings of Galen and Hippocrates returned to circulation, the medical school of Salerno took up the intriguing ideas of pulse music, as did later physicians such as Pietro d'Abano, who connected them with contemporary music and rhythmical theory. In turn, they influenced other medical writers in the fourteenth and fifteenth centuries. Among musical writers, Bartolomé Ramos de Pareia and Franchinus Gaffurius made connections between music, medicine, and astronomy. Above all, Marsilio Ficino drew together these musical, astrological, and medical strands as he presented the first translations of many Platonic and neo-Platonic works, along with commentaries that greatly influenced his contemporaries. Ficino's influence reached across Europe, becoming common currency also in literary works ranging from Shakespeare's plays to Robert Burton's encyclopedic treatment of melancholia, which often referred to the relation between music and medicine.

Johannes Kepler took Platonic ideas about music to new heights throughout his extraordinary body of work. Though best known for his crucial innovations in astronomy, Kepler also reflected on medicine and biology as they related to music and astrology. As innovative as he was, Kepler drew on a Platonic and Aristotelian framework in which physics and biology (as we now call them) together formed what he called "physiology," in which musical harmony was the essence of what he called "soul," animating not only humans and animals but the Earth and the whole cosmos. Chapter 5 treats the ramifications of his harmonic physiology, ranging from his reflections on the symmetries of beehives, pomegranate seeds, and snowflakes to the "spontaneous generation" he envisaged for comets and new stars. Though Kepler dismissed many elements of vulgar astrology as groundless, his fundamental geometrical and musical archetypes mandated what he thought were well-grounded connections unifying physiology throughout the heavens and the Earth. These connections formed the basis of his purified astrology. In his *Harmony of the World* (1619), he used these connections as he tried to persuade James I of England to heal the "cross-shaped wounds" that afflicted Protestants throughout Europe (including Kepler himself) by taking up arms in their defense, thereby engaging the various "hard" and "soft" (major and minor) harmonies of the heavens. In contrast, his astrological work for Albrecht von Wallenstein called that preeminent generalissimo to heed the harmonies indicating that he should turn toward peaceful means.

Tarantism presents a striking, perhaps even unique, example of music as curative agent. Chapter 6 considers how those bitten by spiders would respond to certain kinds of music that excited them to frenzied dancing. This music seemingly exacerbated their distress yet was capable of relieving them if they could dance long enough. In that case, their eventual collapse and subsequent recovery seemed to show that the music had paradoxically purged their malady by making it worse. Tarantism also raised questions about the nature and timing of its crises, which seemed to reoccur on anniversaries of the original bite as well as on the Feast of St. Paul, the patron saint venerated by those afflicted by venomous bites.

As medicine sought to go beyond the visible manifestations of illness, sound offered ways to access the hidden interiority of body and mind, the sonic turn introduced in part II. Chapter 7 considers how the body went from being considered primarily an assemblage of humors to an ensemble of sensitive constituents, fibers, and organs capable of responding sympathetically to sonic vibrations. Sound thus called for and offered a new way of thinking about the body that would explain the responsivity of living organisms without recourse to vital forces that defied rationality and mechanics. Poised between the visible and the invisible, sound offered a new way to mediate between matter and vital activity.

This sonic turn came in several distinct (and somewhat disconnected) waves. In the mid-eighteenth century François-Nicolas Marquet proposed a musical notation of normal pulse that connected its felt pulsation with heard music (a minuet), though without attempting to hear (rather than feel) those rhythms, as described in chapter 8. His minuet reflected contemporary musical styles, as did his notations for various kinds of abnormal pulse, though without relating those rhythms to heart function or disease. Given Marquet's relative obscurity, his ideas received a surprising amount of attention in articles by Joseph-Jacques Ménuret de Chambaud in the famous *Encyclopédie*. Though vitalists such as Ménuret and Théophile de Bordeu rejected Marquet's work, mechanists such as Albrecht von Haller included it in the larger discourse about the physiological manifestations of bodily fluids and fibers.

The decisive phase in the sonic turn included the diagnostic techniques of percussion and mediate auscultation advocated by Leopold von Auenbrugger and René T. H. Laennec. Chapter 9 explores their musical backgrounds in relation to their work. Both were deeply involved with music, Auenbrugger (and his daughter Marianne) with Joseph Haydn and Antonio Salieri, Laennec with Breton folk music. Though neither discussed Marquet's work, they had been trained in the mechanistic physiology of Haller, which was notably open to musical and sonic approaches. Auenbrugger used musical vocabulary to present his work on thoracic percussion; Laennec's musical experience shaped his exploration of the new timbres he discovered using his stethoscope in the emerging practice of mediate auscultation.

Part III turns from physical diagnosis to the search for a new typology of mental illness and the emerging sonic interventions that addressed them. Chapter 10 explores the changing meanings of "mania" and the emergence of what now are called obsessional-compulsive disorders. Yet decades before clinical descriptions of "monomania" and idée fixe, a symphony by Gaetano Brunetti gave a detailed dramatization of a musical obsession he called *manía*. In his Symphony "Il Maniático" (1781), Brunetti's solo cello is fixated on an oscillating semitone. The orchestra around him tries to shake him from his obsession by cheering him up, but the "maniac" cello persists, though to some degree reshaping his fixed notes to accord with the surrounding harmonic context. In the end, the cello seems to give up its obsession by inducing the orchestra to perform equally obsessive figurations of its own, as if all together recognized the obsessional qualities of the classical style itself. Written for the crown prince who later reigned as Carlos IV, the symphony can be read as a message of sympathy regarding the prince's struggles with his obsessional father, then reigning as Carlos III, not to speak of the half century of Spanish royal insanity that had gone before. Then too, the symphony seems to poke fun at the rather obsessional style of Luigi Boccherini, a rival composer at the court.

Where Brunetti offered a new kind of musical insight into psychic maladies, Franz Mesmer used music as a crucial means to induce radical changes in mental and physical states. He was steeped in music, a noted performer on the recently invented glass harmonica, and was involved with the Mozarts and other musicians. Chapter 11 shows how Mesmer's "animal magnetism" relied on music as a powerful modality of treatment as well as providing a basis for his theoretical foundations and self-understanding. Drawing on the procedures by which the classical style manipulated musico-dramatic crises, Mesmer was able to alter the timing and rhythm of therapeutic crises, long considered to be inalterable characteristics of each disease.

Music continued to be an important factor in hypnotic therapy even after Mesmer's successors turned away from his reliance on crises, beginning with his disciple Amand-Marie-Jacques de Castenet, marquis de Puységur, who preferred a deep somnambulistic state to Mesmer's convulsive crises with their violent disorder and possible menace. Chapter 12 relates the ways Jean-Martin Charcot used a tam-tam (a large gong) to induce profound states of hypnotic sleep verging on catalepsy in which patients would be intensely suggestible. Charcot's work came after the tam-tam had become familiar in orchestral practice, beginning with funeral marches during the Revolution in which the tam-tam played an overwhelming role, thereafter becoming a staple of operatic dramaturgy. Though Charcot's work was shadowed by the deliberate imposture practiced by some of his associates and patients, even if it is taken to be suggestion or deception, it operated through musical models of dramatic disruption. The tam-tam's complex mixture of overtones seemed to overwhelm consciousness, overriding normal mental responses, compared to the sequence of conscious (but intensely felt) states Mesmer induced through the harmonica's unfamiliar sounds.

Finally, part IV addresses new frontiers of sound beyond those audible to unassisted human hearing, beginning with the investigation of how bats fly in the dark, pioneered by Lazzaro Spallanzani at the end of the eighteenth century. The bats' "sixth sense" was identified as hearing only in the twentieth century, after their mysterious methods of navigation had been imitated by early developments in sonar and radar. Only after World War II was the electronic technology available to detect the bats' sounds as ultrasonic. That technology ultimately rested on the hetrodyne principle, developed from the musical phenomenon of "Tartini tones." Chapter 13 details the interweaving strains of biological investigations and military technology that grew from the bats' navigation and eventually illuminated these animals' uncanny skills.

Ultrasound went on to have enormous importance for many kinds of clinical diagnosis, as outlined in chapter 14. After two world wars, the development of sonar led to many attempts to ultrasound the human body. These all faced unfamiliar problems of understanding acoustic echoes in the body's complex, reflective internal environment, so different from locating submarines or ships. For a surprisingly long time, experimentation relied on makeshift bricolage of war-surplus electronics and audio equipment. Industrial ultrasonic flaw detectors were repurposed by applying their probes to the human body, but interpreting the resultant images posed many difficulties. Prior practices based on looking at X-rays were misleading when applied to ultrasonic echoes, which required newer modes of visualization than those adapted even from radar and sonar. Building on the work of others and using these new modes, Ian Donald and his collaborators brought ultrasound to wide acceptance for the first time. In so doing, they had to attend to the body acoustically, not just visually, by addressing the new acoustic "slices" of the body that led to tomography and thereby to the three- and four-dimensional imaging that became universal medical practice in both ultrasonic and computerized tomography (CT) scans.

The final four chapters concern the use of sonic techniques to explore the functioning of the nerves. Chapter 15 begins by connecting vibrating nerve fibers to more general concepts of "nervousness" that gained traction in the eighteenth century. Above all, Luigi Galvani's discovery of the connection between electricity and muscular action called physiology to investigate their interrelation, especially the role of the nerves. Much depended on the details of the instruments deployed to observe that electrical activity, particularly the galvanometer. Using it, Emil du Bois-Reymond demonstrated that muscular contraction led to electrical activity. Hermann von Helmholtz measured the speed of those nerve currents and used tuning forks to regulate electrical stimulation of the nerves as well as to time their response. He also listened to the sounds made by muscles as they contract, investigations Julius Bernstein carried further as he developed a chemical theory by which membranes mediate nerve action.

The new technology of the telephone proved very useful in pursuing these ever more detailed studies of the precise details of how nerves and muscles work, as chapter 16 recounts. Du Bois-Reymond was able to make a frog's leg twitch or lie still, seemingly by commanding it through a telephone attached to the nerve. Other researchers were able to use electromagnetic tuning forks to stimulate a nerve whose response they heard through a telephone receiver. Bernstein and others conducted ever more elaborate experiments that converted the rattling sounds they heard in the telephone into moving light beams that could make permanent traces along photographic plates. In 1900, Nicolas Wedensky considered the telephone "virtually irreplaceable" in tracing any arbitrary point in a nerve without disturbing its connection to the muscle, able to hear the action of poison as it acted on the nerve. Likewise, Rudolf Höber used the telephone with new "valve amplifiers" to catalogue a whole range of sounds he evoked from muscles stimulated at various frequencies. He and Ferdinand Scheminzky began using loudspeakers to make the sounds of nerves and muscles audible through an entire room.

The scene then shifts in chapter 17 from Europe to Britain and the United States. The quest to measure the firing of a single neuron taxed the different limitations of the available instruments. Using great ingenuity and skill, Keith Lucas proposed and argued for the all-or-nothing action of muscle fibers by gradually paring them apart and noting the quantized "steps" in their contraction. To establish the truth of this principle in general, he realized, would require measuring the electrical activity of a single nerve cell. After his untimely death, his student Edgar Adrian fulfilled this quest by finding the output of a single nerve through listening to its amplified signal, at first through a telephone and later through a loudspeaker. His student Bryan Matthews turned the loudspeaker into an even better recording device than contemporary galvanometers or oscilloscopes, the "Matthews oscillograph" in use until after World War II. These sonic devices did not merely render the nerve action audible; they were important in separating a single neuron from others and locating its signal. So significant was this technique that Adrian played gramophone records of his nerve experiments during his Nobel lecture in 1932.

Chapter 18 reflects on the epistemic significance of hearing versus seeing or analyzing purely numerical data. Episodes from the later work of Adrian exemplify *rhythmic knowledge*, in which the "analog" perception of bodily sounds led to hearing "binary" neural rhythms that characterized critical changes in living processes. Registering such rhythmic knowledge can bypass the complex dialectic of objectivity versus subjectivity as it has been theorized by historians of science for the visual analysis of scientific phenomena. Rhythmic knowldge gives direct access to living processes in real time, through the vivid aural experience of their changing rhythms. Thus, the thrill of hearing the "machine gun" firing of a cat's visual neuron helped David Hubel and Torsten Wiesel in their landmark studies of vision in the late 1950s. Indeed, audio monitoring is still used in neurophysiology labs around the world, such as Leslie M. Kay's group studying the olfactory bulb at the University of Chicago. There and in many other labs and classrooms students are initiated into the special kind of hearing needed to find a single neuron. In the process, they learn to hear how different parts of the brain *sound* different.

Many of the other sonic techniques discussed here have entered into new phases, as chapter 19 relates. Diagnostic ultrasound now can be run through a smartphone, allowing practitioners far from medical centers to use this powerful tool inexpensively. Even though the venerable practice of mediate auscultation has fallen into neglect as a core skill of physicians, computer-assisted auscultation (CAA) allows greatly enhanced stethoscopes that facilitate powerful analysis of the sonic signal past the limitations of human ears, easily transmitted for remote consultation. Handheld ultrasound devices allow a similar ease and portability of this crucial diagnostic tool. Likewise, handheld "brain stethoscopes" enable quick and ready diagnosis of silent seizures through sonifying the electroencephalogram. Moreover, sonifications of genomic data open the possibility of understanding the genomes of human and coronavirus in new ways through hearing their detailed structure.

Though these developments echo and update earlier discoveries, the biomedical sciences continue to rely on sonic and rhythmic knowledge. If the evidence of the past is any guide, these modes of knowledge will continue to be a valuable resource to overcome the limitations of the human gaze by finding ways to hear critical changes in the deluge of numerical data. We are only at the beginning of sounding bodies.

Throughout the book, when I refer to various "♪ sound examples," see http://mitpress.mit .edu/sounding-bodies. That link will give further information on purchasing enhanced digital editions that will be available in a variety of formats, in which the text and examples are easily and seamlessly available; in them, you need merely touch an example to hear and see it. Readers with special needs will also find instructions for accessing the audio and visual examples in alternate forms.

I

Musical Origins

1

Pythagorean Medicine

We begin part I by considering the Pythagorean approach to medicine as part of the Pythagoreans' larger project to understand everything in terms of number and music. In response, the ancient Greeks developed a rational medicine closely allied with the development of natural philosophy. The Pythagoreans' quest influenced the development of rational (rather than supernatural) understandings of living organisms as well as of the cosmos in general. Drawing on Pythagorean ideas, Plato applied music and mathematics in his influential account of living beings. These ideas reverberated throughout Roman, Muslim, and Western medicine and eventually led to new sonic understandings of bodies. According to the Pythagoreans, music reflects the changeless beauty of disembodied numbers and hence can cure mortal disharmonies. But music was also restless vibration, which gave medicine and biology new ways to understand living bodies through sound.

Though Egyptian medicine used an extensive array of drugs, it ascribed illness to evil spirits or the wrath of the gods; likewise, Babylonian medicine required patients to atone for their sins and placate angry deities.[1] The same basic attitude characterized the earliest Greek texts: Homer's *Iliad* begins by attributing a plague afflicting the Greek army to the anger of Apollo.[2] As remedy, they offered music and dance to propitiate the angry god: "All day long these young Greeks propitiated the god with dancing, singing to Apollo a paean as they danced, and the god was pleased."[3]

Greek medicine often looked back to Egypt; for instance, in the *Odyssey*, Helen of Troy obtains from Egypt a drug (*pharmakon*), "heartsease, free of gall, to make one forget all sorrows," so potent that it would leave one unmoved "even if his mother died and his father died, even if men murdered a brother or a beloved son in his presence." Homer notes that Egypt's "fertile earth produces the greatest number of medicines, many good when mixed with wine, many deadly poisons. Every man is a healer there and more skilled than any other men on earth—Egyptians born of the healing god himself."[4]

Despite their admiration for Egypt, Greek physicians turned to the rational explanation of disease rather than invoking gods or supernatural forces. In this, they were akin to the early Greek thinkers who sought to give a rational account of the cosmos; they called

Box 1.1
Pythagorean Discoveries about Musical Ratios

> The mythical scene illustrated in figure 1.1 showed the original context in which Pythagoras was said to have noticed that two hammers in the weight ratio of 2:1 would sound a musical octave (such as from one C to the c an octave higher), likewise 3:2 a fifth (C–G) and 4:3 a fourth (C–F), while other ratios that were not made from simple whole numbers would be discordant. Though even in ancient times it was realized that this is highly problematic (if not impossible) using hammer weights alone, these ratios do work when applied to string lengths held at constant tension (for instance). The Pythagoreans took this discovery as the touchstone connecting the realm of numbers to physical reality.

themselves *physikoi* or *physiologoi*—those who seek the *logos* (word, reason, logic) in *physis*, the realm of growth and change, as opposed to *theologoi*, those who seek *logos* in the divine. At first, these *physiologoi* tried to understand everything in terms of a single element: Thales asserted that everything was fundamentally water, while Heraclitus chose fire (signifying change) as the central unifying principle. Parmenides, who argued for the unchangeableness of true reality, was honored as the founder of a medico-religious group (*pholeon*).[5] Most significant for our story, the followers of Pythagoras held that number was the basis of understanding everything, for which they found crucial confirmation in the connection between simple whole-number ratios and musical consonance (figure 1.1 and box 1.1).[6] This had profound and long-reverberating influences on the physical sciences, as my book *Music and the Making of Modern Science* details.[7]

The Pythagoreans also applied their general approach to medicine; one might say provocatively that in its beginnings, Greek rational medicine was a Pythagorean project because it "laid the foundation for the most important theories of ancient medicine, or at least decisively promoted its advance."[8] Because of the paucity of writings from the earliest Pythagoreans, we must often rely on accounts of later writers. Thus, in the fourth century BCE the musician and theorist Aristoxenus (who studied with Aristotle and knew the last Pythagoreans) recorded that they "used medicine for the purification [*catharsis*] of the body, and music for that of the soul."[9] Among the neo-Platonists, Iamblichus noted in about 300 CE that Pythagoras

held that music too made a great contribution to health, if properly used: he took this form of purification very seriously, calling it "healing by music." . . . [He] used music as a kind of medicine. There were songs designed for afflictions of the soul, to counter depression and anguish of mind (some of Pythagoras's most helpful inventions); others to deal with anger and bursts of indignation and every disturbance of that kind of soul; and yet another kind of music devised to counter desires. They also used dancing.[10]

Writing in the following century, Porphyry added that Pythagoras

Figure 1.1
The blacksmith shop in which Pythagoras was said to have discovered the relation between simple whole-number ratios and consonant musical intervals sounded by the hammers, as depicted in Franchinus Gaffurius, *Theoriae musicae* (*Theory of Music*, 1462). The weight of each hammer is indicated above it (see box 1.1).

paid attention to his disciples' health, he cured those who labored under physical diseases and comforted those suffering from diseases of the soul, just as we said, the former by means of sung spells and magic, the latter by means of music. Actually, he had melodies that succeeded in healing the diseases of the body, and the sick got up as he sang them. He also had melodies that allowed people to forget pain, and he soothed wrath and removed wicked desires.[11]

According to the Roman rhetorician Aelian, Pythagoras wandered around cities "not to teach, but to heal."[12] His followers ruled Croton in southern Italy about 500 BCE, during a period when that city was famous for its physicians. Indeed, all three extant ancient lives of Pythagoras note that he practiced medicine.[13]

To be sure, the Pythagorean movement also took on certain religious aspects. Porphyry noted that "marvelous and divine things were said about this man," and the secrecy of his brotherhood projected the aura of a religious cult guarding esoteric teachings.[14] The numbers to which they ascribed such power also acquired a certain numinous quality, beginning with the One, the primal monad out of which all the numbers flowed. Nevertheless, the Pythagoreans did not personify numbers as divinities or connect them with the worship of the Olympian gods. If the realm of number was higher than the human world, it was also higher than the gods. When (according to legend) Pythagoras sacrificed oxen to celebrate a great theorem, he did not offer them to any particular god but seemingly to the divine realm of mathematics itself. The Pythagoreans' sense of sacred awe went hand in hand with their mathematics, whose rationality they also applied to music and cosmology.[15]

Those who heard Pythagoras also included pioneers of the new rational medicine, beginning with Alcmaeon of Croton, the first to leave medical writings in Greek. According to Aristotle, Alcmaeon "was young in the old age of Pythagoras and gave a similar exposition" to the Pythagoreans.[16] Alcmaeon's theory of health and disease was "the first rational theory in Greek medicine known to us"; indeed, some scholars even call him the "father of medicine."[17] He considered health to be "the equality of the powers—moist and dry, cold and hot, bitter and sweet and the rest," so that disease was an unbalanced supremacy (*monarchia*) of one power over the rest, compared to "a harmonious blending of the qualities" characteristic of health.[18] This "dynamic concept of illness" constituted the basis for the physiology of humors that was later developed in the Hippocratic corpus and canonized by Galen.[19] Alcmaeon also was the first to dissect the eye (perhaps also the body as a whole) and introduced the concept of female (as well as male) human seed; he was the first to formulate and address basic questions of embryology and heredity. He had considerable influence on other pre-Socratic thinkers outside the Pythagorean movement, leading them into a new physiological direction toward problems of the human organism, not just the cosmos.[20]

Though Alcmaeon is mainly known to us through references by later contemporaries, the medical dimension of Pythagoreanism becomes clearer in the writings of Philolaus, about fifty years after Pythagoras's death. Though there is some controversy about whether Alcmaeon was really a Pythagorean, Philolaus unquestionably was, bearing in mind that the Pythagorean movement was by no means dogmatically unified but rather a loose collection of thinkers who shared certain common convictions (particularly about the centrality of number). Philolaus was the first in the movement to write a book, *On Nature* (extant now only in fragments), perhaps even the first who made public Pythagorean teachings that were previously held secret.[21] His book began by asserting that "Nature [*physis*] in the world-order [*kosmos*] was fitted together both out of things which are unlimited and out of things which are limiting."[22] Here, "fitted together" is literally "harmonized" (using a verb derived from *harmonia*), while "unlimited" and "limited" seem to denote even and odd numbers, respectively. Thus, this opening sentence already highlights the intimate relation between number, harmony, and nature (terms to which we shall return). In particular, Philolaus thought the human body was structurally determined by the "unlimited" number four, referring to the head (intellect), heart (life and sensation), navel (first growth), and genitals (generation).[23] Each of these he called the "origin" or "principle" (*archē*), respectively of humans, animals, plants, and all living things (because "all things both flourish and grow from seeds"). Further, Philolaus called bile, blood, and phlegm the principles (*archai*) of disease, explicable in terms of disharmonies between those three substances. Here Philolaus noticeably changed the connotations and usage of the word *archē*, which in Homer had meant "beginning" or "first cause," such as the dispute between Agamemnon and Achilles that sets the *Iliad* in motion. More pointedly, Philolaus used *archē* to mean a rational explanatory principle or origin, such as an explanation of a disease in terms of disbalanced natural causes.[24] This usage also is found in Philolaus's Pythagorean contemporary Hippocrates of Chios, who used *archē* to mean a mathematical hypothesis that then serves as a starting point for a proof.[25] Thus, Philolaus's use of this term seems to reflect its prior mathematical meaning in Pythagorean circles and shows again how the rational approach to medicine arguably grew from—and used some of the same terms as—mathematical arguments.[26]

During Philolaus's lifetime, the earliest Hippocratic texts began to use *archē* to denote a starting point or premise in the medical sense of the inferred origin or cause of some condition—its etiology, as Galen later called it. The Hippocratic corpus was a body of sixty-odd treatises (probably by different authors) that shared a commitment to rational explanations of disease.[27] To be sure, Hippocratic healers (*iatroi*) were surrounded by other practitioners who relied on supernatural intervention or traditional remedies.[28] Nor can we consider the Hippocratics "scientific" in later senses of that word, for they did not perform experiments or clinical trials (as they later came to be known). Still, they presented the first case histories, the individualized, detailed narratives that became central to later medicine.[29]

Hippocratic medicine primarily relied on diet and regulation of activity or regimen, which had been greatly emphasized in Pythagorean healing.[30] Such regimens could include "exercise of the voice, whether speech, reading, or singing, all these move the soul. And as it moves it grows warm and dry, and consumes the moisture."[31] In Greek, *pharmakon* meant both a drug as well as a poison, an ambiguity whose importance remains to this day: every known drug has some degree of toxicity.[32] Thus, playing with this double meaning, as Athens executed its sentence on him, Socrates called the cup of hemlock his *pharmakon*, which would kill him but also "cure" him of life, which he likened to the imprisonment of an immortal soul in a mortal body.[33]

In order to give a rational account for their various therapies, the Hippocratics relied on ideas of the constitution of the body they shared with various natural philosophers who were speculating about the cosmos and its constituents. Yet the Hippocratic treatise *On the Nature of Human Beings* turned away from the view that the body is "air, or fire, or water, or earth, or anything else that is not an obvious constituent of a human being." In contrast to Philolaus's three principles, the Hippocratics argued that "the human body has in itself blood, phlegm, yellow bile, and black bile," the four humors (*humoi*), as they came to be called, which became the canonical principles of physiology for the succeeding millennium and a half.[34] Though analogous to the four elements (fire, air, earth, and water) invoked by the natural philosophers, some Hippocratic texts considered the four humors to be manifest in medical experience and therefore rightful "elements" or constituents of the body. Perfect health results "when these elements are duly proportioned to one another in respect of compounding, power, and bulk, and when they are perfectly mingled. Pain is felt when one of these elements is in defect or excess, or is isolated in the body without being compounded with all the others."[35] Though there is only one mention of them in a fifth-century Hippocratic text, the four humors later became standard, especially through Galen's advocacy.[36] For over a thousand years, they provided a basic typology of human constitutions still in common use: "sanguine" individuals (those dominated by blood), "phlegmatic" (phlegm), "bilious" or "choleric" (yellow bile), and especially "melancholic" (black bile, *melan cholē*), to which we will return in subsequent chapters.

Hippocratic medicine was qualitative for the most part, varying its prescriptions according to the typology of various patients, their relative states of health and habits, without assigning fixed or definite numerical proportions. In that regard, the Hippocratics did not show the influence of Pythagorean ideas.[37] Still, the writings in the Hippocratic corpus showed considerable interest in counting various aspects of medical conditions, such as the periodicities of pregnancy and childbirth as well as the turning points of diseases, their "crisis" or "critical day." This was not necessarily a grave emergency or crisis, as we might put it, but a day of decision on which a marked change of symptoms occured, indicating "a significant moment in the evolution of the disease, for better or for worse."[38] Ancient Greek had three words for "time," personified as *Aion*, eternity and cosmic recurrence (a man walking through the circle of the zodiac); *Chronos* (an old man, sometimes veiled), embodying the terrestrial passage of past, present, and future; and *Kairos*, a fleeting moment of opportunity (a young man with winged heels and shoulders, holding a

Figure 1.2
Kairos (*Occasio*, opportunity or fortune), a young man whose winged shoulders and heels suggest speed. The scales balanced on a knife's edge weigh the fleeting balance of the moment. His hair (full in front but bald behind) alludes to the maxim about seizing opportunity by the forelock: the critical moment, once passed, cannot be seized. From a Roman sarcophagus (160–180 CE) in the Museum of Antiquities, Turin, after an original sculpture by Lysippos (ca. 350–330 BCE).

balance on knife's-edge; figure 1.2).[39] As a Hippocratic precept put it, "*chronos* is that in which there is *kairos*, and *kairos* is that in which there is not much *chronos*."[40]

A *krisis* is a decisive moment (related, like *kairos*, to the verb *krinein*, to decide or judge) at which a fever may "break," indicating hope of recovery, begin another cycle of fever, or make a turn for the worse. The whole impulse to count the days of illness probably emerged in keeping track of the periodicity of the critical days, moments of *kairos* that physicians needed to recognize in order to gauge their treatment and assess the prognosis.[41] This search for exactness overlapped (and may well have emerged from) the Pythagorean concern with number and its connection with music. Indeed, the Roman physician Celsus referred to critical days as "Pythagorean numbers."[42]

For instance, tertian fevers have a crisis every other day (on the third day, as the Greeks counted) when the fever spikes. In some cases, fever "begins gently without producing obvious signs, increasing day by day paroxysmally until at the crisis it fairly shines out. In other fevers, the start is mild but the fever increases in paroxysms to its height and then persists until the crisis be reached and passed." Beside quartan and other periods, others without obvious periodicity were called "disorderly" or "wandering." Among them, "the most severe, serious, troublesome and fatal maladies produce continued fever," whose lack of periodicity reflects the disease's intractability. The Pythagorean emphasis on the fundamental distinction between odd and even numbers carried over into the timing of crises: "Fevers attended by paroxysms at even numbers of days, reach their crisis also in an even number; if the paroxysms are on odd days, so is the crisis. . . . It must be noted that if a crisis occurs on any other day than those mentioned, there will be a relapse and also it may prove a fatal sign."[43]

To be sure, Hippocratic healers were very restricted in what they could do even were they to recognize the crisis correctly: "When a disease has attained the crisis, or when a crisis has just passed, do not disturb the patient with innovation in treatment either by the administration of drugs or by giving stimulants. Let them be."[44] This was part of a general sense of restraint that remains an important medical principle: "either help or do not harm the patient."[45] Then too, Jacques Jouanna notes that the Hippocratic treatise *On Ancient Medicine* "asserted that number is of great importance in the evolution of disease and yet at the same time challenged the existence of number as the criterion for evaluating the proper adaptation of regimen to man. Arithmology did not penetrate as far as treatment. Health was not defined by number."[46]

To this day, medical practitioners distinguish tertian fever from quartan, whose different insect-borne malarial parasites require different treatments. Current medical textbooks continue to use the term *crisis* in essentially the same sense as did the Hippocratics: the turning point of a disease for better or worse. Even their numerical emphasis on odd versus even critical days long persisted: as late as 1947, William Osler's *The Principles and Practice of Medicine*, a widely used text by one of the most eminent physicians of his time, taught that "from the time of Hippocrates it [the crisis in pneumonia] has been thought to be more frequent on the uneven days, particularly the fifth and seventh."[47]

Number and ratio also informed the Hippocratic approach to embryology. *On the Nature of the Child* holds that male embryos first move in the womb at thirty days, compared to females at forty, perhaps reflecting the traditional associations of 3 with males and 4 with females (as well as gendered assumptions about the greater vigor of the male). But some scholars read this as having musical implications: if a "tone" (a whole step such as C–D) is taken to be twelve days, then the male thirty days is a fourth (two and half tones, C–F), while the female forty is a fifth (three and a half tones, C–G).[48] In the *Regimen*, musical ratios described how children develop in the womb:

If, on changing position, they achieve a correct attunement, which has three harmonic intervals, covering altogether the octave, they live and grow. . . . But if they do not achieve the attunement, and the low do not harmonize with the high in the interval of the fourth, of the fifth, or in the octave, then the failure of one makes the whole scale of no value, as there can be no consonance.[49]

The treatise *Eight Months' Child* also asserted that conception and birth follow numerical periods, following the same logic as crises, "on the basis of the principle of harmony, to be the true and perfect number system, for reasons it would be too long to go into on this occasion."[50]

The famous Hippocratic Oath included several views that seem peculiarly Pythagorean and specific to their way of life.[51] According to Iamblichus, the Pythagoreans "absolutely rejected surgery," while the Oath stipulated that "I will not use the knife, not even for sufferers from stone, but will withdraw in favor of such men as are engaged in this work."[52] Here the Pythagoreans were exceptions to the general acceptance of surgery in Greek medicine, particularly for kidney stones.[53] Thus, the subsequent separation of medicine from surgery, maintained for many centuries, arguably goes back to the Pythagoreans.

Plato further elevated the importance of Hippocratic medicine by connecting it to a cosmology based on Pythagorean music and mathematics. About thirty years younger than the historical Hippocrates, Plato referred respectfully to that "famous physician" for having shown "the way to think systematically" about the nature of the body, as opposed to the merely "empirical and artless practice" of many healers.[54] During his three visits to Sicily, Plato came into direct contact with the Pythagoreans as well as with Sicilian medicine, on which he drew heavily.[55] This Sicilian medical tradition went back to Empedocles, a natural philosopher who held that the body and its illnesses should be understood in terms of the four elements common to the whole cosmos, in contrast to the four humors of the Hippocratics. Sometimes called "empiricists," the Sicilians "relied on experience," as the Roman natural historian Pliny put it.[56] In Sicily, Plato also came to know Archytas, a student of Philolaus and a famous mathematician and general who saved Plato from likely death at the hands of the Sicilian tyrant Dionysius II.[57] Plato's Pythagorean inclinations affected his work deeply. In his dialogue depicting his teacher's final hours, Socrates defends Philolaus's view that it is not right to kill oneself and also devotes much discussion to the Pythagorean teaching that the soul is a kind of harmony of the body.[58]

Though generally critical of artisans who merely rely on habit, without reasoning more deeply, Plato made an exception for the art of medicine. In his dialogues, he constantly brought forward the metaphor of the philosopher as a wise physician, whose treatments are informed by study of the cosmic order and its musical proportions, clearly following the Pythagoreans. Plato's *Timaeus* stressed the deep parallels between the macrocosm and the human microcosm. The Sicilian Timaeus tells Socrates and other listeners what he calls a "likely story [*eikos mythos*]" about how the world we see around us came to be. In his story, a divine maker or craftsman—*demiourgos*, the word also used to designate an ordinary medical practitioner—shaped the unformed matter of the macrocosm to reflect the ideal Platonic Forms of numbers and their musical interrelations. Rather than creating from nothing, this craftsman dealt with preexisting raw material called *hyle*, the word used for timber that might then be shaped into furniture or houses. Curiously, Plato does not call his divine craftsman an *architektonikos*, the term for a master physician or architect, as opposed to an ordinary practitioner, perhaps to emphasize that this workman did not presume to prescribe the architecture of the cosmos on his own but relied on using the higher musical (and mathematical) forms that lie beyond human art.[59] Using these musical ratios, he shaped the cosmos as "a single living thing that contains within itself all living things, mortal or immortal."[60] This was necessary because visible matter "moved unmusically [*plēmmelōs*, literally out of tune] and without order."[61] To harmonize it, the demiurge formed the elements out of triangles, whose innate proportions then were incorporated into five fundamental forms—the "Platonic solids" (figure 1.3) that correspond to fire, air, earth, water, plus a fifth element worthy to constitute the stars.[62] Timaeus offers arguments why each geometric solid was chosen to constitute each element as we perceive it, arguments that depend on the particular combination of proportions and angles, of rational and irrational lines, peculiar to each solid.

Timaeus notes that the divine craftsman or physician "himself fashioned those [beings] that were divine, but assigned to his own progeny the task of fashioning the generation of those that were mortal."[63] Though Timaeus does not emphasize it, this transition was momentous: human beings—and the whole of the mortal, changeable world—are the product not of the divine craftsman directly but of his children. Are we, then, derivative, secondary, perhaps even student work? Though these subsidiary makers were indeed gods, compared to the demiurge, they operated at one further remove from the transcendent Forms that had been the original template of creation. Timaeus seems much less given to humor than Socrates, but still one wonders about the possible irony of "outsourcing" the creation of the human world. Could there be a more powerful (and subtly humorous) explanation of all that is awry in the world? In one possible reading, we are second-hand beings, which may explain many troubling shortcomings in our makeup. Worse still, Timaeus does not explain why this came to be. Why didn't the demiurge

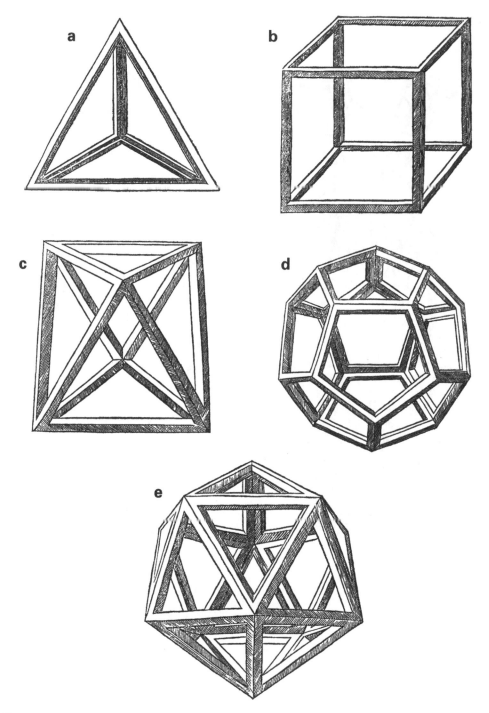

Figure 1.3
The five Platonic regular solids (tetrahedron, cube, octahedron, dodecahedron, and icosahedron) as drawn after Leonardo da Vinci for Luca Pacioli's book *De divina proportione* (*On the Divine Proportion*, 1509).

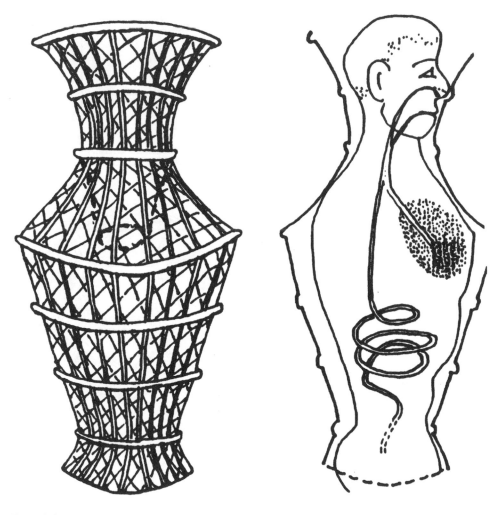

Figure 1.4
Renditions of the fish-trap described in Plato's *Timaeus* 78b–79c. (a) The basic geometric structure of the fish-trap, showing two funnels, in the top and bottom sections. (b) A fish-trap superimposed over a human body, showing the salient resemblances. The top filter corresponds to the processes of breathing and respiration, the bottom to digestion. From James Longrigg, *Greek Rational Medicine* (1993).

create us himself? Did he find the task boring or distasteful? After having shaped the heavens, did he feel he could not sully himself with our mortal clay? Or did he merely feel that it was a good exercise for his apprentices, something a bit challenging but appropriate for their level?

In response to these deep questions, Timaeus's account emphasizes that we were formed according to the same mathematical and musical Forms as the rest of the cosmos, even though applied at one remove from the heavens. After all, we, unlike the stars, are subject to corruption and mortality. Given the worst things humans can do, the creator could have disclaimed all responsibility or relegated us to a far lower order of creation. Instead, Timaeus says that the younger gods fashioned our bodies from a "mixing bowl" in which the maker had blended soul-stuff from the leftovers of creation "and assigned to each soul a star." This beautiful detail indicates that the human world is a microcosm that imitates, as best it can, the macrocosm.[64] Timaeus gives a detailed presentation of the shaping of the various parts of our bodies and the ideal Forms they embody. He also gives an overview of various diseases, whose cure flows from restoring the original harmony of body and soul, for "we are plants grown not from the earth but from heaven."[65]

Timaeus presents the young gods applying to the human body a mathematical-musical harmonization based on the same patterns the demiurge used to shape the heavens. For example, to address the whole process of respiration, Timaeus offers a detailed physical model that he calls a "fish-trap," structured like the external shape of the body (figure 1.4).[66] The fish-trap resembles a common lobster pot and offers a physical structure capable of the funneling and filtering Timaeus considered important for respiration. By using more complicated figures than the elemental solids, Timaeus applied geometry to provide physical structures capable of accomplishing physiological tasks. This unification of structural geometry with anatomic function remained a preoccupation for later anatomists as well.

According to Timaeus, we need to consider the harmonious ordering of the whole body and soul in order to redress any pathological imbalances. Those whose soul is more powerful than their body need to exercise the body, and vice versa for those in whom the body predominates, lest they fall victim to "the gravest disease of all: stupidity," *amathia*, the inability to learn.[67] Rather than drugs (which affect the body through external means), we should seek the curative motions that have their source in ourselves, for "this is the motion that bears the greatest kinship to understanding and to the motion of the universe."[68] More generally, Plato called for the philosopher to act as a physician especially for the human commonwealth, which (in his *Republic*) he treated as if it were a single individual, more or less healthy, as shown in its political life. As part of this supreme medical and political art, Socrates emphasized that "styles of music are nowhere disturbed without disturbing the most important laws and customs of public order."[69] Thus, music had paramount importance for the health of both the individual and the city.

2

The Controversial Project

Despite the power and compelling beauty of Plato's arguments, his Pythagorean project to understand the cosmos and living things in terms of music and mathematics became highly controversial. His student Aristotle argued that neither physics nor biology should be understood in mathematical terms. Aristotle's writings exerted enormous influence, including the first treatises on biology (as we now call it) as well as physics. Nonetheless, some of his successors began to revive the Pythagorean project. Born about the time Aristotle died, Herophilus applied ratios to describe the rhythms of the pulse in musical terms, a provocative idea that will reverberate throughout this book.

As the son of a physician, Aristotle grew up with medicine. Although he did not enter that profession, he remained interested in animals after he began (at age eighteen) his twenty years of study in Plato's Academy. Aristotle agreed with Plato that the principles of medicine ought to come from general philosophical principles. Yet Aristotle was probably more versed in dissection and practical studies of animals, to which he devoted important works. Most of all, Aristotle differed by considering mathematics—including music—to have no relevance to the study of *physis*, the realm of growth and change that includes what we now call physics and biology, studies that have their origin in Aristotle's *Physics* and his writings about animals. He argued that "perceptible things require perceptible principles, eternal things eternal principles, corruptible things corruptible principles: and, in general, every subject matter principles homogeneous with itself."[1] Thus, it made sense for astronomy to invoke eternal principles, including numbers, because it dealt with the eternal and incorruptible heavens. There, the numerical measures seem evident: the number of days between various recurrent positions of the planets, their oppositions and conjunctions in the sky.

Aristotle devoted considerable attention to the Pythagoreans, who "devoted themselves to mathematics; they were the first to advance this study, and having been brought up in it they thought its principles were the principles of all things . . . because the attributes of numbers are present in a musical scale and in the heavens and in many other things."[2] Nevertheless, he argued that one cannot "construct natural bodies out of numbers, things that have lightness and weight out of things that have not weight or lightness," for "the phenomena show that nature is not a series of episodes, like a bad tragedy" lacking all intelligible coherence.[3]

Further, we see around us matter moving and changing, including animals and plants growing and decaying. How could these things be governed by numbers, which never grow or die? Nor are numbers available to our senses that would describe earthly phenomena, compared to the numbers describing musical ratios or the planets' regular courses. Thus, for Aristotle to speak of "mathematical physics"—even less of "mathematical biology"—was a contradiction in terms. His view remained prevalent until about 1600, when natural philosophy began its search for a mathematical physics, drawing inspiration from Plato's mathematical and musical vision of the cosmos, which had remained a powerful alternative account. Indeed, the new natural philosophy required a new kind of mathematics, adequate to dealing with changing quantities, as Galileo, Kepler, and Newton gradually discovered. The life sciences, however, remained apart from this new mathematical stream until the growth of biophysics and mathematical biology, which developed far later than mathematical physics itself. The possibilities of a mathematical understanding of biology, compared to physics, remain to be fully explored.

Aristotle's students opened a new realm of observation and speculation in their collection of *Problems* when they included the first description of what we call "sympathetic vibration": they asked why if you pluck a string and then dampen it, you can hear a faint resonance in another string an octave higher. They began with a general surmise that this happens "because the sound from this string is particularly close in nature to that [original] note, because it is concordant with it" and thus somehow can become an "answering-song [*antōidē*]" to it. Then too, they speculated that the original string, while fading, might somehow also sound the note an octave higher, which then leaves that pitch as an echo picked up by the other string.[4] This phenomenon of sympathetic vibration, its possible meanings and causes, will come back at many points.

Aristotle emphasized that "every realm of nature is marvelous . . . so we should venture on the study of every kind of animal without distaste; for each and all will reveal to us something natural and something beautiful," namely, the "conduciveness of everything to an end," the purposiveness of every part.[5] His emphasis on detailed empirical investigation deeply affected subsequent developments in biology and medicine. When Greek medicine traveled to Alexandria in the third century BCE, a century after Aristotle, it brought its rationalistic tradition and built on it. E. M. Forster's observation that Alexandria, "then as now, . . . belonged not so much to Egypt as to the Mediterranean" applied to Greek medicine in those early days of Alexander's new city, in which "the Greek community remained remarkably insulated from the Egyptian population."[6] Indeed, native Egyptians were for long legally excluded from holding Alexandrian citizenship.[7] Yet for a short time in Alexandria, dissection of the human body was permitted as part of medical training, perhaps as a result of contact with Egyptian practices of mummification and embalming, which involved removing the internal organs.[8]

Into this complex milieu entered Herophilus, born in provincial Chalcedon. Educated in Cos, long a center of medical training associated with the Hippocratics, he studied with Praxagoras, who had discovered that arteries pulsate though veins do not.[9] In Alexandria, Herophilus made important anatomical discoveries for which Andreas Vesalius called him "prince of anatomists."[10] He described what he called a "winepress" (*lenos*) in the skull (figure 2.1, the confluence of sinuses later anatomists called *torcular Herophili*, "Herophilus's winepress") and the ventricles of the brain, as well as the coats of the eye, the structure of the spermatic ducts and fallopian tubes.[11] This "winepress" seems to follow the example of Timaeus's fish-trap by giving a homely but also geometrical description of an important anatomical feature. In this case, Herophilus modeled the distribution of blood among the ventricles of the brain by analogy with a winepress diverting wine into various directions. As with Timaeus's fish-trap, Herophilus sought not just a visual

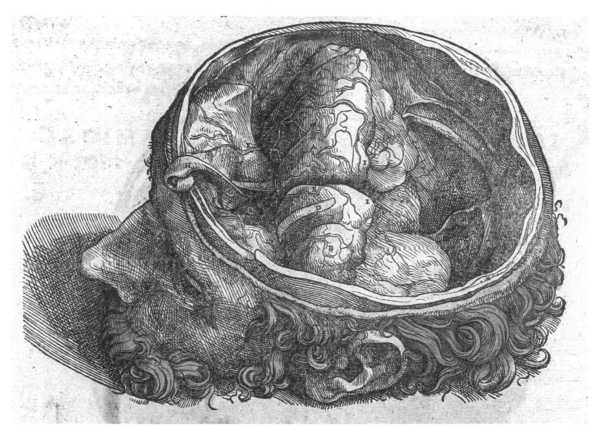

Figure 2.1
Torcular Herophili, "Herophilus's winepress," the confluence of sinuses inside the skull, as depicted in Andreas Vesalius, *De humani corporis fabrica* (*On the Fabric of the Human Body*, 1543).

resemblance but also a model for the underlying engineering of the structure, whose geometry underlies its functionality.

Herophilus may also have followed Plato's radical ideas advocating the education of women. According to one account, Herophilus was the first physician to give a medical apprenticeship to a woman, Hagnodice, who had disguised herself as a man to study with him (perhaps motivated by his anatomical work on the reproductive system). No evidence remains whether Herophilus knew his apprentice's secret. She went on to attend Athenian women in childbirth. Later, accused of having abused her female patients, she revealed her sex to the court to refute the presumption that she was a man and received so much support from her patients that "the Athenians amended the law so that free-born women could learn the science of medicine."[12]

Herophilus took Praxagoras's discoveries about the pulse much further. First, he argued that the arteries did not have an innate faculty of pulsation; instead, as Aristotle had suggested, the power (*dynamis*) that enables the arteries to dilate and contract flows from the heart. Second, Herophilus differentiated pulse from tremor, spasm, and palpitation, whereas Praxagoras considered all four of these arterial motions as essentially similar, differing only in size.[13] Third, to make these distinctions concrete and quantitative, Herophilus applied musical concepts to describe the pulse. Scholars have speculated that he derived his conception of rhythm as "a motion which has a defined regulation in time" from the musical writings of his contemporary, Aristoxenus.[14] Though Herophilus's own texts have not survived, Galen noted that

just as musicians establish rhythms according to certain defined sequences of time-units, comparing the upbeat [*arsis*] and the downbeat [*thesis*] with each other, so too Herophilus supposes that the dilation of the artery is analogous to the upbeat, while the contraction is analogous to the downbeat. He made his observations having started from the newborn child, and, on the supposition that a primary perceptible time-unit is that in which he usually found the artery [of a newborn] dilating, he says that the time-unit of the contraction is also equal to it.[15]

Throughout, he treated pulse as a purely tactile phenomenon (compared to the aural hearing of the pulse we shall consider later). Though it was inaudible, Herophilus described it as musical rhythm, in terms of various kinds of beats, as Galen recounted his views that "the rhythm of the pulse is divided into two time-units in all: (i) a unit of perceptible motion, when the dilating artery beats against our touch; and, (ii) for the rest, the entire time-unit composed of the 'external' pause and of the contraction after this pause, and then again, of the pause following upon this contraction, and of the first parts of the dilation—which themselves, too, are imperceptible."[16]

Thus, Herophilus tracked nuances of pulsation and pause, giving what may be the earliest description of a "rest," many centuries before any musical notation developed capable of recording such measured silences between sounds. He also divided "the pulse into 'beat' [*plēgē*] and 'interval' [*dialeimma*], locating frequency and infrequency in the amount of time of the interval" between beats, "just as they locate speed and slowness in that of the beat."[17] These sophisticated distinctions between speed and frequency had no precedent other than their musical context; many centuries would pass before these concepts would be further elucidated.

In this passage, Herophilus specifically addressed "those to whom the contraction of the artery is imperceptible," which would include many later physicians, who thought that the pulse was only the perceptible dilation of the artery, felt under the fingertips, not the contraction, which they considered imperceptible. Presumably Herophilus's sense of touch was so refined that he could feel a contraction that might escape observation even by another physician; much later, the development of auscultation and the stethoscope amplified these subtle features and made them audible. Over the intervening fourteen centuries, though, everything depended on the physician's fine sense of touch, developed through practice but taught in terms of Herophilus's musical explanations, as Galen passed them down. As we will discuss in chapter 9, the audible hearing of these rhythms (as opposed to feeling them by touch) came only in the nineteenth century.

Herophilus also recognized rhythmic differences in the pulse at different stages of life. To begin,

the pulse of newborn children, then, is completely short and not distinct in its contraction and dilation. Herophilus says this pulse is constituted "without definable ratios" [*a-logos*, irrational]. He calls the pulse which is without a relation to some ratio [*ana-logia*, by analogy] a pulse "without definable ratios," for it has neither a double ratio, nor a ratio of one and a half to one, nor any other proportion [*logos*], but rather is completely short and we observe it to be similar in size to the prick of a needle.[18]

These considerations of ratio emerged directly from established Greek mathematical definitions, in which the unit was not a number and 1:1 was not a ratio, properly speaking (see box 2.1).[19] Hence, Herophilus consistently concluded that the equal pulsation of the newborn was "irrational" in the sense of not having a properly defined ratio. In Greek music, similar arguments excluded the unison (1:1) from being a proper musical interval, compared to ratios such as the octave (2:1) or fifth (3:2) (see box 2.2).

Box 2.1
Greek Concepts of Number and Ratio

In contrast to modern ideas, Greek mathematics consistently treated the unit as the basis of number, though it was not itself a number, which Euclid (an older contemporary of Herophilus and a fellow Alexandrian) defined in his *Elements* as "a *multitude* composed of units." Here, Euclid was recording a long-established consensus: the unit, as monad or the One, was the transcendent basis on which numbers were founded, not simply a number among other numbers. Likewise, Euclid defined a ratio as "a sort of relation in respect of size between two magnitudes of the same kind" so that magnitudes capable of having a ratio to each other "are capable, when multiplied, of exceeding one another." But units related as 1:1 can never exceed each other, thus cannot have a ratio, strictly speaking.

Box 2.2
Musical Intervals and Ratios

> As the Pythagoreans discovered (see box 1.1), the most consonant musical intervals cor-
> responded to simple whole number ratios, such as the octave (2:1), fifth (3:2), and fourth
> (4:3). Greek theorists generally excluded the unison (1:1) from this list on the grounds that
> it was not really an interval at all, the two notes being the same but also because (as box 2.1
> showed) 1:1 is not really a proper ratio.

Accordingly, the newborn pulse corresponds to "the rhythm of a short-syllabled metri-
cal foot . . . consisting of two [short] time-units," later symbolized by the notation | ˘ ˘ | (in
which ˘ denotes a short syllable). This "pyrrhic foot" | ˘ ˘ | did not fit into the established
poetic meters, which required a definite ratio between the successive syllables. Its name
came from a fiery war dance (*pyrrikhē*) in which Spartan children practiced the quick
movements of battle that (it is said) Achilles danced around the pyre of his beloved Patro-
clos.[20] For his part, Edgar Allan Poe thought that the "purely chimerical" character of "so
perplexing a nonentity as a foot of two short syllables, affords, perhaps, the best evidence
of the gross irrationality and subservience to authority which characterize our Prosody."[21]
Thus, poets as well as Herophilus were troubled (in different ways) by the "irrationality" of this rhythm, found in such lines as Andrew Marvell's "*Tŏ ă green thought ĭn ă
green shade.*"[22] Here and in the following examples, we should bear in mind that ancient
Greek poetry (like verse in Latin and other Romance languages) was built on the quantity
(length) of syllables, not on the stress accent so prominent in English and absent in the
ancient languages.[23]

Listening to the recorded sound of a newborn's pulse (♪ sound example 2.1), one hears
the near-equality and rapidity of its beats, as Herophilus described them. He further
observed that as a child grows, the pulse "increases with respect to ratio, getting a dilation which is proportionately more extended than the contraction," making a trochaic
rhythm | ¯ ˘ |, "holding its dilation for two time beats [¯] but its contraction for one [˘]."
(Recall William Blake's trochaic line "*Tȳgĕr, Tȳgĕr, burning bright.*") He described this
rhythm as comprising three metrical beats (compared to the newborn's two) and emphasizes the musical and numerical ratio 2:1. Herophilus then noted that "the pulse of those
in the prime of life is equal in both [dilation and contraction]," the foot called spondaic
|¯ ¯| comprising four beats, each equal but twice as long as the newborn's pulse unit (as
in Gerard Manley Hopkins's line "With swĭft, slōw; swēet, sōur; adāzzle, dīm"). Finally,
"those who are beyond their prime, and almost old" have a pulse that reverses the young
person's trochaic rhythm to become the iambic | ˘ ¯ |, familiar as the basis of English blank
verse (as in Shakespeare's line "Sŏ lōng lĭves thīs, ănd thīs gĭves lĭfe tŏ thēe").[24] In the
pulses of certain old people, Herophilus even noted a contraction ten times longer than
the dilation.[25]

Herophilus also thought that those who are ill have pulses "contrary to nature" so that "great deviations from the natural rhythms . . . signify great harm."[26] Besides five primary kinds of pulse, he described several kinds of abnormal pulses with names reflecting their rhythm, such as "double-hammering" or "dropped."[27] His descriptions became renowned not least for their memorable inventiveness. Thus, "gazelle-like" pulse "consists of uneven beats in one dilation," while "quivering pulse" draws its name from the motion of a membrane "placed around the holes in flutes and a musician then breathes into the flutes" so that the membrane quivers in turn, vibrations he then connects to the uneven motion of the artery. Finally, "mouse-tailed" pulse tends to taper off.[28]

Further, Herophilus "gave the opinion that a person has a fever whenever his pulse becomes more frequent, bigger, and stronger, [and is] accompanied by a high internal temperature. So, if the pulse loses its strength and magnitude, [it is] because the fever is getting [some] relief." As with the ratios he found in pulse rhythms, Herophilus applied number to measure fever, but not through measurement of the temperature, long before any kind of thermometer was developed. Instead,

he constructed a water clock capable of containing a specified amount for the natural pulses of each age. And, upon entering to visit a patient, he would set up his water clock and feel the pulse of the person suffering from fever. By as much as the movements of the pulse exceeded the number that is natural for filling of the water clock, by that much he declared the [patient's] pulse too frequent— that is, that [the patient] had more or less of a fever.[29]

As G. E. R. Lloyd notes, this shows that Herophilus had "the evident *ambition* to make the inquiry an exact one, to construct pulse theory on the model of music, the successful mathematization of harmonics. If the main concords are expressible in terms of simple numerical relationships, why not also the main ratios between the dilations and contractions of the arteries?"[30]

This emphasis on quantitative and numerical measure also carried over into the administration of drugs, whose dosage previously had not been quantitatively exact. Those who followed Herophilus (as well as the Empiricists mentioned above) "became meticulous in their specification of measures—ounce, drachm, obol, ladle, cup—in their prescriptions. . . . Historically, his advance on these predecessors [Praxagoras, Aristotle, the Hippocratic authors] was a stunning achievement," Heinrich von Staden judged, "while Erasistratus's understanding of the heart, and in particular of its intake and outlet valves, was superior to Herophilus's, it was Herophilus's theory of how the music in our arteries is made that became a model for almost all subsequent ancient writers on pulse-lore."[31] As we shall see, that musical approach to cardiology also had a significant following among modern as well as ancient thinkers.

3

Musica humana

Even after Alexandria was eclipsed by the Roman Empire, Herophilus's reputation remained high. Largely through Galen's advocacy, Herophilus's ideas about musical ratios of the pulse came into wider circulation. Writing about music, Aristides Quintilianus reflected the widespread influence of Herophilus, as well as describing musical therapy and its various modes of operation. In the shadow of the fall of Rome, Boethius transmitted the concept of *musica humana* to later Muslim and Christian writers, providing the cornerstone of "liberal education" as well as a basis for further speculation and investigation among medical practitioners. Around the turn of the first millennium, Christian and Muslim scholars began to reconsider the ancient ideas about music and pulse.

Writing in the first century CE, the natural historian Pliny the Younger considered Herophilus to have been "an oracle of medicine" for having advanced his rhythmic theory of pulse, which still "regulates the steering of our life" even though Pliny thought it had been "abandoned on account of its excessive subtlety."[1] Pliny also remarked that this "school [of medicine], too, was abandoned since its members had to have a literary learning," presumably in order to understand the musical theory that Herophilus used.[2] It is not clear why this was so; perhaps at that point in Roman history, physicians lacked the kind of "literary" education in the quadrivium that Herophilus and his Alexandrian contemporaries enjoyed. Then too, Pliny seemed unaware of some schools of medicine that continued to use Herophilus's pulse theories.[3]

To some extent, this changed in the following centuries, particularly through the extraordinary career of Galen, a Greek born in Pergamon who moved to Rome in the second century CE and became the personal physician to several emperors. Among them, Marcus Aurelius described him as "first among physicians and unique among philosophers," for Galen was both. He transmitted and augmented Hippocratic medicine into an authoritative and extensive corpus, so vast that Galen's extant 3 million words represent perhaps only a third of his works, arguably the largest single body of writings from the ancient world. Overall, Galen consolidated the Hippocratic physiology based on the four humors.[4] Through their authority, Galen's teachings dominated Western medicine for thirteen centuries. In his estimation, Herophilus "attained the highest degree of

accuracy in things which became known by dissection, and he obtained the greater part
of his new knowledge not, like the majority [of physicians], from irrational animals, but
from human beings themselves."[5] Further, Galen judged that "Herophilus is a man who
is known by everybody to have surpassed the great majority of the ancients, not only in
width of knowledge but in intellect, and to have advanced the art of medicine in many
ways; as, for instance, by his rational account [*logos*] of the pulsation of arteries, which
one needs more now and finds more useful than any other *logos*, for deriving benefit
therefrom."[6] In comparison to other physicians, Galen thought that Herophilus "alone
understands and the shadows flit around."[7] Indeed, according to Lynn Thorndike, "Galen
estimates that the chief progress made in medical prognostication since Hippocrates is the
gradual development of the art of inferring from the pulse."[8]

Not content merely to respect Herophilus, Galen probed the exact meaning of his various ratios and "just what he means by 'rhythm': is it the ratio of the time of dilation
[*diastolē*] to the time of contraction [*systolē*] only, or does he also attribute to 'rhythm'
the time of the pause which follows upon each of the two motions? This is why there is
no agreement, not even among those who are named 'Herophileans' after him, concerning just what Herophilus really thought about rhythms."[9] In a sense, medicine remains
concerned with these issues to this day: the ratio of systole to diastole remains a central
index of cardiac function, still considered a "vital sign," now stated as the ratio of systolic
(heart contraction) to diastolic (heart resting) blood pressures in the arteries. Then too,
Galen questioned what Herophilus's followers meant by the "size" of a pulse. Compared
to an adult, a child's pulse might be "small" yet be fully adequate and "good-sized" for the
child's age.[10] Galen noted that pulse was always relative to the circumference of the arteries, whereas the primary perceptible time unit was taken as an absolute unit. This was an
important early observation of the relativity of clinical signs with respect to the patient's
age and body, perhaps the first case in which such relativity had been noted because the
quantitative characterization of the pulse allowed (and even invited) such comparisons.

For his part, though Galen rebuked those whom he thought did not pay enough attention to Herophilus, he also asked how those ideas might be put to clinical test. At times,
Galen seemed to want even more from Herophilus, criticizing him because he "omits
what someone who is educated in a manner worthy of the art should have learned from
the musicians, and he discourses with them as though they understand this, taking from
them [the musicians] what is useful for the art of medicine. But whenever they give an
exposition of this 'amazing scientific system' [*technologia*]—as they call it—concerning
the rhythms, they are no longer interested in demonstrating how a person could become
competent at prognosticating or interpreting something through them."[11] Besides his several works on the pulse, Galen even devoted an entire treatise (now lost) to Herophilus's
work. Compared to the controversies and confusions he noted among his contemporaries,
Galen wanted to draw much more diagnostic information from Herophilus's "technology." As we will see, several important developments in subsequent medicine came to

address Galen's wish. Neither he nor Herophilus had a definite view about the exact role of the heart in causing circulation between arteries and veins, though they seemed to understand that some "power" flowed from the heart to cause arterial pulsation. Galen himself demonstrated that blood (not air or breath, *pneuma*) flowed through the arteries, correcting his predecessors, including Herophilus and Erisistratus.

In the process of summarizing and extending Herophilus's work, Galen distinguished at least twenty-seven main types of pulse, whose differentiation required great rhythmic sensitivity and whose rate measurement would seem to require precise timekeeping.[12] He took Herophilus's ratios even further, going past double, triple, and quadruple proportions to include pulse ratios such as 5:2, 7:2, 9:2, 11:2, "and so on." These findings were problematic. First, how could Galen have measured these ratios with anything like their stated precision? Did he use devices such as the water clocks Herophilus used? Even using such devices, these ratios seem exotic because they go past the boundaries of what Herophilus (and indeed ancient music theorists in general) considered to be proper musical intervals, which rely on simple ratios such as 1:2, 2:3, 3:4 for consonances. Galen's exotic ratios were thus highly dissonant, which he may have considered important. He considered the movements of the pulse to be circular, comparable to the motions of the heavenly bodies and hence also describable through numerical proportions.[13] Complex and dissonant intervals would then have corresponded to pathological conditions. In the spirit of Herophilus, Galen may well have wondered whether precise pathologies could be linked to these abnormal pulse ratios. Though his extant works do not broach this question, it recurs several times throughout this book.

Later Roman writers connected Herophilus both with the great healers of the past (such as the semidivine Aesculapius) as well as with Pythagoras. For instance, the third-century rhetorician Censorinus wrote that

before he went to sleep and when he was awakened, Pythagoras used to sing to the lyre, they say, in order always to fill his soul with divinity. And the physician Asclepiades restored the minds of people suffering from delirium [*phrenitis*]—minds agitated by disease—to their own nature through musical harmony. But Herophilus, a practitioner of the same art, says the pulsations of the blood vessels move in musical rhythms. If, therefore, there is a harmony in the movement of both the body and the soul, then doubtless music is not alien to the days of our birth.[14]

The linkage of these three figures indicates the persistence of the Pythagorean strain in medicine, carried forward by Herophilus and his followers. By the turn of the third to fourth centuries CE, in his book *De communi mathematica scientia* (*On the Common Mathematical Science*), Iamblichus argued for the application of the Pythagorean approach to all natural phenomena, not only those in the heavens: "It is also the custom of mathematics sometimes to attack mathematically perceptible things as well, such as the four elements by using geometry or arithmetic or harmonics [*harmonikōs*], and similarly with other matters." Iamblichus clearly had in mind the way Plato's demiurge looked to mathematical forms in shaping the created world, so that "since mathematics is prior in nature, and is derived from principles that are prior to those of natural objects, for that reason it constructs its demonstrative syllogisms from causes that are prior. . . . Thus I think we attack mathematically everything in nature and in the world of coming-to-be."[15] The universality of his argument included biological phenomena no less than the motion of physical objects in general. To be sure, Iamblichus's neo-Platonic vision of thoroughly mathematized sciences had to wait more than a millennium for the "new philosophy" in physics at the hands of Galileo and Kepler and longer still for the mathematical biology of the twentieth century.

Roughly contemporary with Iamblichus, Aristides Quintilianus's survey *De musica* (*On Music*) showed that the connection of music, mathematics, and medicine had become familiar and well established.[16] Aristides noted that "medicine also describes everything in terms of numbers—attacks of palpitations, for instance, or the proportions involved in periodic kinds of disease," those with recurrent crises such as malaria.[17] Aristides made explicit that the musical ratios involved in these diseases "are proportionate to concordant ratios, those occurring every day being proportionate to duple [2:1 ratio], those which occur on the third day to the hemiolic [3:2], and those appearing every fourth day to the epitritic [4:3]," using the familiar terminology for ratios used in musical theory (see box 1.1). Note that some of these ratios (such as 4:3) are more specific to consonant pitches in melody, while others (like 1:2 or 2:3) are common to both melodic and rhythmic theory— for instance, to the rhythms used in poetry. As indicated by their consonant ratios, these diseases "are not necessarily dangerous. Others are serious, but have certain similarities to the first group (for example, those of the semitertians [having strong crises on alternating days with weak crises] that have affinities with those ones), and these are dangerous, but offer some little grounds for hope. Others again are altogether discordant (continuous ones [having no perceptible rhythm of crises], for example), and these are deadly and to be feared," their lack of musical and mathematical structure reflecting the gravity of the illness and the resulting disorder in the body.[18]

In general, Aristides gave an important exposition of rhythm, including its effects on the soul as well as the arteries. He defined rhythm (*rhythmos*) as "a *systēma* of durations put together in some kind of order," ranging from "bodies that do not move, as when we speak of a statue having a 'good rhythm' [*eurhythmos*], to anything that moves, as when we speak of someone walking with 'good rhythm.'"[19] Melody by itself may be "obscure and confuse the mind: it is the elements of rhythm that make clear the character of the melody, moving the mind part by part, but in an ordered way."[20] Besides seeing and hearing, rhythm can also be felt by "touch, by which we perceive, for instance, the pulsations of the arteries."[21] He engaged with Herophilus's theory by noting that rhythms

that remain within a single genus [of rhythms] are less disturbing, while those that modulate to others pull the soul violently in opposite directions, forcing it through their multiplicity to follow and assimilate itself to every variation. Thus it is also true that those movements of the arteries which keep the same [rhythmic] form, though varying slightly in their durations [*chronoi*], are disordered but not dangerous, while those that alter too much in their durations, or go so far as to change from one genus [of rhythms] to another, are frightening and deadly.[22]

In explanation, Aristides noted that "medical men . . . say that the most fundamental parts of the body, which are proportional to the natural masses and do most to hold the body together, the parts in which, so they tell us, even a slight alteration puts the creature at risk, are membranes and tubular vessels: and these are nothing other than sinewy films like spiders' webs shaped into pipes, enclosing breath inside them," here seeming to adhere to the pre-Galenic view that the arteries contain breath (*pneuma*) as well as (or even instead of) blood.[23] Through these pipes, "the soul, not the body, is set in motion, stretching out as they expand, subsiding as they contract. They demonstrate this also from the beats of the pulse, whose orderly movement they identify with the creature's health, and whose disordered and chaotic movement they diagnose as a warning of death: and they maintain that a complete cessation of its movement indicates that the soul has finally departed."[24] Conversely, "in the movements of the pulse the healthiest people are those in whom contraction and dilation answer to one another through movements of these kinds," namely, ratios having "purity of their numbers and perfection of their ratio."[25]

Aristides goes on to use the phenomenon of sympathetic vibration to explain why and how the soul is affected by music. After Aristotle's students first described it, others returned to that phenomenon and broadened its application by remarking that a plucked string, when damped, could excite sympathetic vibrations in another string "in concord" with the first, not necessarily an octave higher (as Aristotle's students had originally specified).[26] For his part, Aristides noted that "if on one of two strings *tuned in unison* you place a light little piece of straw, and then strike the other, which is strung at some distance from it, you will see that the one which carries the straw is quite visibly set in motion."[27] His use of unison evoked the closest possible similarity (and hence sympathy) between the strings; the "little piece of straw" drew attention to the physicality of the phenomenon, much more than the early Aristotelian description of a disembodied "answering-song" or "echo."

Thus, even bits of straw can respond to vibrations, not just strings, so that potentially all matter could vibrate sympathetically. For Aristides, these experiments with strings implied that

it is therefore not surprising that the soul, having acquired through its nature a body similar to things that set instruments in motion, which are sinews and breath, is moved with them when they are moved that when a breath gives out a melodic and rhythmic sound it is affected with it, in the breath which it itself contains; and that when a sinew is harmoniously [*enharmoniōs*] plucked it responds with a similar sound and tension in its own sinews. . . . The divine craftsmanship, it appears, has a marvelous skill to bring things about and to act, even through inanimate objects. How much greater must be the power of similarity to cause activity in things that are moved by soul?[28]

For Aristides, because it shakes the body and hence also the soul, music's medicinal effect comes directly from its physicality, not from ratios as disembodied principles.

Further, Aristides extended the application of musical ratios by arguing that the step from qualitative to quantitative drug prescription took place under the influence of mathematical and musical concepts because medicine cannot "contrive the qualities and capacities of its drugs except through the proportions of the quantities used."[29] This relation goes both ways, from medicine back to music: "Just as in the case of medical drugs, no one substance has the natural capacity to cure the afflictions of the body, but full recovery is brought about by a mixture of several, so also in our own field melody alone makes only a slight contribution toward putting things right, while a complete combination of all the elements is fully sufficient."[30] By Aristides's time, these ideas had become so familiar that he referred to them as well-known lore. These concepts uniting medicine, music, and mathematics had gone from being the peculiar vision of the Pythagoreans to entering the wide stream of general awareness and practice.

Not only should musical ratios govern medical prescriptions, but Aristides considered music itself to be a potent drug: "Just as one and the same drug applied to the same kind of complaint in several bodies does not always work in the same way, depending on the slightness or severity of the condition, but cures some more quickly, others more slowly, so music too arouses those more open to its influence immediately, but takes longer to capture the less susceptible."[31] In particular, "no cure could be found through reason alone" for those suffering (for example) from illnesses induced by extreme grief or those experiencing "inspired ecstasies," hence prey to "superstition and irrational fears."[32] Yet "for each of these there was a fitting style of treatment through music, which brought the sufferer gradually, and without his knowledge, into a proper condition," showing the power of music to act beyond the ability of reason. This is possible because "a person who is under the moderate influence of any of these emotions makes music of his own accord, while one who has succumbed to untempered emotion may be taught through his hearing. For a soul subject to excesses of disorder cannot be benefited except through the means by which it acts itself when it is affected only moderately."[33]

In that vein, Aristides argued that we should speak of music as we speak of medicine because we owe to both our good health as well as recovery from illness, both music and medicine taking us from "idleness and self-indulgence into our original condition" of physical and mental harmony.[34] Above all, "music is a treatment for the passions of the soul" for which Aristides offered a useful categorization. Such therapy can act through "amelioration, which we use to diminish little by little an emotion that we cannot persuade all at once, to the point where it is no longer felt: the other is eradication, used when we achieve a complete transformation from the start."[35] Whether acting gradually or suddenly, this therapy follows a well-established Hippocratic maxim that "the physician should treat disease by the principle of opposition to the cause of the disease."[36] To effect this, Aristides outlined three genera of rhythmic or melodic composition: "expansive"

or "exalting [*diastaltikos*, diastalic]," "contractive" or "depressing [*systaltikos*, sysaltic],"
and "peaceful [*hesychastikos*, hesychastic]."[37] Each of these terms deserves a closer look.
Two of them have close links with Herophilus: in the context of music, *diastaltikos* means
excited or exalted or even expansive (from a verb meaning separate, "divide," sometimes
open or expand) while *diastolē* refers to the dilation or expansion of the pulse. Contrariwise, *systaltikos* means depressed, and *systolē* means contraction.

   Standing apart from these two opposites, peaceful (*hesychastic*) denotes no specific
emotional affect but, on the contrary, a cessation from them. The word *hesychia* meant not
only peacefulness but a profoundly quiet and contemplative state; for instance, the apostle
Paul used this word to describe "the peace [*hesychia*] that passes all understanding." Long
before it became associated with the Hesychasts (Eastern Christian mystics who said the
Jesus Prayer to commune with the divine), this word already had a strong and specific
association with the Pythagoreans. For instance, in the Roman satirist Lucian's *Philosophy for Sale*, representatives of various philosophical schools hawked their respective
views. Among them, a Pythagorean described how he would teach his prospective buyer
music and geometry through a process of "cleansing [*katharsis*]" involving "long silence
[*hesychia*] and speechlessness [*aphōniē*], and for five entire years no word or talk," a
religious silence evidently meant to encourage thoughtfulness and meditation, here comi-
cally exaggerated.[38]

   These widely used terms described ways in which music could affect body and psyche.
Writing about the same time as Aristides, the music theorist Cleonides characterized the
expansive (diastaltic) as suitable for "heroic deeds, the grandeur and loftiness of a manly
soul, and an affection akin to these. It is most used in tragedy" and similarly serious genres.
Contractive (systaltic) music brings the soul "into dejection and an effeminate condition.
Such a state will fit with erotic affections, dirges, expressions of pity," and the like. In contrast, the peaceful (hesychastic) brings about "quietude of soul and a liberal and peaceful
state," fitting for "hymns, paeans, encomia, counsels, and things similar to these."[39]

   These terms fall into two distinct groups: the contrasting pair of expansive (diastaltic)
and contractive (systaltic) describe a spectrum of emotional affects, whereas peaceful
(hesychastic) does not denote any particular emotional affect but rather the transcendence
of all such affects. This corresponds to an important (and longstanding) division in music
itself that I have described as *music as expression* (comprising expansive and contractive
music) versus *music as science* (hesychastic). As I discussed in *Polyphonic Minds*, these
two distinctly different streams of musical tradition seem to go back as far as human
music can be traced.[40] At present, music as expression is far more familiar: compare the
vast number of "happy" or "sad" songs to the alternative stream of music as science, such
as the "music of the spheres," the dispassionate, sublime concert that for Plato (and the
Pythagoreans) shaped the structure of the cosmos.

   Early Christian authors steeped in this Platonic tradition of the liberal arts also showed
deep interest in the relation between numbers and bodies. One of the young Augustine of

Hippo's earliest works was *On Music* (*De musica*).[41] In it, a master and his disciple consider what happens when they pronounce the verse *Deus creator omnium* ("God the creator of everything"). They agree that the line is governed by the numerical ratios of its four iambic feet (*Dĕūs| crĕā|tŏr ōm|nĭŭm*) composed of twelve "time units" (*tempora*), one for each short syllable (˘), two for each long (ˉ equaling ˘ ˘ in length). For Augustine, the word for "rhythm" is *number* (*numerus*), following the long tradition he learned from Aristides and underlining the close relation between rhythm (as well as pitch) and mathematics.[42] The master wants to "move on from the corporeal to the incorporeal": "Should we say that these rhythms [*numeros*] are merely in the sound which is heard, or also in the hearer's sense, which belongs to the ears, or also in the activity of the pronouncing person, or, since the verse is known, also in our memory?"[43] Yet what happens when nothing sounds, when the verse resounds in the mind alone? "Even when we are silent, it is possible for us, by thinking, to produce in ourselves some rhythms [*numeros*] during the same time-span as the one during which they would be produced with the voice. It is obvious that these rhythms exist in some kind of activity of the mind, and since this activity does not emit any sound or produce any reaction in the ears, it shows that this kind can exist without the other two, of which one is in the sound, the other in the hearer, when he hears."[44]

The master seems poised to conclude that this rhythm is created within the mind but then pauses to ask how that might be—specifically whether the rhythm "would exist without the help of memory." Yet here the master pivots away from the mind to the body: "If, however, the soul activates the rhythms [*numeros*] that we find in the pulse of our veins, then the problem is solved" because these rhythms exist in a bodily activity not dependent on memory.[45] The disciple adds that he does not "doubt that the different pulses of the veins and the intervals of breathing are produced according to the constitution [*temperatione*] of the bodies."[46] Thus, both seem to take for granted something like the numerical proportion of vein pulsations handed down from Herophilus, though they do not name him. In the ensuing dialogue, the master extends the concept of rhythm beyond the human pulse to a universal principle of biology: "For there is no kind of plant which does not, in accordance with the time-spaces that have been determined for its own kind of seed, grow strong and put out shoots and run up into the air and unfold its leaves and gain strength, and either bear fruit or give back the power of the seed with the help of the most hidden rhythms of the tree. How much more true is this not of the bodies of animals, in which the intervals of the limbs exhibit even more a rhythmical equality to the view?"[47] Ultimately, these rhythms come "from that supreme and eternal origin of rhythms," *Deus creator omnium*, God the creator of everything.[48]

These themes continued to resonate in the waning years of the Western Roman Empire. Writing in the fifth century (and drawing heavily on Aristides's work on music), Martianus Capella mentions the sack of Rome by the Visigoth Alaric in the course of his encyclopedic work, *On the Marriage of Mercury and Philology*.[49] This extravagant allegory shows the enduring appeal of the neo-Platonic and Pythagorean vision, at whose center lay music with all its connections to the other mathematical arts. Its story concerns the search of Mercury, the messenger of the gods, to find a suitable bride, who turns out to be Philology, meaning the love of letters and learning; the book culminates with an extended account of the various liberal arts (comprising the trivium and quadrivium), ending with an elaborate vision of Music (figure 3.1), in particular the polyphonic music of the spheres that rises far beyond a single melodic line into a dazzling whole. In keeping with this emphasis on the liberal arts, the practical arts of Architecture and Medicine were present at the nuptial feast but were required to keep silent in the company of divinities. Nevertheless, in the course of his narrative (filled with learned references and digressions) Martianus reminds his readers that "Herophilus used to examine the blood vessels of the ill through a comparison of their rhythms," showing the continuing interest in Herophilus's ideas.[50]

Writing in the early sixth century, just after the fall of the Western Roman Empire, Boethius was "the last of the Romans and the first of the scholastic philosophers," in the judgment of the Renaissance humanist Lorenzo Valla. Alongside his philosophical, theological, and mathematical works, Boethius also transmitted ancient musical theory to the West. His *De institutione musica* (*Fundamentals of Music*) "alone forms the historical link between the Pythagorean musical speculation of Greek antiquity and the theoretical musical speculation of the Middle Ages."[51] This book became the principal text that instructed the West in music for the next thousand years and more. Between the exalted music of the macrocosm (*musica mundana*) and the practice of ordinary musical instruments (*musica instrumentalis*), Boethius placed the music of the human being (*musica humana*), for "the whole structure of our soul and body has been joined by means of musical coalescence [*coaptatione*]. For just as one's physical state affects feeling, so also the pulses of the heart [*pulsus cordis*] are increased by disturbed states of mind," seemingly alluding to Herophilus's musical ratios in the pulse.[52] Boethius considered music the key to all questions about the integration of soul and body,

for what unites the incorporeal nature of reason with the body if not a certain harmony [*coaptatio*] and, as it were, a careful tuning of low and high pitches as though producing one consonance? What other than this unites the parts of the soul, which, according to Aristotle, is composed of the rational and the irrational? What is it that intermingles the elements of the body or holds together the parts of the body in an established order?[53]

Boethius also transmitted stories about the healing powers of music. Arguing that "no path to the mind is as open for instruction as the sense of hearing," he noted that even the severe Spartans "were so attentive to music that they thought it even took possession of minds."[54] A number of Boethius's stories became staples of musical lore, endlessly repeated by subsequent authors:

Who does not know that Pythagoras, by performing a spondee, restored a drunk adolescent of Taormina incited by the sound of the Phrygian mode to a calmer and more composed state? One night, when a whore was closeted in the house of a rival, this frenzied youth wanted to set fire to the house. Pythagoras, being a night owl [*nocturnus*], was contemplating the courses of the heavens (as

Figure 3.1
A depiction of Musica by Gherardo di Giovanni del Fora (about 1450–1470), to illustrate Martianus Capella's
Marriage of Mercury and Philology.

was his custom) when he learned that this youth, incited by the Phrygian mode, would not desist from his action in response to the many warnings of his friends; he ordered that the mode be changed, thereby tempering the disposition of the frenzied youth to a state of absolute calm.[55]

Nor was the therapeutic use of music limited to such passing mental or emotional unbalances, for he also relates that "Terpander and Arion of Methymna saved the citizens of Lesbos and Ionia from very serious illness by the aid of song. Moreover, by means of modes, Ismenias the Theban is said to have driven away all the distresses of many Boetians suffering the torments of sciatica."[56]

By the ninth century, toward the end of the "Dark Ages," Boethius's works again appeared in manuscripts, particularly in the north of the Carolingian kingdoms, which became a center of renewed interest in ancient writings. Martianus Capella's book also became "one of the half-dozen most popular and influential writers of the Middle Ages," copied in 241 medieval manuscripts and even becoming a school text, obscure and confusing as it was.[57] Martianus's assertions about Herophilus drew the attention of commentators; for instance, a certain Remigius tried to explain what the name "Erophilus" meant ("strong love," he conjectured), as well as the meaning of the rhythms of the pulse, which he understood more accurately: "For if the pulse of the blood vessels were in accordance with natural—i.e. appropriate—metrical feet, the person would be healthy."[58] Evidently, over the centuries even Herophilus's name had become garbled in the process of manuscript copying, but as these ninth-century commentators began trying to piece together what the original work said, they kept alive Herophilus's intriguing assertions so that future generations could ponder them anew.

Still, in her drama *The Conversion of the Harlot Thais*, the tenth-century Saxon canoness Hrotsvit of Gandersheim had the saintly hermit Pafnutius explain to his disciples that *musica humana* is observed "not only, as I said before, in the union of body and soul, and not only in the emission of high and low sounds, but also in the pulse of our veins and in the measures of our limbs, as in the parts of our fingers, where we find the same mathematical proportions of measure as we mentioned in harmonies, because music is not only the agreement of sounds but also that of dissimilar entities."[59] To this, the disciples object that "had we known beforehand that the solving of the knot of our question [about the nature of music] would pose such a degree of difficulty for us ignorant students, then we would have preferred not knowing anything about the microcosm [human beings] to undergoing such a difficult lesson." To address these difficulties, Pafnutius recounts how he confronted Thais with the discord of her sins "because the sickness of both body and soul must be cured by the medicine of contraries."[60] In order for her to experience those dissonances most acutely, "it follows that she must be sequestered from the tumult of the world, immured in a small cell, so that she may contemplate her sins undisturbed."

Far from these remote Carolingian outposts, Herophilus's ideas were already being studied by Arab and Persian scholars, who were largely responsible for the preservation and continued work on Greek medicine during the long period in which its writings were not available in the West. During the ninth century, translation of Greek writings began in the circle of Abū Yūsuf Ya'qūb ben Isḥāq al-Kindī, "the first philosopher in the Arabic tradition, and his thought was defined in large part by his engagement with the Greek tradition that preceded him."[61] Even so, al-Kindī's "project was not just to recover and explain Greek science but to expand upon it."[62] For instance, he took Galen's idea that drugs affect us because they are in varying degrees hot, cold, dry, or moist, which his *Formulary* elaborated into a complex theory of drug formulation built on arithmetic proportions between the degrees of heat of the various ingredients. Among these proportions, al-Kindī emphasized the relation of doubling (2:1), which he notes that musicians confirm is "the most excellent" because it corresponds to the ratio of an octave.[63] This exemplifies al-Kindī's general view in his work on *Quantity* that "the science of harmony discovers the relation or combination of one number with another, and the knowledge of what is and what is not harmonious."[64] Consequently, he put forward harmony as the most comprehensive mathematical science because it "is composed from arithmetic, geometry, and astronomy. For there is harmony in everything, though it is most obvious in sounds, and the composition of the universe and of human souls." This beautiful and daring insight responded to Plato's vision of cosmic and human music in the *Timaeus* and carried it further still.

Building on this insight, al-Kindī also took up the therapeutic uses of music. Among his five extant treatises on music, *The Informative Parts of Music* develops his thesis that "musical relationships are fundamental to a wide range of physical and psychological phenomena."[65] Al-Kindī thought that playing on the different strings of the *'ūd* (figure 3.2, then as now a widely used four-string instrument, ancestor and namesake of the lute) would evoke particular kinds of actions and bodily states in its listeners because those strings acted specifically on the four bodily humors. Thus, the highest string of the *'ūd* excites the yellow bile and hence evokes cheerfulness, the second string excites blood (aiding digestion), the third phlegm (sometimes associated with despondency and forgetfulness), and the fourth black bile (melancholy).[66] At about the same time (if not somewhat earlier), the Nestorian Christian Ḥunayn ibn Isḥaq, chief physician of the caliph in Bagdad and an important translator of Greek texts, also stated this correspondence between strings and humors, which he attributed to Archytas.[67] Thus, Ḥunayn and al-Kindī took the Greek association of humors and strings on a lyre (each string having a fixed pitch) and transposed it imaginatively to the *'ūd*, a very different kind of instrument whose strings can each play a large range of pitches (as with lutes or guitars). These Muslim and Christian scholars were evidently more concerned with preserving the essential aspects by which music can aid healing than with slavishly copying Greek usages that did not fit Arabic instruments.

An important account of al-Kindī's therapeutic procedures was recorded by his follower Ibn al-Qifṭī about a case of a paralyzed boy. After al-Kindī had diagnosed the boy's condition based on his pulse, he had four students play the *'ūd* to the patient, doubtless

Figure 3.2
An *'ūd* player shown on an Iranian architectural ornament or relief (11th–12th century CE).

using the appropriate strings indicated by al-Kindī's theories. Following a lengthy session of this music therapy, al-Kindī again took the boy's pulse to check that its music had regained its proper rhythm. The boy became fully conscious and began to move again.[68] Al-Kindī's theories and this story in particular influenced later Western practitioners, as we shall see. Within the Muslim world, the great encyclopedia compiled in the tenth century by the Brethren of Purity (Ikhwān Al-Ṣafā) also gave considerable attention to the correspondence between *'ūd* strings and humors.[69]

Probably as a result of Al-Kindī's activities, most of Galen's treatises on the pulse were available in Arabic translation by the end of the ninth century.[70] During that century, Abū Bakr Muhammad ibn Zakariyyā al-Rāzī (known in the West as Rhazes) introduced the meter ("harmony") of the pulse, following Galen, with whom he disagreed on other matters.[71] In the tenth century, the oldest extant treatise on medicine in Persian, Al-Akhawayni Bokhari's *Learner's Guide to Medicine* (*Hidayat al-Mutaallemin fi al-Tibb*), gave a lengthy description of pulse following Galen that included a description of a tremulant pulse (*pulsus tremulus*) in which "the artery vibrates like the string of a musical instrument."[72]

By 1025, Abū' Alī Ibn Sīnā, the Persian scholar known in the West as Avicenna, completed his extensive *Canon of Medicine* (*al-Qānūn fī aṭ-Ṭibb*), an overview of contemporary medical knowledge that became a standard textbook used in Europe as late as the eighteenth century. Ibn Sīnā emphasized that

> you should know that there is in the pulse a musical nature, for as the art of music is realized in the melody [*ta'līf al-niḡam*, composition of sounds] according to the relation between them as to high pitch and low, and in the rhythmical recurrence of time-intervals between the strikings [of strings], so it is with the pulse: its temporal relation in respect of swiftness [duration of movements] and frequency [duration of pauses] is a rhythmical relation, and its qualitative relation as to strength or weakness [of the impact on the finger] and size [of dilation] is a relation like that of melody. And as the time-units of rhythm and the durations of notes may be concordant or inconcordant, so the variations of the pulse may be orderly and may be disordered, and the qualitative relation of the pulse as to strength or weakness and size may be concordant or inconcordant, indeed uneven.[73]

Ibn Sīnā related that "Galen holds that the observed values of metrical proportions are in accordance with one of these musical relations mentioned," specifically the octave (2:1), fifth (3:2), fourth (4:3), third (5:4), and the compound ratio of octave plus fifth (3:1 = 2:1 × 3:2), concluding that "no further ratios can be observed." If Ibn Sīnā was aware of Galen's more exotic pulse rhythms (such as 5:2, 7:2, 9:2, 11:2), he here seemed to ignore or perhaps even reject them. Also, Ibn Sīnā phrased these ratios in terms of *melodic* intervals, not just pure rhythms, as Galen had. Some scholars suggest that Ibn Sīnā was interpreting Galen's rhythmic patterns within the music theory (*al-mūsīqī*) of his own time, developed from ancient musical theory but arguably related to contemporary practices.[74]

Ibn Sīnā already registered important questions about the meaning and observability of various pulse rhythms, even while he affirmed their importance: "I am amazed that these relations should be detected by touch; but it is easier for one who executes the progression of rhythm and the proportion of sounds in [practicing] the art, and then has the opportunity to become acquainted with music theory [*al-mūsīqī*], so as to correlate the art with the discipline. Such a man, should he turn his attention to the pulse, would be enabled to identify these relations by touch."[75] Ibn Sīnā's candid admission was only the beginning of a controversy about the "music" of the pulse that (as we shall see) would extend for centuries and eventually overlap with the beginnings of modern cardiology.

As Pythagorean teachings were transmitted from Rome to the Middle East, Herophilus's ideas about musical rhythms in the pulse spread ever wider. Galen was largely responsible for this wide transmission, but the interest of Roman and Muslim readers went far beyond mere respect for ancient authorities. Herophilus offered a radically new approach to the body based not on the qualitative balance of humors but on the hidden operation of mathematical and musical principles. This promised a whole new level of understanding of the body that thoughtful physicians and scholars did not ignore.

4

Homage to Herophilus

As they were rediscovered in the West, Greek teachings about the music of the pulse and musical healing found new resonance and new advocates. This revival of learning directed attention to ancient accounts of the power of music to affect body and psyche, power that new generations sought to use. In particular, Boethius's concept of *musica humana* and his stories of its profound effects had a wide influence on the medicinal use of music. In combination with astrology, music became a familiar component of medical interventions, advocated by Marsilio Ficino and practiced by many others.

Though certain elements of ancient ideas about music and medicine were registered in some Carolingian texts, it took centuries for these same ideas to circulate fully in the West, thanks to Muslim scholars like al-Kindī and Ibn Sīnā. This slow recovery of ancient Greek medicine shaped the various periods of medieval Western practice and theory. Between the tenth and twelfth centuries, medical practice was centered in monasteries. There, music was central to the celebration of the Mass as well as of the Divine Office, in which the entire community sang the chants and psalms prescribed for each day, in different services eight times daily (according to the Benedictine Rule). Appropriately, such omnipresent music also marked the care of the dying. For instance, after receiving last rites, a monk at Cluny was constantly surrounded by the singing of psalms.[1] Death was not solitary but shared by the community through its singing. Though little evidence survives, it would seem logical that a similar therapeutic use of music would have extended as well to those sick in the monastery's infirmary.

The writings of Hildegard von Bingen, the remarkable twelfth-century abbess, visionary, and composer, give some insight into her views on medicine and music. Among her many accomplishments, she also was a medical practitioner who left two substantial treatises on the subject, *Physica* (*Book of Simple Medicine*) and *Causae et curae* (*Causes and Cures*). In terms of the transmission of ancient medicine, she seems not to have had access to full texts by Galen or Hippocrates but instead relied on her own training and experience, along with some accounts that transmitted ancient ideas in various compilations or summaries, as well as her own speculative constructs. These emerge, for instance, when she speaks of "humors," for which she does not name the traditional foursome (blood, phlegm, yellow bile, and black bile) but uses her own individual terminology: dryness, wetness, foam, and warmth.[2] Her writings indicate that she had some knowledge of the four "qualities" as described by Galen (dry and wet, hot and cold) but reconceived them in her own way, whether because she lacked his detailed descriptions or departed from his ideas based on her own clinical experience.

Hildegard was also the most prolific composer of monophonic song in the entire Middle Ages, though she said that she did so "without being taught by anyone, since I had never studied neumes [musical notation] or any chant at all."[3] Yet during her monastic life, she surely sang a great deal of chant, just as she must have accumulated considerable medical and curative experience. Drawing both on what she did know as well as on her own speculations, in *Causes and Cures* she argued that the various humors should be related by ratios:

But the humor that is highest is greater than the one which follows by a fourth part and half of the third part. And the next should temper two parts and the rest of the third [part]. . . . And when it is this way a person is tranquil. But when some humor exceeds its measure, then there is danger.[4]

Her commentators have some difficulty interpreting what exactly these ratios meant, but one of them thought she was using "a garbled version of the cosmic proportions" from Plato's *Timaeus*, proportions "repeatedly invoked by medical writers."[5] Another thought that such ratios between humors "is unique, without historical precedent and without a clear physiologic correlation."[6] In so doing, Hildegard implicitly proposed musical ratios between the humors, ratios that (as we noted earlier) stem from the Pythagorean musical intervals. This may have been her own speculative extension of that tradition.

In the larger framework of her views, this is not surprising because she considered music essential to human activity, particularly to the religious life. In her major visionary work, *Scivias* (*Know the Ways of the Lord*, written between 1141 and 1151), song is omnipresent; she noted that "the song of rejoicing, sung in consonance and in concord, tells of the glory and honor of the citizens of Heaven, and lifts on high what the Word has shown. . . . And so the words symbolize the body, and the jubilant music indicates the spirit; and the celestial harmony shows the Divinity, and the words the Humanity of the Son of God."[7] If this music were interrupted, she thought grave physical and psychic consequences would follow. When she refused to disinter an excommunicated man who she thought had been reconciled to God, the ecclesiastical authorities had forbidden her monastery to sing the Divine Office (though still allowed to read it). Hildegard wrote to her superiors that this caused her nuns great distress because "the body is the garment of the soul and it is the soul which gives life to the voice. That is why the body must raise its voice in harmony with the soul for the praise of God." She extended this also to "all the other musical instruments that clever and industrious people have produced," noting that the devil himself tried "to bring to disharmony, and even to ban, God's praise and the beauty of spiritual songs."[8] Indeed, she considered the human body itself to be a musical instrument, "one limb fitted to another in such a way that no limb should exceed its proper proportion . . . just as God made the mighty instrument of the universe according to careful standards."[9] Here and elsewhere she seems to take what she may have heard of Plato's cosmic harmonies even further than the accounts transmitted from antiquity (to whatever extent she may have had access to them).[10]

Hildegard's understanding of the humors led her to argue that all humans, not just cloistered monastics, are endangered by melancholy if they lack music. She held that blood is the only positive humor, representing Edenic purity, whereas the other humors entered into the body as a result of sin.[11] Melancholy in particular developed "in Adam and in all his posterity" because "banishment from Eden severed him from the heavenly choirs, with whose voices he had hitherto been in harmony."[12] In this understanding, melancholy was not the result of saturnine character, planetary influence, or genius but a symptom of separation from heavenly music. Accordingly, reimmersion in that music represented not just a cure but a return to the proper human condition. Hildegard probably applied this insight whenever melancholy troubled her nuns.

The ancient opinions about the curative powers of music were also reflected by Robert Grosseteste, whom A. C. Crombie described as "the real founder of the tradition of scientific thought in medieval Oxford . . . Grosseteste united in his own work the experimental and the rational traditions of the twelfth century and he set forth a systematic theory of experimental science. He seems to have studied medicine as well as mathematics and philosophy."[13] Versed also in law, Grosseteste eventually became bishop of Lincoln. His treatise *On the Liberal Arts* (*De artibus liberalibus*, written about 1220 to 1235) noted that "they say that even wounds and deafness can be curable by music; for since the soul follows the body in its passions and the body follows its soul in its actions," knowledge of the due proportion of the elements in the human body and of the concord of soul with body can restore any lack of harmony in body or soul.[14] At the same time, Grosseteste emphasized that astronomy or astrology was the supreme science, whose aid natural philosophy needed more than any other study.[15]

By the end of the twelfth century, about forty treatises attributed to Hippocrates (and even more than that ascribed to Galen) were available in Latin. These were still incomplete; in the West, large portions of Galen's anatomical works were unknown until the fourteenth century, while his and Hippocrates's full corpus of extant works became available only in the sixteenth century.[16] During this gradual recovery of Greek medicine, ancient ideas about the music of the pulse reentered the stream of Western medical discourse; in the process, the image of Herophilus (often paired with his contemporary, the great anatomist Erisistratus) was enshrined among the pantheon of wise physicians as an inspiration for the rising generations increasingly aware of these pioneers of Greek medicine and its importance (see figure 4.1).

Figure 4.1
Detail from a woodcut by Lorenz Fries (1532) depicting Herophilus and Erisistratus as medical colleagues and friends.

As these translations and the Arabic commentaries on Galen's works began to circulate in the West, his influential endorsement of musical pulse theories naturally spurred his new readers to apply them as well.[17] Studying these texts, the scholars of Salerno in southern Italy made a new synthesis that connected medicine with *physica*, which they presented as a natural philosophy connected integrally to the liberal arts and hence to music as well as mathematics.[18] At the end of the eleventh century, Alfanus I of Salerno wrote a *Treatise on Pulse* (*Tractatus de pulsibus*) that included the musical pulse ratios.[19] These ideas then traveled farther afield, especially via Gilles of Corbeil, who studied at Salerno and then practiced at Paris, eventually becoming physician to King Philip II Augustus. Referring specifically to Alfanus, in about 1200, Gilles wrote a poem *On Pulse*

(*De pulsibus*) that versified ancient pulse lore (including its metrical rhythms) so that medical students might more easily learn these concepts, rather than spending their time poring over "the lascivious verses of Ovid."[20] Closely associated with the circle of Peter the Chanter at Notre Dame in Paris, Gilles was so involved in musical matters that he fulminated against the use of the "couplings of voices" (*copula*) in polyphony, which he compared to sodomitic couplings "against nature."[21]

Going forward, Herophilus's musical theory of pulse received careful attention in Pietro d'Abano's compendious *Conciliator* (1303), which sought to reconcile the major medical and astrological texts then in circulation. Claude Palisca noted that "as a natural scientist and medical doctor by profession, d'Abano was attracted most to the analysis of physical and medical problems" included in the Aristotelian *Problems*, on which he was the first commentator in the West.[22] Indeed, Palisca counted d'Abano among the earliest musical humanists, "the first commentator on an ancient Greek work dealing (though not exclusively) with music. . . . [He] transmitted questions and interpretations concerning acoustics that were to occupy scientists and musicians for several centuries afterwards . . . [and] gave the first impetus at the University of Padua to the scientific movement that was eventually to lead to the discoveries of Galileo."[23] Further, d'Abano gave "a vigorous defense of the proposition that music is found in pulse."[24] He even referred to contemporary polyphonic music to express the changes in human pulse music over the life cycle.[25]

Recall that Herophilus (and those who followed him) had used the terminology of poetic meters to describe how the rhythms of the pulse change over the human life cycle, from the tripping iambs of infancy to the slow spondees of old age. To present this idea, d'Abano created a new subcategory of *musica humana* that he called *musica organica*, as if to signal that he considered the music of the pulse to be a notably distinct part of more general analogies between the human microcosm and the larger cosmos. In that connection, d'Abano referred his readers to the contemporary terminology of rhythmic modes, which applied the ancient classification of poetic meters to the rhythmic patterns used in the music of his time, the many-voiced motets and conductus (see figure 4.2 and ♪ sound example 4.1).[26] Though d'Abano does not underline it, his readers probably noticed that he drew on both ancient and modern ideas to explain these connections between medicine and music.

D'Abano also translated a text on astrological medicine then (falsely) ascribed to Hippocrates.[27] D'Abano argued that the light coming from the heavens could effect changes in the four qualities—heat, cold, dryness, and moisture—and hence in the four bodily humors corresponding to them. He concluded that physicians must know the characteristics of the planets and their particular dominion over the humors in order to identify the rhythm of crises. Because "man is a perfect representation of the vault of the heavens," the same musical laws of proportion that govern the cosmos also regulate the humors in

Mode I	 ¶ ■	long-short	_ ◡	trochaic	 ♩ ♪
Mode II	■ ¶	short-long	◡ _	iambic	 ♪ ♩
Mode III	¶ ■ ■	long-short-short	_ ◡ ◡	dactylic	 ♩ ♪ ♩
Mode IV	■ ■ ¶	short-short-long	◡ ◡ _	anapestic	 ♪ ♩ ♩
Mode V	¶ ¶ ¶	long-long-long	_ _ _	spondaic	 ♩ ♩ ♩
Mode VI	■ ■ ■	short-short-short	◡ ◡ ◡	tribrachic	 ♪ ♪ ♪

Figure 4.2
The basics of rhythmic notation according to Johannes de Garlandia (about 1280). Each of the rhythmic modes is shown in contemporary notation (neumes), as well as modern (on the far right). They correspond to ancient poetic meters, such as the first mode (trochaic) and the second (iambic), illustrated in ♪ sound example 4.1, "Breves dies humanis."

the human microcosm.[28] In so arguing, d'Abano set forth a thoroughgoing analogy connecting among music, astrology, and medicine.

D'Abano's writings on the music of pulse were cited by several other medical writers throughout the fourteenth and fifteenth centuries, including Gentile da Foglino, Jacopo da Forli, and Ugo of Siena. Each of them formally endorsed the concept of musical proportions in pulse, though they devoted different amounts of attention to it—some more, others less.[29] They cited Galen's ratio of 5:2, as well as more standard (and simpler) musical ratios such as 2:1, 3:2, 4:3, and 5:4, but they also registered some perplexity, as had Galen himself. Should the ratios be considered between systole and diastole or vice versa? Should the rest between systole and diastole be counted as part of these ratios or not? Though these concerns were already present in the ancient sources, the Italian physicians went further with the musical analogy. Following d'Abano's references to the rhythms of poetry, in 1547 Jacopo da Forli compared "the circle of chances and times," meaning the fixed and recurrent patterns manifest in the pulse, to the "times of mensurated voices," explicitly referring to the rhythmic patterns of contemporary polyphonic music. Jacopo explained that "singers proceed in a circular manner, now lowering, now raising the voice, now cutting short, now drawing out the time." As if to underline the comparison, he concluded, "Concerning this, you may inquire more fully from musicians if you don't believe me."[30]

Those writing primarily about music also made connections with medicine. Likely drawing on Muslim sources circulating in his native Spain, Bartolomé Ramos de Pareia's *Practica musica* (1482) published detailed correlations of musical modes, notes, planets, muses, bodily humors, and affect or ethos and made them available to Western readers.[31] He included these correspondences in a complex diagram (figure 4.3) whose spiral design challenged visual imaginations familiar with circular representations of cosmic order. Instead of nested circles or spheres, Ramos wrapped all the planets, modes, and notes along one large spiral, as if to bring forward the ways they were unified rather than emphasizing their hierarchy or subordination. We earlier mentioned some of the ways astrology had long been used in medical practice, which were not separate from the connections between music and medicine. By correlating music to the ways in which the planets (and their associated divinities) could affect human health, Ramos made a three-way connection of astrology, music, and medicine. To take one of his examples, "The moon increases phlegm and moisture in man while the sun dries them up. In this way, these two planets, since they are foremost and light-giving, rule the first and second modes," namely, the Dorian and Hypodorian modes of contemporary practice.[32]

Though Ramos set out a theoretical framework for a thorough unification of music, medicine, and astrology, his treatise did not go further into the practical than this kind of general advice about the physiological effects of the various modes. Ramos apparently spent time during the 1470s in Florence, where he probably circulated his treatise and may even have encountered Marsilio Ficino, whose own work included similar connections joining music, medicine, and the magical arts, though no surviving evidence documents any direct interaction between them. A Florentine priest, philosopher, scholar, physician, and musician, Ficino provided the earliest translations of the whole Platonic corpus into Latin, thereby giving access to many works previously unavailable, which deeply influenced the revival of learning. Until that time, only Plato's *Timaeus* had been generally available in the West, shaping its impression of Plato as primarily a cosmological thinker who interwove music, astronomy, and medicine. Ficino's influential commentary on the *Timaeus* certainly emphasized those connections. As a physician (and the son of a physician), Ficino studied Galen closely and was also much influenced by the medical writings of al-Kindī, an example of how Muslim scholars affected the reception of Greek medicine in the West.[33] In particular, he took note of how al-Kindī used music to heal a paralyzed boy, as related in the previous chapter.[34]

His close study of Plato's works led Ficino to become the prime exponent of the therapeutic use of music. Ficino's *Three Books on Life* (*De vita libri tres*, 1489) drew also on many neo-Platonic sources that he translated for the first time at the behest of Lorenzo de' Medici, "the magnificent," patron of the Platonic Academy in Florence as well as the city's virtual ruler. Ficino's translations and commentaries were important parts of the Medici effort to station Florence as a new Athens, the center of the educated world and the revival of ancient learning.

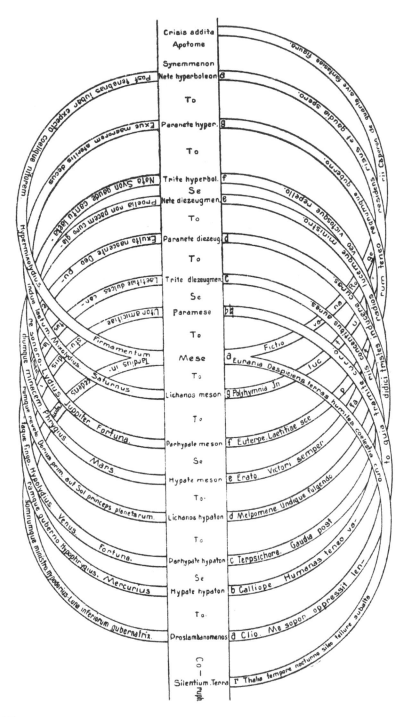

Figure 4.3
Bartolomé Ramos's diagram of the relation of modes, notes, and planets from his *Practica musica* (1482).

Primarily addressing those suffering from melancholy, Ficino noted that song is "altogether purer and more similar to the heavens than is the matter of medicine," meaning material drugs, so that music, if rightly used, could be more powerful than drugs:

Hence it is no wonder at all that by means of song certain diseases, both mental and physical, can sometimes be cured or brought on, especially since a musical spirit [*spiritus*] of this kind properly touches and acts on the spirit which is the mean between body and soul, and immediately affects both the one and the other with its influence. You will allow to the Pythagoreans and Platonists that the heavens are a spirit and that they order all things through their motions and tones.[35]

Ficino is careful to warn that we are "not to think that we are speaking here of worshipping the stars, but rather of imitating them and thereby trying to capture them" through "a natural influence."[36] He was evidently aware of the risk of advocating magic or idolatry, both condemned by the Church; at one point, he himself was accused of practicing magic, though the formal charges were eventually dropped. In Penelope Gouk's judgment, "although Ficino firmly denied consorting with demons, he nevertheless believed himself to be practicing a form of natural, or spiritual, magic" that he considered licit.[37] She and other scholars over the years have considered the complex relations between such "natural magic" and the emerging "new philosophy" that eventually led to what we now call "science." Whatever one concludes about this larger issue, the art of medicine arguably had a different relation to "magic" than did other fields of natural philosophy. At least that seemed to be Ficino's assumption as he used music to invoke Jupiter and Venus, the appropriate planets that would help him transmute the metaphorical iron of scholarly melancholy into the gold of poetic inspiration.

His concern centered on the *spiritus*, then considered a subtle and airy substance generated in the heart from the thinnest part of the blood: "Medicine heals the body, music the spirit [*spiritus*], and theology the soul" because "song is the most powerful imitator of all things. . . . When the heavens are imitated with [its] power, it marvelously rouses our spirit to the heavenly influx, and indeed the influx to our spirit."[38] Thus, if we use the therapeutic effects of music, we implicitly accept its seemingly magical powers.

To explain what is "natural" about the power of music, Ficino turned to sympathetic vibrations as understood in the neo-Platonic writings he studied and translated. In particular, Plotinus had argued that prayer and magic spells operate on basic principles of musical sympathy "by the mere fact that part and other part are wrought to one tone like a musical string which, plucked at one end, vibrates at the other also," just as "the vibration in a lyre affects another by virtue of the sympathy existing between them." Further, Plotinus emphasized that such effects work involuntarily because "it is the reasonless soul, not the will or wisdom, that is beguiled by music," just as "a human being fascinated by a snake has neither perception nor sensation of what is happening; he knows only after he has been caught, and his highest mind is never caught." Despite what might seem questionable or even alarming in the possibilities of influencing others without their consent, Plotinus argued that such "sorcery raises no question, whose enchantment, indeed, is welcomed, though not demanded, from the performers."[39] Following his logic, love potions or other magical charms—like music itself—operate as naturally as vibrating strings and hence could be justified post facto: when they worked, they showed themselves to be natural and hence blameless.

With that in mind, Ficino explained that "all music proceeds from Apollo [the Sun]; that Jupiter is musical to the extent that he is consonant with Apollo; and that Venus and Mercury claim music by their proximity to Apollo."[40] In his view, these different planets correspond to different kinds of song; to Venus, "songs voluptuous with wantonness and softness"; to Jupiter, music "deep, earnest, sweet, and joyful with stability"; to Apollo, "grace and smoothness . . . reverential, simple, and earnest"; to Mercury, songs "more relaxed, along with their gaiety, but vigorous and complex." In contrast, "the other three planets [Mars, Saturn, the Moon] have voices but not songs": Saturn's voice, "slow, deep, harsh, and plaintive"; Mars's "quick, sharp, fierce, and menacing"; and the Moon with "the voices in between" these.[41] Thus, Ficino's astrology corresponded to a typology of songs and voices according to their character and affect, so that singing a voluptuous song implicitly invokes Venus, as indicated by that planet's long-established connotations: "Accordingly, you will win over one of these four [planets] to yourself by using their songs, especially if you supply musical notes that fit their songs." Further, Ficino associated with each singing planet a definite musical interval, presumably the range of their characteristic song: to Jupiter, an octave and a fifth; to the Sun, an octave; to Venus, a fifth; to Mercury, a fourth.[42] Here, and at many points to follow, the astrological dimension of such musical interventions was ever-present.

Thus, Ficino asserted that if "at the right astrological hour you declaim aloud by singing and playing in the manners we have specified for the four gods [Apollo, Jupiter, Mercury, Venus], they seem to be just about to answer you like an echo or like a string in a lute trembling to the vibration of another which has been similarly tuned. And this will happen to you from heaven as naturally, say Plotinus and Iamblichus, as a tremor reechoes from a lute or an echo arises from an opposite wall."[43] In short, the response of the soul to these songs is as natural as the sympathetic vibrations of strings; in this way, an Apollonian song evokes the corresponding vibration of that same part of the listening soul, which had been primordially tuned to that same note. The singer who understands this can thereby act in whatever manner is needed because "from your spirit influenced within, you have a similar influence on your soul and body."

Ficino also compared this sympathetic musical action to other bodily actions, digestive and sexual:

For just as the natural power and spirit, when it is strongest, not only immediately softens and dissolves the hardest food and soon renders harsh food sweet but also generates offspring outside of itself by the emission of the seminal spirit, so the vital and animal power, when it is most efficacious, not only acts powerfully on its own body when its spirit undergoes a very intense conception and agitation through song but soon also moves a neighboring body by emanation.[44]

In Ficino's bold account, the emission of semen and the outpouring of inspired song both evoke conception physically and musically because of the thorough parallelism between the processes underlying sexuality, sympathetic vibrations, and passionate music. Just as the emission of semen can lead to physical conception, Ficino's singer can generate conceptions within the listener that would enable physical and psychic healing as well as emotional and sexual arousal. Based on this passage, Ficino may have found in music the essence of immaculate conception that he worshipped in the Christian mysteries as a priest.

Ficino's book reappeared in over thirty editions and translations over the next century and a half, inspiring others to respond sympathetically and try his methods for themselves. Among his readers in Milan, Franchinus Gaffurius's *Practica musica* (1496) presented a beautiful diagram that correlated planets, musical modes and notes, and muses (figure 4.4), synthesizing the work of Ramos and Ficino. Beside studying Ficino's rendition of Plato, Gaffurius also commissioned Latin translations of Aristides's treatise *On Music* and Ptolemy's *Harmonics*, important ancient works heretofore untranslated and unknown. As we saw in the previous chapter, Aristides offered provocative connections between musical and medical concepts, which Gaffurius then studied and incorporated (mainly as quotations) in his own work. In particular, Gaffurius presented more details of pulse music than had Ramos. In his *Practica musica*, Gaffurius noted that "physicians agree that the correct measurement of a short unit of time ought to be matched to the equal beat of the pulse, establishing arsis [upbeat] and thesis [downbeat] as equal to that which they call diastole and systole in the measurement of each pulse."[45] In the context of long and short syllables in poetry, Gaffurius thereby introduced the medical measure of pulse to his humanist readers, evidently needing to define diastole and systole for an audience not familiar with these terms. He also noted that "modern [musicians] have assigned to the regular semibreve a measure of one unit of time, including diastole and systole in the sound of each individual semibreve," here going beyond poetic meters to the semibreves used to notate rhythms in contemporary music.[46] Indeed, in his *De harmonia musicorum instrumentorum opus* (*On the Harmony of Musical Instruments*, 1518), Gaffurius included a musical setting of a poem illustrating the characters of the various modes and planets through the various poetic meters (♪ sound example 4.2).[47]

For instance, between 1628 and 1630, Tommaso Campanella (a utopian poet later condemned by the Inquisition) and none less than Pope Urban VIII (Galileo Galilei's sometime friend and eventual nemesis, Maffeo Barberini) closeted themselves to perform astrological magic. Surrounded by auspicious gems, colors, and plants, they lighted tapers, drank special liqueurs, and performed music chosen to avert the malignant astral influences that astrologers had predicted would menace the pontiff's life.[48] Gary Tomlinson has emphasized the degree to which Renaissance music (and also the graphic and poetic arts that joined its efforts) had implications of occult magic, using the powers of music to draw down beneficent astral influences for the worldly benefit of its patrons.

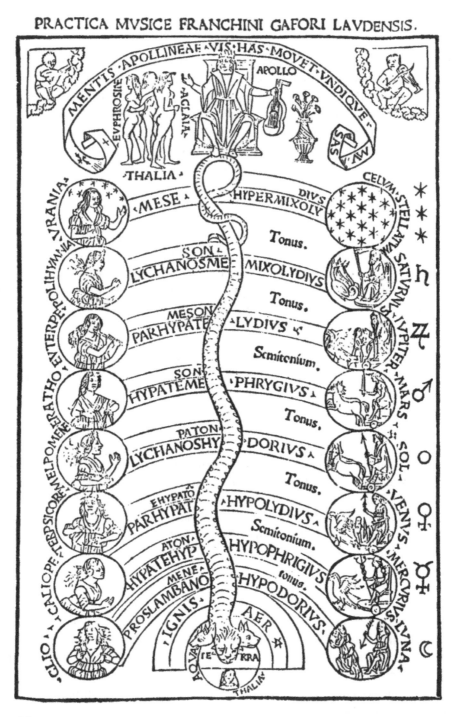

Figure 4.4
Franchinus Gaffurius's diagram correlating the Muses (far left), musical notes (middle left), musical modes (middle right), and planets (far right), from his *Practica musica* (1496). In the highest heaven, Apollo puts his feet over the tail of a three-headed snake, whose body stretches across the celestial spheres to the Earth at the bottom of the diagram.

Ficino's ideas influenced theorists and practitioners far from Italy. Writers from Germany, the Low Countries, and Poland were struck by his teaching that the voice stirs the *spiritus* and thereby affects soul and body.[49] In his *Margarita philosophica* (*Philosophic Pearl*, 1496), Gregor Reisch, a Carthusian monk and confessor to the Holy Roman Emperor, Maximilian I, quoted with evident relish Ficino's description of how musical sound

excites the aerial spirit [*spiritus*], which is the node between soul and body. Through its emotion it influences the sense and the discursive intelligence [*animus*] simultaneously. Through its meaning it works upon the intuitive intelligence [*mens*]. Finally, through the very motion of the subtle air it penetrates vehemently. Through contemplation it licks (or moves) [the mind] sweetly . . . and at the same time it seizes and takes possession of the individual. . . . Apollo takes possession of the soul more forcefully with melody than Bacchus takes hold of the sense of taste with wine, or Venus customarily besieges the sense of touch with her itch.[50]

Especially widely received was Ficino's advocacy of music's special power to alleviate melancholia, then considered to be a surfeit of black bile (often associated with the planet Saturn) that caused this characteristic malady of scholars and artists. In the early decades of the sixteenth century, some Swiss preachers and German physicians echoed Ficino's views.[51] Nevertheless, others continued to take the negative view of melancholia prevalent in the Middle Ages and resisted Ficino's musical approach, which associated melancholia with those who were talented or even divinely inspired.[52]

Ficino's ideas particularly influenced the German polymath, physician, and occultist Heinrich Cornelius Agrippa von Nettesheim whose book *On Occult Philosophy* (*De occulta philosophia*, first published in 1533) considered musical sound to be "one of the most efficient ways of transferring the imagination [*phantasia*], emotion [*affectus*], and mind [*animus*] of a singer to those listening, by means of the subtle spirit."[53] Even more efficacious than instrumental music, according to Agrippa, singing "proceeds from the conception of the intellection and the ruling emotion of the imagination and heart. Having broken up and tempered the air simultaneously, through its motion it easily penetrates the hearer's aerial *spiritus*, which is the bond between soul and body."[54] In this way, singing moves the listener "as far as the innermost parts of the mind, and by degrees infuses the moral as well. Moreover, it moves the members and restricts the humors of the body." Agrippa also stipulated that the various humors in a healthy man should follow musical proportions, so that the weights of blood:phlegm:choler:melancholy (black bile) should be proportioned 8:4:2:1, and all the bodily measurements should follow a complex canon of ratios.[55]

Agrippa and others connected the various planets and humors with the eight ecclesiastical modes.[56] Thus, in 1540, the Silesian composer and theorist Jerzy Liban associated the four humors with the four final notes of those modes (D, E, F, and G); he also connected the seven planets and the firmament with the eight modes.[57] For instance, Liban argued that the seventh mode (Mixolydian) corresponded to Saturn, the lord of melancholy. Through his influence, this mode "makes people sad negligent, slow, sluggish, and difficult," hence suitable for "hard and harsh words with vehement signification."[58] To illustrate this, Liban composed eight settings of the beginning of the Magnificat (the canticle Mary sang in response to the angel's annunciation) in the eight modes, based on the appropriate Gregorian melodies for each mode. You can hear for yourself how Liban presented the differences between what he considered the subdued melancholia of mode seven, the sanguine hopefulness of mode four, and the choleric brightness of mode six (figure 4.5, ♪ sound examples 4.3, 4.4, 4.5).

To be sure, Ficino's ideas were extreme versions of commonly held beliefs about the healing powers of music that went back to biblical accounts of David using his harp to heal the madness of King Saul as much as to the ancient pagan sages. Music was a familiar part of therapeutic regimens that became staples of literary depiction. Shakespeare had King Lear regain his senses amid the curative strains of sweet music, as the stage directions specify. Prospero calls for solemn music "to work mine end upon their senses / That these airy charms are for"; even his beastly servant, Caliban, is a connoisseur of "sounds and sweet airs, that give delight, and hurt not."[59]

These literary, biblical, and classical streams converged in *The Anatomy of Melancholy* (1621), in which Robert Burton used music as a central remedy in his encyclopedic treatment of melancholia (figure 4.6), "Philosophically, Medicinally, Historically Opened and Cut Up."[60] Among Burton's vast list of therapeutic approaches, he considered "none so present, none so powerful, none so apposite as a cup of strong drink, mirth, music, and merry company." Armed with a barrage of classical quotations and sources, Burton argued that music is

a roaring-meg against melancholy, to rear and revive the languishing soul; "affecting not only the ears but the very arteries, the vital and animal spirits, it erects the mind, and makes it nimble." . . . In a word, it is so powerful a thing that it ravisheth the soul, *regina sensuum*, the queen of the senses, by sweet pleasure (which is a happy cure), and corporal tunes pacify our incorporeal soul, *sine ore loquens dominatum in animam exercet* [speaking without a mouth, it exercises dominion over the soul] and carries it beyond itself, helps, elevates, extends it.[61]

Beyond the "excellent power it hath to expel many other diseases," Burton singles music out as "a sovereign remedy against despair and melancholy, and will drive away the devil himself."

Even so, Burton notes that by his time, music's place in the world had changed, for "the Greeks, Romans, have graced music, and made it one of the liberal sciences, though it be now become mercenary." By this, he seems to mean that music was losing its connection with the other "liberal sciences," the mathematical arts of the quadrivium, in the process of becoming what much later would be called a "fine art." Burton's comment reflected his critical sense that music was beginning to be considered more as a rhetorical art, used as a means to move the passions, rather than as an end in itself that was connected with the deepest mathematical structures of the cosmos. He was living through a larger intellectual

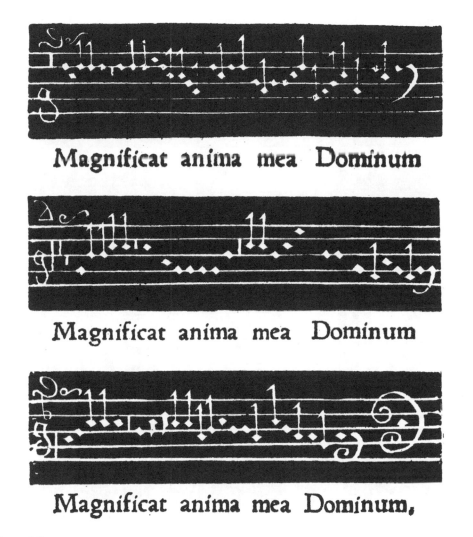

Figure 4.5
Jerzy Liban's versions of the Magnificat in modes (a) seven, (b) four, and (c) six, from his *De musicae laudibus oratio* (1540, ♪ sound examples 4.3, 4.4, 4.5).

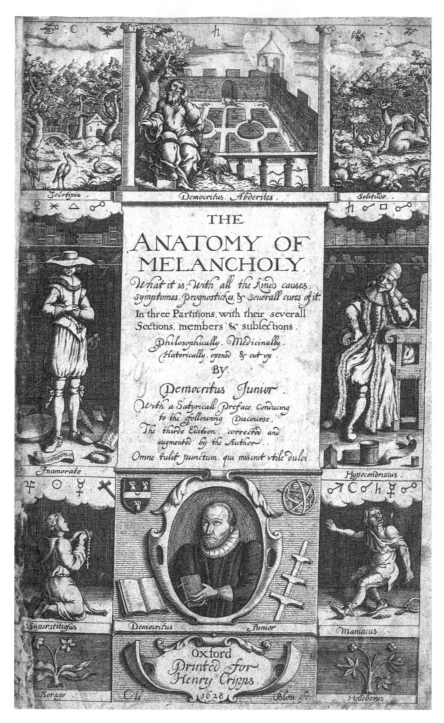

Figure 4.6
The title page of Robert Burton, *The Anatomy of Melancholy* (first published 1621). The images illustrate different kinds of melancholy, including love sickness (middle left), hypochondria (middle right), superstition (lower left), and mania (lower right). Note the lute at the feet of the love-sick youth, denoting his use of music to allay his melancholy.

and artistic movement that emphasized the power of expressive music, as compared with the dispassionate beauty of music as a "liberal science."[62]

Burton still assessed music to be so potent a medicine that it could have dangerous side effects if misused. For instance, he strongly recommended against the use of music to cure young, lovesick sufferers because "it will make such melancholy persons mad, and the sound of those jigs and hornpipes will not be removed out of the ears a week after."[63] For such cases, he followed Plato's advice to omit wine and music, "lest one fire increase another." Yet it is important to note that Burton was a clergyman, not a physician. Indeed, melancholy was not considered a medical condition and hence was not treated by physicians until about 1800.

5

Kepler's Harmonic Physiology

In the wake of Ficino's translations and commentaries, growing interest in Platonic and Pythagorean ideas supported various musical treatments for the ills not only of the individual but of the larger human community. Even Johannes Kepler's groundbreaking harmonic discoveries in astronomy had important implications for his theory that the cosmos spontaneously generated new stars and comets just as he thought worms, eels, and whales were spontaneously generated on Earth. Throughout this chapter, we need to suspend later assumptions that neatly separated physical from biological sciences. In his concept of physiology (*physiologia*), Kepler still looked to Aristotelian physics, which included all kinds of growth and change, biological as well as physical (as we now call them). At the same time, he extended those concepts from the Earth (to which Aristotle had limited them) throughout the cosmos: Kepler's "universal physics was a study in life."[1] His fervent Platonic and Pythagorean convictions made him the most radical exponent of the harmony of the physical world. In turn, his blending of music and physiology affected his astrological attempts to heal the political and religious wounds he experienced firsthand and then presented in terms of medical imagery. Using music and astrology, Kepler tried to sway King James I and Albrecht von Wallenstein in directions he thought might alleviate the widespread suffering surrounding him.

In contrast to Burton, Kepler considered music still to be a living part of the mathematical and liberal arts. Ficino's revival of Platonic views was deeply appealing to Kepler, who drew on it in his bold account, *The Harmony of the World* (*Harmonices mundi libri V*, 1619). Searching for new mathematical harmonies in the celestial data obtained in Tycho Brahe's exacting observations, Kepler found new relations between the planets' orbits and periods, most notably what came to be called his "Third Law" of planetary motion, which became a cornerstone on which Isaac Newton built a new physics and celestial mechanics. To be sure, Kepler's involvement in medicine was not direct; unlike Ficino and many others we have been considering, he was not a physician, though quite a few of his associates and correspondents were.

Still, he worked within the Platonic tradition in which the cosmic macrocosm and biological microcosm were continuously connected. As Max Caspar noted, "at that time, medicine and astronomy, or rather astrology, were not so very far apart: even Copernicus and Tycho Brahe practiced the art of healing and busied themselves with the preparation of medicaments."[2] At a critical point in his youthful career, the city councilors of Graz advised Kepler to take up medicine.[3] In 1599, having difficulties finding a position, Kepler announced to his teacher Michael Mästlin that "I shall enter upon a medical career," though in the end he went to work with Tycho.[4] Still, as Jonathan Regier has argued, "medical ideas deeply inform Kepler's theories of light and solar force (*virtus motrix*)."[5] In his *Dream* (*Somnium*, 1634), Kepler noted that "medicine and astronomy are related studies, arising from the same source, the desire to understand natural phenomena."[6] A number of scholars have noticed the methodological closeness of medicine to physics and astronomy, both of which frequently "mingle the probable with the necessary."[7] As Patrick Boner put it, "Kepler often compared astrology to the imprecise art of the physician, advanced by the accumulation of knowledge about physical qualities"; both were highly empirical practices, the physician gathering information about the therapeutic value of "the many and various herbs" just as the astrologer developed "a discrete science [*scientia*] and body of knowledge" about the effects of the planets.[8]

At the same time, Kepler described astrology as "a silent music" whose effects were made possible by the existence of the soul created to "dance to the tune of the [celestial] aspects."[9] In a letter of 1599, he specified that

the analogy with music and astronomy is absolute. I show that the analogy must necessarily be seen in this way, since the origins of all things are derived from geometry. Nature confirms these principles in the creation of a single species and employs these principles in everything that is capable of them. This occurs in music, the motions of the planets, the operation of the plants [on Earth], the measure of musical notes according to time, the dances of men, and the composition of songs. For although these things are the discoveries of men, nevertheless man is the image of the creator.[10]

With that in mind, Kepler moved back and forth between considerations about the harmony of the whole cosmos and its constituent bodies to the relative harmony—hence, health or illness—of the bodies of humans and animals, even of the Earth itself.

Though certainly interested in mechanistic models of the cosmos, he considered the Earth to have a soul that mediated between the heavens and meteorological phenomena. Here and throughout Kepler's writings, "soul" needs to be understood in terms of harmony. For him, "soul" did not merely mean a generic animating principle or a disembodied consciousness that merely contemplates or contains harmonies. As Jorge Escobar has argued, Kepler's "soul" was "not a bucketlike thing containing innate or inborn (ideas of) harmonies"; rather, "the soul is its contents, the contents are the soul," namely, "the harmony itself."[11]

Kepler consistently applied anatomy and biology "to the ether and every other part of the cosmos," as Boner has pointed out.[12] Thus, in the *Harmony of the World*, Kepler's climactic Book V begins with an epigraph from Galen's treatise *On the Function of Parts*, praising the "adornment" and divine "wisdom" shown in the parts of the human body, which Kepler used as justification for his own search for the "greatest possible adornment" found in the cosmos at large.[13] This indicates that Kepler was familiar enough with Galen's writing to make use of them, less surprising because he considered Galen a fellow Platonist. Kepler was always seeking to extend and perfect Plato's Pythagorean vision, always assuming that music is integrally related to mathematics and astronomy, which in their unity shaped the whole cosmos and everything in it.

For Kepler as for Plato, therefore, the same geometric and musical archetypes that informed the architecture of the cosmos also shaped the bodies of living things. As Kepler put it in his *Harmony of the World*, "Geometry, which before the origin of things was coeternal with the divine mind and is God himself (for what could there be in God which would not be God himself?), supplied God with patterns for the creation of the world, and passed over to Man along with the image of God."[14] Thus, Kepler explicitly stated that geometry "is God himself." His creation was not limited to an initial act but, in Kepler's view, continued to operate throughout the cosmos in order to compensate for the natural forces of decay. After all, one of Kepler's greatest innovations was "to envision astronomy as a part of physics," thereby extending the meaning of *physis* (the realm of growth and change) to include the whole cosmos, not only, as Aristotle had held, the sublunary world, from the Earth to its moon.[15]

Kepler applied this insight in his wonderful short work *On the Six-Cornered Snowflake* (*De nive sexangula*, 1611). Standing on the Charles Bridge in Prague, he was struck by the six-fold symmetry of the snowflakes falling around him, each of which "descends from the sky and bears a likeness to the stars."[16] Kepler wondered what might give the snowflakes this particular symmetry, compared to the forms of beehives (six-sided), pomegranate seeds (five-sided), and the foliage and fruit of apples and pears (also five-sided). He then asked: "What does an animal [in the sense of a living thing] have in common with snow?"[17] Despite their differences, he believed that "the cause behind the six-cornered shape of the snowflake is no other than the one responsible for the regular shapes and the constant numbers that appear in plants."[18] Each bears the imprint of a geometrical archetype, such as the regular polygons (the hexagon and pentagon) or the "Platonic solids" (figure 5.1). Each symmetry emerges from "a formative faculty in the body of the Earth," whose "vehicle is vapor" in the case of the snowflake; in their various ways, bees and plants also respond to that faculty and its "constructive heat." That faculty "distributes itself among bodies and by means of bodies, growing within them and building one thing or another according to the internal conditions of each form of matter, as well as external circumstances. In vapor, too, which had been entirely pervaded by a soul, it is no wonder that as the cold begins to break up the continuous whole, this same soul should occupy itself in shaping the parts as they contract, just as it was busy earlier shaping the whole."[19] Though he could not answer his own question about the rationale for the six-cornered snowflake, Kepler raised a host of important questions and "knocked on the doors of chemistry," which responded only centuries later.[20] Indeed, we are only beginning to traverse the boundary between "living" and "nonliving" matter.

Figure 5.1
Kepler's depiction of the five Platonic solids in his *Harmonices mundi* (1619), illustrating Plato's identification of each of them with an element: cube (earth), octahedron (air), tetrahedron (fire), dodecahedron (Sun and heavenly bodies), and icosahedron (water).

Throughout the Earth, Kepler considered this generative faculty to be "the source of flies and swarms of bees and beetles, hornets, and wasps and extraordinary flutters of butterflies," here agreeing with Aristotle's ideas about spontaneous generation.[21] Kepler judged that "whenever this generative faculty encounters any superfluous matter, it converts it into a living being that serves the nature of things, either by adding or taking away. . . . The philosophers call this putrefaction, which is in fact nothing other than the instauration of old material from an expiring form, as if dying, to some new one, which is produced by this industrious architect, whoever he may be."[22] Still, Regier notes that "the biological side of [Kepler's] natural philosophy is not naively Aristotelian. Instead, he is up to date with contemporary discussions in medically flavored natural philosophy."[23]

Where Aristotle had opined that the vital heat in semen is analogous to "the element of the stars," Kepler postulated that this generative faculty operated throughout the cosmos, far beyond the Earth.[24] Thus, Kepler argued that the new star of 1604 had emerged from superfluous matter in the heavens in the same way that eels were born from bogs, the new star emerging from the "oily and impure vapors" emitted by the other stars.[25] As Boner put it, the new star "served as a sort of cleansing agent, preserving the transparency of the ether and contributing to cosmic renewal."[26] Similar processes were at work in the formation of comets. In *On Comets* (*De cometis libelli tres*, 1619), Kepler proposed what he called "a new and extraordinary physiology of comets" (figure 5.2).[27] He compared the motion of comets to a whale "that strays and is washed up on shore, left struggling to breathe on dry land." Just as he considered whales to be extraordinary "sea marvels" spontaneously generated by the ocean, comets too were "beyond the ordinary nature of the heavens" but not contrary to them or miraculous, as Tycho had thought.[28] Throughout, Kepler unified the physiology of comets and whales in ways that challenge our current categories and manifest the consequences of his original unification of mathematics, music, and astronomy.

These cosmic acts of generation could also have important implications for human health. For instance, Kepler argued that comets could cause the Earth "to perspire many moist vapors," resulting in "heavy rain and flooding," which then led to "headaches, dizziness, catarrh, as in [the comet of] the year 1582, and even pestilence [plague], as in the year 1596."[29] The divine geometry was transmitted by the angles between the rays from various planets to the Earth. In *A Third Man Intervenes, That Is, Warning to Some Theologians, Medics, and Philosophers, Especially D. Philip Feselius, That They in Cheap Condemnation of the Stargazer's Superstition Do Not Throw Out the Child with the Bath and Hereby Unknowingly Act Contrary to their Profession* (*Tertius interveniens*, 1610), Kepler tried to point out what he thought was the real physics behind astrology: each "person in the first igniting of his life . . . receives a character and pattern of all the constellations of the heavens or of the form of the rays flowing onto the Earth, which he retains until he enters his grave: This character afterwards leaves noticeable traces in the formation of the countenance and of the remaining shape of the body" as well as of the ensuing personal and psychological qualities.[30] To do so, Kepler revised astrology according to "harmonic doctrine," rejecting the traditional lore about the division of the zodiac into twelve equal parts, the signs or "houses" used by vulgar astrology, which he considered a form of "fortune telling" he found "not only wrong, but even childish and vain."[31]

For Kepler, astrological aspects and musical consonances were "different peoples, as it were," who came from "the same country of geometry": "the consonances in music and the aspects in astrology spring from the same geometrical source of the division of the circle."[32] Until 1608, he had thought that there were only as many aspects as there were "harmonic sections in music"; after that, he extended the aspects from the eight consonances to include five more that did not correspond to the standard musical consonances. Because aspects are derived purely from circular arcs, he thought they could include some cases not included in the consonances, which he derived from the straight-line sides of polygons inscribed in a circle.[33] Consonances corresponded to polygons that could be constructed with straightedge and compass; aspects might include other intervals. By including other intervals as aspects, Kepler confessed that he was moved by "experience alone . . . against all reason," prompted by the need to reconcile certain weather

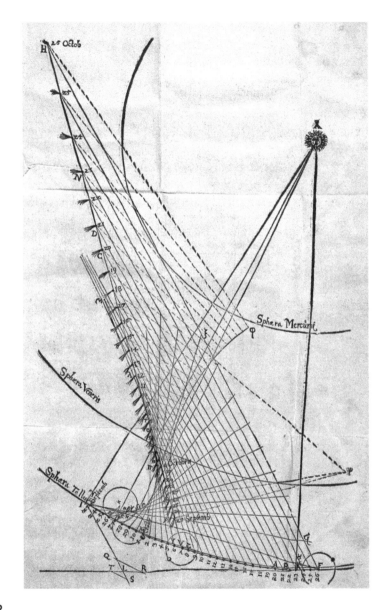

Figure 5.2
Kepler's diagram of the "physiology" of Halley's comet, from *De cometis* (1619). The Sun is shown at the upper right (*X*); the Earth's path (*E*) is contrasted with what Kepler thought was the comet's straight-line path.

observations with the astrological aspects presumably affecting them.[34] He emphasized that "an aspect is a geometrical harmony [*geometrica concinnitas*] between the light rays of two planets here on Earth. A geometrical harmony is a relation, a thing of reason. A relation is not an effect, and so it follows that the aspect can produce no rain on its own."[35]

Thus, Kepler's stance about the nature of such astrological effects reflected what he considered the musical quality of terrestrial weather and human health, which Boner calls "a symphony orchestrated by the aspects, which produced a palpable, if often unpredictable, effect on Earth."[36] Though ultimately basing both music and astrology on the objectivity of geometry, Kepler emphasized that geometry "cannot act objectively on anything except for the animate faculties, as music does on the hearing of a dancing farmer."[37] As Boner interprets this, "Just as the farmer could dance to a melody without fully knowing the mathematics of music, so the Earth could move the air in response to the geometry it sensed in the aspects instinctually. . . . As an animate being that became aware of these proportions instinctually, the soul of the Earth could distinguish their momentary manifestation from an otherwise constant cacophony of celestial configurations." Indeed, Kepler had noted the prevalent dissonance of the cosmic music, in which planetary conjunctions, oppositions, and other important aspects represented rare moments of relative consonance. Kepler had decoded the "song of the Earth" as "*MIseria et FAmes*," misery and famine, expressed in the doleful semitone *MI FA*—notes also called the *diabolus in musica*—all too well explained by the almost continual dissonance raining down on us from the heavens.[38]

On the other hand, Kepler envisaged the body of the Earth as capable of corporeal pleasures that he compared to animal sexuality. He argued that the Earth could emit vapors that were analogous to the "seminal excretion" of animals:

It is not so absurd, really that we should attach some pleasure to this excretion since the Earth shares so many other things with animals. In fact, physicians tell us that if the blood vessels swell with seminal fluid when a pleasant image is presented during sleep, an erection occurs at once without any contact taking place. And is there anything more similar to such a thing than what is known to happen with that faculty in the Earth, the perceiver of celestial aspects that exudes rainy vapors when stimulated by one of these aspects? On the one hand, the immaterial image, the object of the imagination, rouses the reproductive substance, so what, on the other hand, prevents the existence of pleasure in the expulsion of that substance from the perception of this image?[39]

For Kepler, Earth was not passive matter but ensouled and even endowed with a kind of sexuality that acts as freely as human sexuality; whereas vulgar astrologers treat the planets as determining everything, "I attribute nothing to the heavens except for the venereal image, the aspect, provoking the outflow of this seminal fluid from the Earth."[40] Kepler also invoked an analogous generative faculty in the heavens to explain the generation of new stars and comets. Nevertheless, he judged that this natural faculty worked greater wonders in the human body and "the infinite arrangement of parts of living beings" on Earth than it did in the heavens; though humans are tiny compared to the heavenly bodies, our generative faculties are "far more noble" than cosmic generation, even though it is able "to roll up celestial refuse into the single form of a new star."[41]

Kepler's *Harmony of the World* set out a new understanding of that harmony that had implications for astrological medicine. This theme already appears in his dedicatory letter to King James I of England, describing what Kepler called a "civil war" during which his patron, the Emperor Rudolf II, had been forced to abdicate in favor of his brother Matthias.[42] In 1618, during these dynastic struggles, Kepler was trying to complete his book and had been invited to England by Sir Henry Wotton, King James's ambassador, to escape the political struggles convulsing the empire. Already in 1590, James had visited Tycho at his island observatory, showing his evident interest in the new astronomy; in his dedication, Kepler noted that he had planned for almost twenty years to dedicate his world harmony to James.[43]

Taking a musical view, Kepler referred to the political and religious struggles surrounding him as "manifold dissonance in human affairs" that postpone and put in doubt "the promise of the pleasing consonance which is to succeed . . . for what else is a kingdom but a harmony?"[44] Shifting from musical to medical terms, Kepler felt that these harsh dissonances wounded him and many others afflicted by civic and ecclesiastical strife: "What wounds to my person treated by what harmonies, by what physician?"[45] Because he united "two extremely hostile nations" and "produced one kingdom and one harmony" from England and Scotland, Kepler referred to "your Davidic harp, glorious king," so that (like David treating Saul) James might heal Kepler's wounds and, by extension, those of many Protestants suffering religious persecution on the Continent. "What more appropriate patron could I choose for a work on the harmony of the heavens, with its savor of Pythagoras and Plato, than that king who has borne witness to his study of Platonic learning by domestic tokens," referring both to King James's well-known intellectual interests and to the marriage of James's daughter Elizabeth to the Elector Frederick of the Palatine, the leader of the German Protestants. Thus, Kepler portrayed King James as a Platonic ruler who also was the natural protector of the Protestant cause, reminding the king of his title as "Defender of the Faith among faithful followers of Christ." As Kepler (and the king) well knew, the pope had given that title to Henry VIII before he embraced the Protestant faith; by continuing to use it, the king implicitly claimed to defend that faith when Catholics besieged Protestants.

In many ways, Kepler's dedication drew on Ficino's vision of musical medicine, now applied on a grand scale to the dissonances wracking the empire and menacing Kepler's search for new cosmic harmonies: "But what use will it be, if in striving for harmony with my private clamor I do not overcome the public roaring, through the weakness of my support, and in addition I increase the annoyance of the absurd chorus to my ears?" Kepler went on to make ever more specific the religious character of his "wound":

For my part I must confess—ah! what sorrow—that the criss-cross wound is still swollen, or if we prefer a more sacred and more felicitous word, the cross-shaped wound, is swollen, I say, with its multiple lips, and though none of them winks at it, the medicine has so far been useless, and jeered at from all sides, because the physician, to force a deceptive medicine on a crazy patient, makes many pretenses and many embellishments, which seem to stray far from sensible reasoning.[46]

For Kepler himself, the "cross-shaped wound" was his exclusion from the Lutheran communion because of his refusal to accept the doctrines in the Formula of Concord, even more painful because he (at different times) had been expelled from both Catholic and Protestant lands. In the midst of this uncertain political and religious situation, Kepler considered himself "invigorated by the very thought that the supreme Healer of our wounds is sure in his art, and applies no remedy in vain." In Kepler's estimation, King James's "work at home [ruling both England and Scotland] seemed to me to contain a not untrustworthy omen" that he "would perform some greater and more excellent, and also more lasting work." Therefore, Kepler exhorted James to act as the earthly physician who would bring peace and concord to Europe by supporting its Protestants. To that end, Kepler presented the king with these newly discovered cosmic harmonies to "strengthen and stir up in yourself by the examples of the brilliance of concord in the visible works of God [the planets] the zeal for concord and for peace in church and state."[47]

Kepler's subtle rhetoric embodied a two-fold use of Ficino's musical medicine. First, Kepler "set himself all the more resolutely to chant the cosmic harmonies" in order to enchant the king, who then could apply those harmonies to the cure of those suffering on Earth. After all, in Kepler's book the king would learn the doleful song of the Earth that should alert the royal physician to his patients' malady, imploring his therapeutic intervention as well as signaling the underlying causes of their suffering.

This political subtext becomes explicit in a curious "Political Digression" Kepler appended to his Book III, at the very center of his *Harmony of the World*.[48] Having in the previous books explained "the regular figures which give rise to harmonic proportions" (Book I) and the congruence of those figures "as they are 'conceived in the mind'" (Book II), in Book III Kepler showed how those figures lead to "the origins of the harmonic proportions" and hence of "the nature and differences of those things which are concerned with melody." Each of these books had opened with an epigram from Proclus's commentary to Euclid's *Elements*, underlining the Platonic roots of Kepler's endeavor. The epigram for Book III "makes plain the proportions of number which are associated with the virtues, some in arithmetic, some in geometry and others in harmony" in order to "perfect us in moral philosophy, implanting in our behavior order, propriety, and harmony in social relations" so as to avoid "the excesses and deficiencies of the vices" and guide us "to the middle way in behavior and in morals."

At the end of Book III, Kepler's digression explicitly addressed the French philosopher Jean Bodin, whose *Six Books on the Republic* (*Les six livres de la république*, 1576) had applied mathematical proportions to political theory.[49] This digression would have particularly interested King James, who had cited Bodin's writings and knew his works on demonology as well as his *République*.[50] Writing after the St. Bartholomew Day's

Massacre in 1572, Bodin was interested in presenting a "middle way" that would avoid the bloody extremes of Christian sectarianism; he thereby sought a new justification of monarchical sovereignty, topics of deep interest to James as a Protestant monarch who appealed to ideas of divine right, though the pope had denied him any such right and even advocated the violent overthrow of Protestant monarchs.

Kepler pointed out errors in Bodin's understanding of the different kinds of mathematical means—arithmetic, geometric, and harmonic (detailed in box 5.1)—in order to show that the harmonic is the best mean for political purposes. Though these distinctions may seem obscure or unimportant, they actually embodied crucial political tensions: Kepler noted that popular rule rests on arithmetic proportions (such as "one person, one vote"), while aristocratic rule appeals to geometrical proportions, assigning to the aristocrats a greater proportion of power than their sheer number, on account of their presumably greater merit. Thus, the conflict between arithmetic and geometric means paralleled debates between rule by the many versus rule by the few. Such debates had begun in antiquity, especially in Plato's dialogues, and became pressing issues for a king whose "divine right" was no longer reaffirmed by a pope or the Roman Church. Indeed, only twenty-one years after Kepler's dedication, James's son Charles I lost his head as well as his crown during violent struggles about the nature and character of the commonwealth he claimed to rule by divine right.

Writing within the classical tradition in which King James had been educated, Kepler offered tactful help on these vexing issues of sovereignty. To the disturbing conflict between the many and the few, a king needed to offer something more than the naked face of tyranny, the unjustified rule of a single individual that Plato and Aristotle (and many medieval thinkers as well) considered the worst form of polity. Kepler emphasized that beyond the stark contrast between arithmetic and geometry lies the higher ground of harmony: the harmonic mean synthesizes and reconciles the tensions between arithmetic and geometry (as box 5.1 explains). For Kepler, "To this regent, whether he be king, or the aristocracy or the entire people I should recommend harmonic proportions."[51] Though here he phrased this recommendation in the most general terms of political theory, at other points Kepler underlined "the tempering of both patterns of government [popular and aristocratic] in the royal state of affairs," as if only a king could wisely harmonize their conflicting claims.[52] Kepler recommended that "the direction of the commonwealth [*respublica*], unfettered by such great compulsions [between the many and the few], should be adapted to the general well-being at the will of the ruler, according to the circumstances, without a great commotion."[53] Whether out of prudence or from a real preference for monarchy, Kepler invited James to present himself as such a harmonic ruler. Implicitly, James would thereby have asserted a new kind of divine right, one not dependent on a pope but on the even more ancient Platonic tradition of which Kepler was the latest champion.

Such divine right would cover not only James's kingdoms but also dominions in Europe that in 1619 his extended family was on the verge of acquiring. Kepler's dedication was

Box 5.1
Arithmetic, Geometric, and Harmonic Means

In *The Fundamentals of Music* (*De institutione musica*), Boethius gives the following examples: "That mean in which the differences are equal is called 'arithmetic,'" such as 1:2:3, in which the differences between 2 and 1 and between 3 and 2 are equal. Next, "That in which the ratios are equal is called 'geometric,'" such as 1:2:4 because the ratio 1:2 is equal to the ratio 2:4. Finally, in the harmonic mean, "The largest term is related to the smallest in the same way that the difference of the larger terms is related to the difference between the lesser terms," such as 3:4:6, in which 6:3 is the same ratio as that of the differences between the terms, namely 2:1 (from 6–4:4–3).[54]

dated February 13 of that year, in which James's son-in-law Frederick and his daughter Elizabeth were crowned as king and queen of Bohemia on November 4 and 7, respectively. In so doing, the Bohemian nobles had rejected the claim of the Hapsburg heir apparent (Archduke Ferdinand) to their throne because he was a fierce Catholic who persecuted Protestants. Defying the imperial power, the nobles chose Frederick and Elizabeth, who persuaded her doubtful husband to accept the Bohemian crown. Their reign ended precipitously when Ferdinand, by then elected Holy Roman Emperor, defeated Frederick at the Battle of the White Mountain (November 8, 1620). The "Winter King and Queen" lasted scarcely a year on their thrones; Frederick died two years later. His defeat and Ferdinand's subsequent forcible suppression of Protestantism began the Thirty Years' War that overshadowed the remainder of Kepler's life.

All this lay in the future, though, as Kepler wrote his dedication, probably knowing that James (notoriously averse to armed conflict) was weighing whether and to what extent to commit his forces to defend his children and the Protestants who dared to crown them. At stake, then, was not only James's status as a ruler but also how he should deploy his power abroad as "Defender of the Faith." Kepler may well have hoped that James would especially take heed of the parts of his book that explained how "hard" and "soft" harmonies—those based on major and minor thirds, as we now call them—were associated respectively with masculine and feminine and were both essential components of cosmic harmony: "This is the impediment which the marriage of the Earth and Venus has, as male and female, and they are the two planets which distinguish the kinds of harmonies, that is to say, into hard and masculine, and soft and feminine" (figure 5.3).[55]

Kepler noted that Venus may try to beguile the Earth through her "soft" harmonies "and of her embraces, so to speak, so as to make love, laying aside for a little while his shield and arms, and those tasks which are proper for a man." Kepler even invited us to imagine what would happen if we were to "command this antagonistic lady, Venus, to be silent" (figure 5.4). This scene of cosmic seduction clearly recalls the scene in Homer's *Iliad* in which Hera beguiles Zeus with lovemaking in order to distract him from

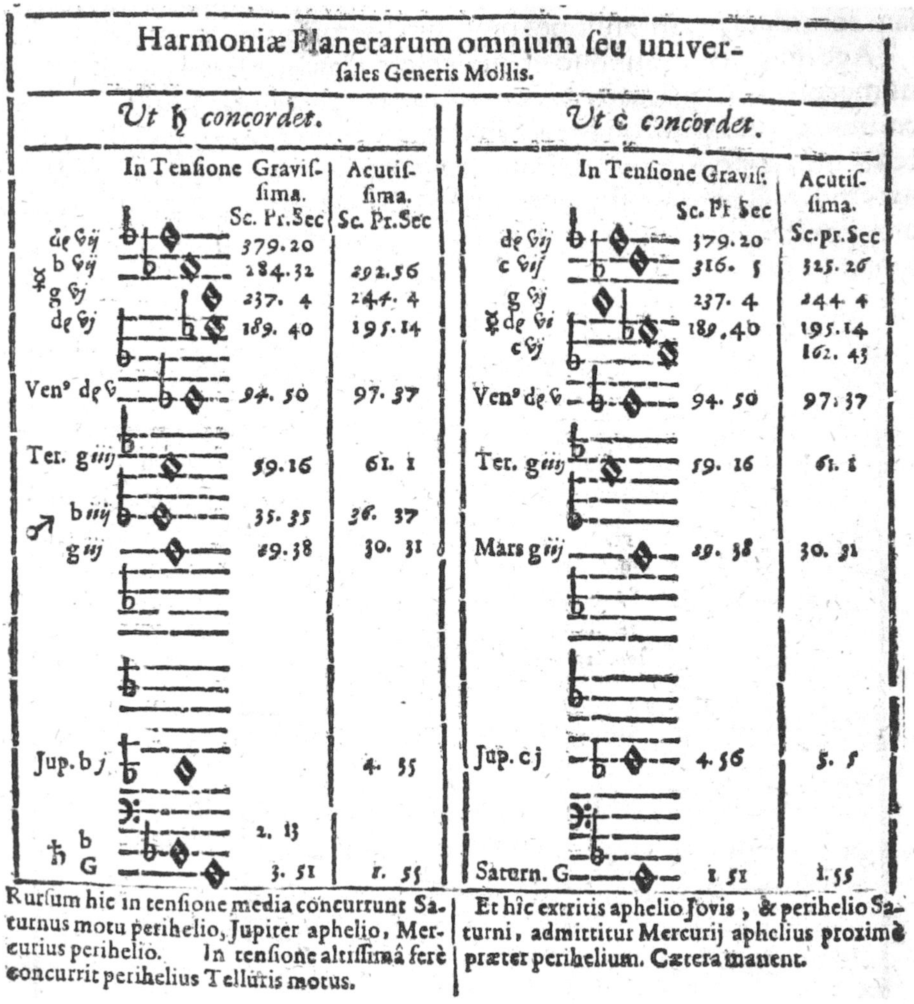

Figure 5.3
"Harmonies of All the Planets or Universal Harmonies of the Soft Kind" (from Book V of *Harmony of the World*). The left column assumes that the note B-flat is in harmony (correcting a misprint that reads B-natural); the right column assumes C. Each column has two subcolumns showing the lowest and highest tunings; each row corresponds to a different planet. The left-hand chord (G, B-flat, E-flat) is the second inversion of an E-flat major triad (using modern terminology); the right-hand chord (G, C, E-flat) is a second inversion of a C minor triad.

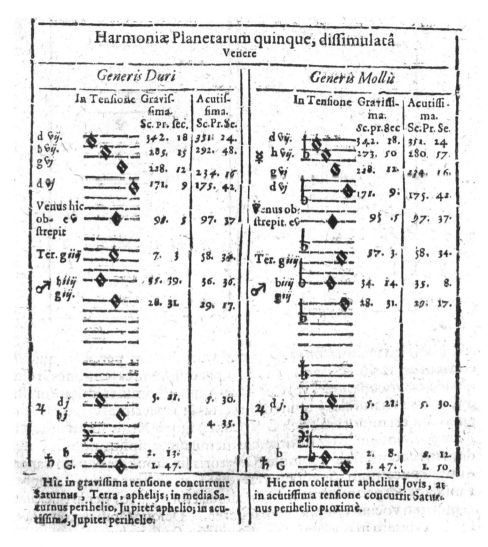

Figure 5.4
"Harmonies of Five Planets, Disregarding Venus" yield more consonant chords. On the left, the "hard" (G major) triad G, B, D; on the right, the "soft" (G minor) G, B-flat, D.

warmaking, thereby giving an advantageous opportunity to the Greeks. It also brings to mind the seduction of Mars by Venus, leading to their being discovered in flagrante, thereupon trapped and enchained by her angry husband, Hephaestus.

Though here Kepler was ostensibly discussing "dissonances" between planetary motions, he read those harmonies as implying that the king should not be deterred by the blandishments of love from taking up his shield at the crucial moment. Kepler seems to have urged this conflict-averse king to use force when necessary through showing him how the "hard" harmonies of combat were necessary for cosmic resolution. After all, Kepler had long identified himself as "a disciple of Mars," referring as much to his battle to find the elliptical orbit of that planet as to his opinion that Mars is "the Master of the Horoscope."[56] Given the situation facing Frederick and Elizabeth in Bohemia and the military threat hanging over the Protestants, it is hard to imagine how Kepler could have advocated their cause without also arguing for war in their defense.

Stepping even further back, Kepler's radical presentation of the sexual significance of music may well have been crafted to pique James's interest, whether or not Kepler was aware of James's predilection for male favorites, widely known even at the time. As I have discussed elsewhere, Kepler argued that "hard" and "soft" cadences (as in ♪ sound examples 5.1 and 5.2) explicitly depicted male and female orgasms, respectively.[57] Kepler's graphic description of the musical and sexual details included the climactic emission of semen, which parallels Ficino's description discussed above, in which he compared seminal emission to the "emanation" by which music projects "vital or animal spirits" onto another soul and body. Kepler's allusions to Ficino suggest that he may have been the main source behind Kepler's extraordinary passage, in which crucial terms were given only in Greek, to spare (or perhaps excite) the prurient (see figure 5.5).[58] But Kepler takes this sexual understanding of music much further, analyzing the dissonances between planets in terms of "marital difficulties" between Earth and Venus, in particular.

As matters later turned out, James's refusal to support Frederick militarily directly affected Kepler. In 1625, agents of the Catholic Counterreformation placed Kepler's library under seal; he had to leave Linz, under siege, and move to Ulm, where he paid out of his own pocket for the publication of his *Rudolphine Tables* (*Tabulae Rudofinae*, 1627), which presented the detailed planetary predictions derived from his long work with Tycho's data. To salvage his situation, Kepler ended his days as an advisor to General Albrecht von Wallenstein, who had been largely responsible for Frederick's defeat at the White Mountain, later becoming a generalissimo for the emperor, even a rival for his power.[59]

The ironies of this situation were many. Rather than persuading a conflict-averse king to take up arms, now Kepler advised a man of arms to seek peace—in fact, the very general responsible for the Protestants' reverses. Indeed, in 1608 a learned physician named Stromair had asked Kepler to cast the nativity of an unnamed nobleman, who in fact was Wallenstein, then a twenty-five-year-old soldier who was neither distinguished nor rich (figure 5.6). In a private cipher, Kepler noted the actual identity of his unnamed

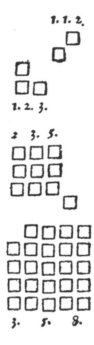

Figure 5.5
Kepler's diagram of the sexual relations between male numbers (2 and 10) and female (3 and 24), from Book III of his *Harmony of the World*. In each case, he depicts the male number as having a diagonal projection that can penetrate the female.

subject, indicating that did not actually work without some knowledge of the person in question. Here again medical and astrological functions overlap, even though Kepler had praised King James's censure of "the excesses of astrology"; elsewhere, Kepler compared judicial astrology to "the stupid daughter who, thanks to her incantations, clothes and feeds her mother," astronomy.[60] In the case of his Wallenstein horoscope, to the physician Stromair, Kepler referred to Galen as evidence "that my work [is] for one who understands philosophy and is infected with none of the contradictory superstition that an astronomer should have been able to predict coming particular things and future contingencies from the heavens."[61] Instead, Kepler himself drew on the powers of music to argue that

it is a property of human souls that at the time of heavenly aspects they get extraordinary impulses to carry through the business which they have in hand. For what a goad is to an ox, what a spur or a jockey is to a horse, what the drum and bugle are to a soldier, what a stirring oration is to its audience, what a measure on flute, bagpipes, and pandura is to a crowd of rustics, that a heavenly configuration of appropriate planets is to all of them, especially when they are assembled together. Thus they are both stirred up individually in their thoughts and deeds, and collectively made more ready to conspire together and to join hands.[62]

Figure 5.6
Kepler's draft of his astrological chart of Wallenstein's nativity, drawn up in 1608 and revised in 1624 at Wallenstein's request. At the center, Kepler recorded his subject's birthdate, which in 1624 he corrected to be six and a half minutes later.

Having expressed his caveats, Kepler presented the young soldier with a horoscope that was far from flattering: "Saturn in ascendency makes deep, melancholy, always wakeful thoughts, brings inclination for alchemy, magic, sorcery, communion with the spirits, scorn and lack of respect of human law and custom, also of all religions, makes everything suspect and to be distrusted which God or humans do, as though it all were pure fraud and there was much more hidden behind than was generally assumed."[63] According to Kepler's horoscope, Wallenstein was "unmerciful, without brotherly or conjugal love, esteeming no one, surrendering only to himself and his lusts, hard on his subjects, grasping, avaricious, deceptive, . . . [showing] great thirst for glory and striving for temporal honors and power, by which he would make many enemies for himself but also would mostly overcome and conquer these." In that regard, Kepler compared this chart to that of "the English queen" (Elizabeth I) and predicted that Wallenstein would acquire high honors and wealth if he paid attention to the course of the world.

Wallenstein pondered this extraordinarily unflattering diagnosis (called a *Judicium*) for sixteen years, then returned it to Kepler in 1624 (again under the cloak of anonymity), asking him to explain it further and extend it into future years. By this time, Wallenstein had become a preeminent warlord wanting to know the specific details that Kepler had insisted were impossible for astrology to provide: Would he die of a stroke? When should he end his military career? Who would be his enemies? Though he accepted this assignment (and the honorarium he badly needed), Kepler objected to his subject's "quite visibly erroneous delusion" with scathing directness:

Whatever person, educated or uneducated, astrologer or philosopher, in the discussion of these questions turns his eyes from the individual's own temperament, or otherwise does not consider his behavior and qualities in relation to the political circumstances, and wants to derive the answers solely from the heavens, be it because of an inner compulsion or only from inclination and disposition, he truly never really went to school, and never yet properly cleaned the light of reason which God ignited for him.[64]

That said, Kepler then answered Wallenstein's illicit questions, though with an ironic twist. After discussing the constellations in the astrological chart for a certain year, he noted that "a ruler who thought as much of astrology as the person in question [Wallenstein], and knew all this, would without doubt send such a commander with such an impressive constellation against his current foreign enemies."[65] Thus, the very belief in astrology (however misguided, in Kepler's estimation) could in fact lead to fulfillment of its predictions, more through credulity than any supposed power of the stars. Kepler did, however, feel that comparing the nativities of two persons could allow certain deductions about their mutual sympathy or antipathy. On those grounds, he predicted that between Ferdinand and Wallenstein there would be "not particularly affection or inclination toward each other, but rather frictions." That indeed proved to be the case. Here again, Kepler's astute personal observations of both men may have helped him more than the stars. Finally, Kepler predicted "horrible disorder" for March 1634. Wallenstein was killed on February 25 of that year, baring his breast to his assassins' weapons as if he regarded their assault as predestined. Kepler himself had died four years earlier; in his personal horoscope for 1630, he noted that the planets stood at the same positions as at his birth.[66] Might Wallenstein have been able to elude assassination had he been less superstitious about Kepler's prediction? Then too, one wonders whether Kepler's own limited belief in astrology actually affected his response to the fateful recurrence in the heavens. Might they have moved him to relax his vigilance against sickness in order to accede implicitly to the celestial design he had already discerned?[67]

Though it may seem that we have departed far from medicine to consider these strange predictions, for Kepler and other thinkers of his time, the matters we have considered were connected in ways that need specifying for us who look at the world so differently. At least since Plotinus and other neo-Platonists, if not already for Plato or Boethius, the cure of the human microcosm depended on what they thought was its many analogies to the macrocosm, the planets and firmament. Thus, throughout this chapter, we have noted the interplay between medicine and astrology, as well as music. Kepler sought to establish the harmony of the cosmos on new foundations, which accordingly would have affected the understanding of the human body and soul, as well as of the formation of comets and new stars. Ficino and his successors used music to draw down cosmic influence; Kepler sought to understand that cosmic music in new detail, using new and more accurate observations. That led him to consider how those harmonies vary in dissonance over time, reflecting the complex interrelation between planetary positions and velocities. Thus Kepler found himself acting as a physician to the whole cosmos, observing its changing harmonies in the fluctuating symptoms of misery and famine around him. Like Galen, he could not alter the fundamental rhythms of those harmonies, but he could try to predict their crises and even to apply that knowledge to specific individual cases, such as Wallenstein and himself, trying to fathom the hard and soft harmonies that imperiled the Protestants and finally extinguished his own life.

6

The Musical Disease

From the Middle Ages onward, medical practitioners and natural philosophers devoted special attention to a unique disease: tarantism, a deep melancholy that seemed to be caused by a spider's bite but responded only to music and dance. The phenomenon of tarantism brought into question both the physiological reality of this disease as well as the surprising effects music seemed to have on its sufferers. Indeed, its recognition as a "disease" was highly contested, given that most mental conditions were not considered medical matters or treated by physicians until the eighteenth century. The peculiar "disabilities" of those subject to tarantism responded intensely to music and dance. Because their purview included poisons, physicians undertook to treat those presumably bitten by spiders, but their attention then turned to the role of music, especially whether it operated through cosmic harmonies or physical vibrations. The growing emphasis on music's vibratory effects accorded with the rising view of the body as an ensemble of fibers, essentially a musical instrument that naturally would respond sympathetically to other instruments.

To this day, tarantism is associated with southern Italy, especially the regions of Apulia and Calabria near the sole of the Italian boot. Consider this account from a British traveler in 1969:

On the way to Taranto I stopped in a small town to look at a church, and on my way back to the car I heard the sound of music. It was a quick kind of jig tune played on a fiddle, a guitar, a drum, and, I think, a tambourine. Looking round for the source of this sound, I saw a crowd standing in a side street. Glancing over the heads of the spectators, I saw a countrywoman dancing alone with a curiously entranced expression on her face, her eyes closed. She held a red cotton handkerchief in her hand which she waved as she undulated around the circle with more grace than I should have expected. I was surprised by the gravity of the crowd. There was not a smile. There was something strange about this. I wondered whether the dancer was mad, or perhaps—unusual as this would be—drunk. Glancing round at the set faces, I did not like to ask any questions, and, not wishing to intrude upon what was obviously a painful scene, I turned away.

Some days later, an Italian friend in Taranto informed him that "the woman had been 'taken' by a tarantula spider and she was dancing, and might dance for days until completely exhausted, to expel the poison." According to his friend, the traveler had seen something rarely witnessed at the time. When he objected that he thought the tarantula bite was not poisonous, his friend shrugged and replied that "this has been going on for centuries. Who can say what is at the back of it?"[1]

Because that part of Italy had already been colonized by Greeks in the eighth century BCE (near the site of Sparta's only colony), some scholars speculate that tarantism goes back to ancient rituals of Dionysian possession, but no secure evidence remains.[2] Ironically, this region also became a stronghold of Pythagoreanism; both Archytas and Eudoxus were from Taranto, which Archytas ruled as general (*strategos*) for seven years. One wonders about the possible relation between the Pythagoreans, so devoted to mathematics and music, and archaic rites that used curative dance, but again no definite evidence remains. Still, in what follows, we will have many occasions to recall the description of musical possession given by Plato in his *Ion*: "That's how it is with the Corybantes, who have sharp ears only for the specific song that belongs to whatever god possesses them; they have plenty of words and movements to go with *that* song; but they are quite lost if the music is different."[3]

The extant documentary trail begins in the Middle Ages in writings about poisons such as the *Papal Garland on Poisons* (*Sertum papale de venenis*, 1362) by the Paduan physician Gulielmus de Marra, who analyzed tarantism as a form of melancholia:

> For what reason do those bitten by the tarantula find miraculous recovery in various songs and melodies? It must be said that, music and song being a reason for cheerfulness, both are thought to be useful for almost every poison: and since the bite of the tarantula produces a melancholic disease, and because the melancholy is treated in the most suitable way with cheerfulness, it follows that songs and music are quite healthy for those who have suffered such a bite.[4]

He noted that "the common people and the uneducated claim that the tarantula emits a music at the moment in which it bites, and that when the sufferer hears melodies or songs similar to this aforementioned music, he receives great benefit from it, but in my opinion this is not so." Instead, Gulielmus referred to the individual taste of each sufferer, noting that "when they hear sounds or songs which are pleasing to them, their souls are joyous, and since cheerfulness is an excellent medicine for that disease, they recover and return to life" because those songs attract the spirits "from the inside of the body to the periphery. This movement prevents the poison from penetrating the interior," thereby sparing the vital organs there.

Though at the time clerics, rather than physicians, treated melancholia, one infers that Gulielmus justified his professional involvement as physician because these were presumably victims of poison, not just ordinary melancholiacs. His discussion of tarantism appeared between passages on the poison of the basilisk and the bite of mad dogs.[5] Nevertheless, this may have been one of the first recorded instances in which a physician treated melancholia, at least in this form. Gulielmus's rejection of the "popular" account of the spider emitting music before biting indicates doubts whether music could really be both cause and cure, poison and remedy.

Yet Ficino used tarantism as a crucial instance confirming his argument "that there is indeed in certain sounds a Phoeban and medical power." He recounted that "in Puglia everyone who is stung by the phalangium [commonly called a tarantula] becomes stunned and lies half-dead until each hears a certain sound proper to him. For then he dances along with the sound, works up a sweat, and gets well. And if ten years later he hears a similar sound, he feels a sudden urge to dance. I gather from the evidence that this sound is Solar and Jovial."[6] Ficino could scarcely have missed the resemblance between this specificity of curative sound and Plato's Corybantic dancers, who await the unique song associated with their tutelary deity. Then too, as Tomlinson points out, this description "also places tarantism clearly in the context of Ficino's natural-celestial magic: the iatromusic [healing music] exemplifies the natural power of sound but, like any other natural power, its proximate source lies in the stars (here Jupiter and the Sun). Thus the musical cure of the spider bite is for Ficino one more instance of the magical interconnectedness of mundane and celestial things."[7]

Others tried to understand the ways in which music might operate beyond its ordinary application to dispel melancholy. For instance, in about 1510, Leonardo da Vinci noted in one of his manuscripts that "the bite of the tarantula maintains a man in his intention, that is, whatever he was thinking at the time he was bitten."[8] Accordingly, if the spider emitted music, the one bitten (called the *tarentato* or *tarentata* by gender) presumably would remain fixated on that music. Here da Vinci echoed an idea found already in texts on poisons written about 1425. In 1521, the physician Ferdinando Ponzetti speculated that the spider's venom passed through the skin to the nerves and brain, thereby causing the blocking action that results in the fixation.[9]

Then too, a persistent tradition linked poison to sorcery. As late as 1706, the Neapolitan physician Ludovico Valletta thought that many *tarentati* were "possessed by the devil." Yet canonical exorcism was strangely ineffective, as he discovered in the case of a woman stung by a tarantula whose mind Valletta judged to be "occupied by tremendous phantasms, or rather assailed by a whole army of violent demons":

There were two other priests together with me to attend the performance, and we all concurred that it was a case of possession, therefore we began to practice exorcisms. But while we were busy doing so, a musician chanced to pass by playing his instrument. Then, suddenly, the woman got out from under our control, and as if seized by an insane frenzy, she ran out of the house dancing in an unseemly manner and followed the musician. Having realized what had happened, the latter stopped in his tracks and interrupted his playing. Then the woman, as if struck down, fell to the ground lifeless. In the meantime, her family had left the house, and they asked the musician to begin playing his music again. He complied, and the woman rose to her feet immediately, restored to her full vigor, and began a frenetic dance which lasted two hours without interruption. After having danced for two days in a row, she recovered completely.[10]

At least in this case, music and dance seemed to have powers greater than those of exorcism.

As this story implies, the Catholic Church struggled to deal with tarantism, which evoked ancient rituals of demoniacal possession, perhaps linked to immemorial pagan rites. Dance was deeply alarming to a church that strove to turn its back on all pagan practices; as St. John Chrysostom put it, "Where there is a dance, there also is the Devil."[11] Still, the Roman Church placed the dancing *tarentati* along with others afflicted with bites under the protection of St. Paul, who had shown his power over venomous serpents in Malta, where "a viper came out because of the heat and fastened itself on his hand" without harming him.[12] His chapel in Galatina became the center toward which *tarentati* gravitated, including women called "the brides of St. Paul."

Physicians also struggled to understand tarantism in relation to other forms of dancing mania, such as "St. Vitus's dance," which they found especially perplexing in relation to contemporary epidemics that included the Black Plague, leprosy, and many others. In the earliest recorded instance (1528), a large number of people both young and old began to dance in the marketplace of Strasbourg, continuing to dance for a night and a day until they collapsed. They were taken to the chapel of St. Vitus near Zabern, where they were give "small crosses and red shoes, and a mass was held for them," according to an account written in 1587. As if to sanctify their dancing, "On the uppers and soles of the shoes crosses were drawn in consecrated oil and in the name of St. Vitus they were sprinkled with holy water. This helped all of them."[13] The saint's name thus became associated with their malady, which the reformist Swiss physician Paracelsus grouped with other dancing plagues he called *choreomania* or, alternatively, *chorea lascivia*. In his view, mania was "a transformation of reason and not of the senses" characterized by "frantic behavior, unreasonableness, constant restlessness, and mischievousness." Still, he thought that those seized by the "voluptuous urge to dance" were largely "whores and scoundrels," their malady befitting their "recklessness and disgraceful living in which there is neither reason nor sense," which he thought occurred more frequently among women than men.[14]

Later writers discriminated between these various "dancing plagues." In 1679, the physician Hermann Grüber noted that both the victims of St. Vitus's dance and tarantism "become so engrossed in dancing and running about that they ultimately become quite exhausted and fall to the ground." Still, he concluded that St. Vitus's dance did not seem to be caused by a poisonous bite, noting that its sufferers are "seized with that peculiar madness, and, without any particular music, do tire themselves out by their movements." For Grüber, the role of music as a causative factor was essential to diagnose these conditions because the rather spasmodic and paroxysmal character of the movements in St. Vitus's dance (and of a similar malady called St. John's dance) operates "independently of the music," as he put it.[15] In contrast, a 1549 account described a young *tarentato* who "danced with a certain elegance, with the movements of his body and mime in keeping with the rhythm of the tambourine." So crucial was the music that when it stopped, the young dancer collapsed on the ground but rose and danced again when it recommenced.[16] Indeed, like the Corybantes Plato described, *tarentati* did not respond indiscriminately to music, each attuned only to the particular songs and dances capable of rousing and curing them.

Over time, certain commentators drew attention to what they considered erotic elements in tarantism, perhaps manifest first of all in the insistent energy of the dancing. This was often coupled with gendering the *tarentati* (the generic plural) specifically as female (in that case called *tarentate*), though in fact the earliest accounts do not bear this out.[17] Already in 1491, in his satirical dialogue *Antonius*, the Neapolitan humanist Giovanni Pontano argued sarcastically that Apulians should be considered to be exceedingly happy because they

have a good and ready excuse to justify their insanity, attributing its cause to the spider they call the tarantula, the bite of which causes men to go mad. And this constitutes their greatest happiness, since, if someone wanted to, he could obtain—quite licitly—the desired fruit of their madness. . . . Women are wont very often to be bitten by this spider; and then, since the poison cannot be extinguished in any other way, it is licit for them to unite with men, freely and with impunity. In this way, what for others would be a shameful act, for Apulian women is a remedy.[18]

Pontano seized on the women to taint all the *tarentati* with deception and immorality in a process of "feminization" that scholars have noted in the derisive treatment and marginalization of other contested cultural practices by powerful males, such as Pontano at the court and Academy of Naples. Similar innuendos also reemerged in later phases of controversies about tarantism.

Though tarantism increasingly became the concern of physicians, they disagreed about the nature of this peculiar condition and its reality even as they medicalized it and included it in medical texts. Probably the earliest detailed case history was given by Epifanio Ferdinando in *One Hundred Histories or Observations and Medical Cases* (*Centum historiae seu observationes, & casus medici*, 1621). A native of Mesagne in Apulia, Ferdinando studied medicine in Naples; he had a distinguished career, called to professorships in Padua, Parma, and Rome before going back to Mesagne, where he returned to practice and write. There, he claimed to have experienced tarantism directly, not merely through second-hand accounts. Educated in the skeptical medical school of Naples, his interests included philosophy, chemistry, mathematics, astronomy, and astrology. His treatise is filled with learned references to the classical authors, in many respects closer to a kind of literary critique than pure clinical description. From the beginning, he presented tarantism as highly contested:

If there is anything in the universe or in the theater of the world, a very complicated problem or most beautiful theorem, more pleasing, curious, difficult, and admirable than the controversy about the bite of the tarantula, I have always thought that that would be the observation of its nature, of its properties, of its extraordinary ability to weave its web, of the multiplicity of forms with which it reproduces its species, of its birth, of its sex, of the fecundity and number of its offspring, of the amazing variety of symptoms its bite produces, and finally of the unusual and almost unbelievable cure found through music, unknown to the ancients.[19]

Ferdinando instanced the case of Pietro Simeone, a nineteen-year-old "of hot and dry temperament" bitten on the left upper abdomen by a "tarantula," the generic term then used to encompass a variety of venomous spiders and insects.[20] Suddenly the young man "felt a violent pain in the bitten part and immediately, in the space of three Ave Marias, fell to the ground with cold throughout his body, shivering, pain in the pubic region, tension in the penis; his legs were weak, he sighed and wept, felt suffocated, wanted to scream but could not." The next morning, he was taken to the city, and musicians were summoned. Pietro responded above all "to that kind of music called *Catena*; the young man jumped up and suddenly began to dance with abundant sweat. . . . For a week he danced and sweated: he was perfectly cured by music."[21]

Drawing on such cases, Ferdinando argued against those who think that "this bite is by no means real but fictitious, chimerical, a kind of melancholy or dementia."[22] On the contrary, he considered the bite to be real and toxic, the curative power of music incontrovertible: "Not a few are dead because of the bite, especially because they lacked the support of the antidote and above all because there was no music."[23] To be sure, Ferdinando noted that "the first thing to do in the cure is to hunt down the venom as rapidly as possible and reinforce the natural heat," to counteract the deathly cold that overwhelms the victim. "Second, it is necessary to extract the venom from the bitten part."[24] To that end, he advised suction of the wound by mouth to remove the venom, drinking good wine and aquavitae, and ingesting an antidote he called *antiphalangium*.[25] In the end, he emphasized that "all these therapies are certainly worthwhile in order to expel the venom, but beyond that, in all of Puglia no other remedy is used that is more effective than music."[26] To demonstrate this, besides Pietro he brought forward many other cases, including "an old man of ninety-four who could not walk without canes and moreover was helpless; bitten by a tarantula, as soon as he heard the music he began to jump and dance like a wild goat."[27]

Saddest of all he considered the cases in which the use of music was denied out of mistaken pride: "Because in many cases the terrible power of the venom is cast out by the power of musical instruments and dance, those sick people who refuse to dance do poorly and almost go crazy. . . . I have observed many religious men and women [clerics and monastics] who did not wish to dance out of shame and dignity and would have been dead, if (whether they wished it or not) music had not intervened. Those who had entirely renounced music and all other remedies fell into malignant illnesses."[28] Ferdinando recounted that the bishop of Polignano "as an experiment in jest had himself bitten in the summer by a tarantula and I call on God as witness that, had he not been treated with music and other antidotes, he would have been dead."[29]

Ferdinando acknowledged various erotic aspects of tarantism. His main case history concerned a young man whose immediate symptoms include pain in the pubic area and "tension in the penis." More generally, Ferdinando inquired, "Why do some have an erection of the member immediately after the bite, as if they had been affected by priapism?"

He considered this a fact "on which there is no doubt," attested by other authors as well as himself. Ferdinando hypothesized that "a strong flatulence being present in the body [as he observed], an erection could occur, above all through the concourse of many *spiriti*," using the term for the "subtle air" we encountered in the previous chapter.[30] He also noted that sometimes young girl *tarentate* "would exhibit their intimate parts, tear their hair, scream, howl."[31] He treated all these as purely physiological effects of the venom affecting both men and women. Indeed, Ferdinando noted that "in some regions there are more women *tarantate* and in others more men," depending on their relative degree of exposure to being bitten: in some places women spent more time in the locales frequented by tarantulas; in others, men (including Mesagne, his own village).[32] Though he did not mention the innuendo that women simulated tarantism to give them erotic license, Ferdinando implicitly contradicted that view.

Throughout, Ferdinando pointed to the general power of the venom as being somehow sufficient to cause many diverse symptoms. Thus, in order to explain why some *tarentate* he knows "have danced for ten, fifteen, seventeen, twenty, and thirty years," he considered that "this kind of venom, hidden and unknown in what it does, operates also through its viscosity, density, and resistance. Because of this, it can persist for many years, like the French disease [syphilis]."[33] He noted that the *tarentati* dance every year (or sometimes even twice a year), "almost always at the same time of year," usually summer, "because the venom, or its residues, are not eliminated from the body; indeed, such residues, fomites [disease-transmitting objects], and impurities can be carried for many long years. . . . Whatever is the residue of the malady generally causes a recurrence."[34] The summer heat presumably triggered the relapse, as it did the initial effect of the venom, which did not seem to be potent at other times of year or other localities than the hottest parts of Apulia. Still, it seems hard to understand why this condition should be so restricted geographically or why recurrences should not occur in other hot seasons beside the yearly midsummer feast of St. Paul.

His explanation why exactly music should be so effective against the venom responded to his question, "Why are all [*tarentati*] not pleased by the same music or melody but some are conquered by this one or by that? All this I can derive from the variety of tarantulas, indeed each tarantula loves one particular melody." His authority was not the popular belief mentioned (and dismissed) by Gulielmus but instead the testimony of Sazio Lupo, "an excellent surgeon and musician" who lived in Mesagne and told Ferdinando the "extraordinary fact" that "he knew well that this tarantula was attracted by one type of music, that one by another. It was just like this: playing in one mode this [tarantula] danced, but playing in another it did not; every day he did so with each kind. This comes from the different complexions of the tarantulas," their different natures as expressed in their own musical preferences.[35] Likewise, *tarentati* have a particular love for a certain color (red in the case of Pietro and of the woman in the story at the beginning of this chapter), corresponding to the color of the tarantula that bit them.[36]

Thus, Ferdinando gave a learned basis for the popular belief that the spider sings as it stings. He argued that "so numerous and almost miraculous are the properties of this little animal" that the tarantulas are "the genius of nature, as Aristotle exclaimed, . . . worthy of admiration that exceeds human comprehension" because "they clearly are learned in the mathematical disciplines [of the quadrivium], namely arithmetic, geometry, astrology, and music."[37] This claim seems less whimsical when one reflects that according to Plato's account in the *Timaeus*, animals (including spiders as well as humans) were formed according to the same mathematical and musical archetypes that shaped the whole cosmos, as studied in the quadrivium. Still, Ferdinando found it remarkable how much these exalted subjects seem familiar to the little tarantula.

These arachnoids know arithmetic because "their web consists of threads numbered ordinately"; geometry, because in a web "it is possible to see how many and how long is the distance from the center to the circumference, which seems measured with a compass." Moreover, if a web is damaged, "they are most able to repair the work" presumably because of their knowledge of geometry. If they build their web high up, "it is a sign that there will be much rain or overflowing rivers, according to Pliny." Such foresight indicated their knowledge of weather and the seasons, hence of astrology (which included those subjects). Because of the tarantula's inherently cold temperament, the construction of its web could also predict "a very severe winter or a very strong atmospheric disturbance or contrariwise a storm." Their knowledge of music was demonstrated by Lupo's experiments, especially the way different species of tarantula responded to different musical modes. Accordingly, Ferdinando considered it reasonable that the *tarentati* likewise respond to the specific mode associated with the kind of tarantula that bit them. He asserted that the "almost incredible" effects of music on the *tarentati*—enabled precisely by the musical affinities of the tarantula that bit them—represented an important discovery "of which no memory can be found among the ancients," as much as they knew about the connections between music and the natural order. This discovery by the moderns was "the fruit of science, of extraordinary commitment and great curiosity."[38]

Indeed, when he listed the different kinds of music, Ferdinando included *musica mundana*, which he defined as consisting "in the connection of diverse elements and in harmony in the order and proportion of all things," in contrast to the earthly practice of *musica instrumentalis*, ordinary playing and singing. Despite the exalted stature of the cosmic music, what heals the *tarantati* "is above all instrumental and vocal music, indeed this double music elicits unbridled dancing and movement of the body and thereby mitigates the deadly force of the venom, with a great deal of sweating of the whole body."[39] Ferdinando thought this also applied to animals, having seen a tarantula bite a wasp; "by chance, a guitarist happened to be there who began to play and in the same tempo both the tarantula began to move rhythmically and the wasp to fly around. We know also of a rooster who had been bitten by a tarantula and so started dancing and jumping beyond his usual habits."[40]

In the end, though Ferdinando gave respectful attention to the musical theories of the "divine Plato" and other ancients, he did not base the therapeutic effect of music on the disembodied numerical ratios that were its basis for Ficino. Instead, Ferdinando emphasized the physical movement of the body that music induced through dance, for "the fact is sufficient that music pushes you to dance, so that with that much shaking of the body, the dormant forces of the poison, kept safe and quiet, are removed and chased out by sweat. Nature, restored by music, wins so great a duel."[41]

Ferdinando noted that "those deaf from birth in no way sense the music and therefore cannot be excited by it," so that they "run the risk of dying" if they are bitten unless the venom can be removed. Then he speculated that "if instead they have only been deaf a short time and see musical instruments, then (I believe) they can dance, above all because the venom stimulates them from inside and furthermore the imagination is urged through the sight." But he then returned to his own experience in which "I have seen one who was hard of hearing and been bitten by a tarantula and who always held his ears close to the musical instruments and thus jumped and danced; if he were a bit further off, he held his ears as close as possible to the trumpeter."[42] These clinical experiences implied that music's effect depended on the *tarentato*'s ability to feel its vibrations physically and therefore to dance.

The famous Jesuit polymath Athanasius Kircher gave perhaps the most widely circulated and wide-ranging account of tarantism, which he treated as a crucial phenomenon. Long a professor of mathematics and Oriental languages in Rome, Kircher was at the center of various intersecting networks of informants that spanned the globe, including learned persons in many different fields. Among other things, he knew twelve languages, claimed to have solved the riddle of Egyptian hieroglyphs, proposed bold views on subterranean geology, gave an ambitious description of China, and wrote a voluminous work on the theory and implications of music; as he paraphrased Plato, "nothing is more divine than to know everything."[43] Throughout his vast body of works, "Kircher regarded magnetic attraction and repulsion as the *lingua franca* of all creation."[44] He considered music as a particularly important example of this magnetism, reaching back to the description Plato gave of how poetic—hence musical—inspiration is transmitted from person to person, just as a magnet not only pulls iron rings but "it also puts power in the rings, so that they in turn can do just what the stone does—pull other rings—so that there's sometimes a very long chain of iron pieces and rings hanging from one another. And the power in all of them depends on this stone. In the same way, the Muse makes some people inspired herself, and then through those who are inspired a chain of other enthusiasts is suspended."[45]

Kircher considered tarantism a prime example of "the magnetic tie among all the natural things in this universe" and presented it in three of his major works, beginning with the chapter "On Tarantism and the Apulian Spider Tarantula, Its Magnetism and Strange Sympathy with Music" in his book *Magnet, or on the Magnetic Art* (*Magnes, sive de arte magnetica*, 1641).[46] He did not have any firsthand experience but (characteristically) recounted information he received from two Jesuit informants in Apulia that described the same basic phenomena as Ferdinando, though without his medical contextualization and analysis. Kircher emphasized "that for a certain resemblance of nature [tarantula poison] arouses the same reaction in man and the tarantula. . . . As the poison stimulated by the music drives the men to dance through the continuous excitation of the muscles, it does likewise the tarantula."[47]

In confirmation, Kircher mentioned experiments in which music was played to a spider hanging in a jar of water. According to his Jesuit informants, a certain melody would induce the spider to leap and "dance"; if that music ceased, the spider would stop moving.[48] Kircher interpreted this as reflecting the similar behavior of a *tarentato*, who would dance when hearing a certain melody but would stop dancing and collapse were the music to cease. He asked his informants to ship him some live tarantulas in glass vials so that he could experiment with them himself, but unfortunately the animals died in transit.[49] Kircher also transmitted certain interesting details about the social milieus of the *tarentati*, which included not just peasants but sometimes nobles and even Capuchin monks. So important were their ministrations, Kircher recounted, that in Taranto, musicians were paid a wage by city officials so that poor *tarentati* also could hear the music that would enable them to dance and recover. Above all, Kircher included the texts and music of a number of therapeutic dances his informants gathered (figures 6.1 and 6.2).

These tarantellas (as they came to be called) give further insight into the relation between music, dance, and the *tarentati*. The music often came with an erotic text, such as: "Where did the tarantula bite you? Under the fringe of my dress."[50] Kircher included a tarantella that compared the body of a lovesick man to a musical instrument (figure 6.2; ♪ sound example 6.2): "This breast has become a harpsichord of love / Keys, the senses, feeling and ready / Strings, the tears, sighs, and pains / Sound hole is my heart, mortally wounded / Rod is a point, my ardors are wounds / Hammer is my thought and my fate / Conductor is my lady, who, singing without end, / Joyfully sings my death."[51] Indeed, the connection between love and death winds through a number of the tarantellas Kircher presented, a feeling he called "joyful misery."[52]

Though Kircher's descriptions were sometimes plagiarized in later accounts of tarantism, his reception by those promoting the new sciences showed their dissatisfaction with his ideas of cosmic sympathy and magnetic attraction. After reading *Magnes*, René Descartes wrote to a friend that Kircher "is more charlatan than scholar."[53] Evangelista Toricelli wrote Galileo Galilei describing that book's "beautiful engravings" and "extravagant words"; though praising the inclusion of "the score of that music that is said to be an antidote to the tarantula's venom," Toricelli noted that he and his friends "laughed for quite some time."[54]

The next detailed clinical discussion of tarantism came forty-five years after Kircher's *Magnes* and eighty years after Ferdinando's treatise, showing how the study of these

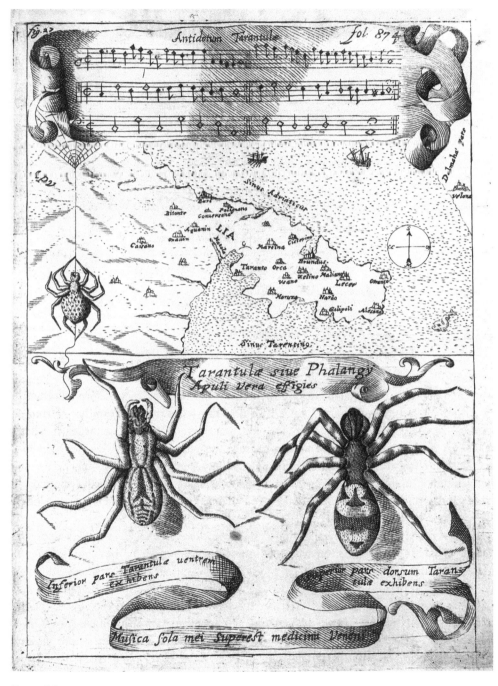

Figure 6.1
Athanasius Kircher's illustration of the map of Apulia and its tarantulas, along with a musical "antidote for the tarantula" (♪ sound example 6.1), from Kircher's *Magnes, sive de arte magnetica* (1641).

Figure 6.2
A tarantella with lyrics given both in Latin and Apulian dialect (♪ sound example 6.2), from Kircher's *Magnes, sive de arte magnetica* (1641). Text: "This breast has become a harpsichord of love / Keys, the senses, feeling and ready / Strings, the tears, sighs, and pains / Sound hole is my heart, mortally wounded / Rod is a point, my ardors are wounds / Hammer is my thought and my fate / Conductor is my lady, who, singing without end, / Joyfully sings my death."

physical vibrations had moved into a new phase reflecting a new concept of the human body and of medicine. Though Giorgio Baglivi was born in Ragusa (modern Dubrovnik) of Armenian and Italian parents, at age fifteen he moved to Lecce in Apulia; he studied medicine at the great centers in Salerno, Padua, and Bologna. In time, he moved to Rome, where he became an eminent professor, personal physician to Pope Clement XI, and a fellow of the Royal Society. Like Ferdinando, Baglivi had a personal connection to the native land of tarantism as well as an up-to-date medical education, but his discussion of music and tarantism showed the advance of mechanical ideas in medical discourse underway in the century between them.

Baglivi referred to Kircher slightingly as one of those "who have judged upon the Credit of others, rather than on what they themselves have seen," but drew on Ferdinando's treatise as the work of a fellow physician and Apulian.[55] Still, where Ferdinando considered that the "cold and dry temperament" of the tarantula causes the morbid coldness it imparts to its victims, Baglivi treated the venom in terms of its "too active, volatile and rapid Particles," preferring such mechanical explanations to the older language of temperaments.[56] The title of Baglivi's *The Practice of Physick Reduced to the Ancient Way of Observation* (*De praxi medica ad priscam observandi rationem revocanda*, 1696) reflected his overarching project to connect modern Baconian observation with ancient practice; indeed, his book begins by referring to Francis Bacon, whom he quoted often and approvingly. In this book, Baglivi's "Dissertation on the Anatomy, Bite, and Effects of the Tarantula" takes a central place, the longest and most extended of the clinical studies he included, meant to "take in a great many Things relating to the Mechanical Doctrine of Musick, Poison, and Dancing."[57]

Baglivi incorporated Ferdinando's account but took it much further, giving a more detailed description of the natural history of tarantulas and several other case histories of tarantism. He noted that "all the Authors that write of the Tarantula have either gone upon Hearsay or coin'd several Things at Pleasure. None of 'em ever was in Apulia," except for Ferdinando.[58] Thus, based on "the infallible Experience of the People of this Country," Baglivi dismissed the "idle Story" that the effect of the venom persists only as long as the spider lives.[59] He also gave an insight into the aural world of the tarantula, which seem to hunt by sound, so that "when the Peasants have a mind to catch 'em, they come to their Holes, and, with a small Reed, imitate the murmuring, buzzing Noise of Flies; upon which the *Tarantula* comes forth in quest of the Flies or Bees thus counterfeited, and falls into the Snare."[60] Then too, Baglivi had "seen the Leg of the *Tarantula* pull'd off, dance for almost two Hours together," it being "in the very Nature of the Insect to be inclin'd to Leaping, and other vehement Motions, and it has scarce ever been observ'd to rest."[61] Here again, the spider's own behavior seems to prefigure its effect on its victims. Likewise, Baglivi noted that "those that are bit by a mad Dog will howl, and in other respects act like Dogs, which proceeds perhaps from the Impressions of certain Ideas of the mad Dog upon the Poison."[62]

In accord with his desire to prefer experiment and observation to tradition, while in Naples Baglivi had a tarantula brought from Apulia to bite a rabbit, which developed pronounced discolored swellings, had difficulty breathing, and collapsed to the ground: "The Musician came and us'd great Variety of Tunes, yet it had no effect" on the poor rabbit, who died after five days.[63] Yet in 1693, a Neapolitan physician allowed himself to be stung in the presence of six witnesses and a public notary. He felt pain comparable to a fly sting and experienced some swelling and discolored markings, but nothing like the spectacular symptoms of the *tarantati*, certainly nothing that called for music and dancing.[64]

Rather than inferring from this case that its sting is essentially harmless to humans, Baglivi instead concluded that "a *Tarantula* transported to foreign Countries does not produce fatal Symptoms by stinging; and that the Heat of such Countries is not active enough to elevate the Venom to a due Pitch."[65] He believed the venom could only be activated in "the excessive Blood and Climate of the *Apulians*," who live in the region having "a greater Frequency of melancholick and mad People" than anywhere else in Italy.[66] As further confirmation, he noted "the great Frequency of mad Dogs, whose Madness is justly attributed to the scorching Heat of the Air" in Apulia.

Baglivi considered the notable "Period and Revival" of tarantism to reflect the behavior of poisons in general. He noted that "the Poison of the *Tarantula* renews it self infallibly every Year, especially about the same time when the Patient receiv'd it," usually midsummer. Like Ferdinando, Baglivi compared this to "the impressions of the vene-real Distemper" that "stick to the blood for thirty years together and better, without any Trouble or Injury to the Patient's Health; but then there is as vigorous a Return of the Symptoms, as if it were upon the first Onset." Neither of them seem aware that syphilitic relapses do not have the annual regularity of tarantism, which they both ascribed to the seasonal heat (evidently considering the Feast of St. Paul on June 29 to be merely coincidental). Baglivi compared the tarantula's venom to the "Poison of a Dog . . . having its stated Seasons of recovering its Vigor," as in the case of "a Matron, who being bit by a mad Dog, found the Poison renew'd its Vigor every seven years, for almost thirty Years together, till at length she died."[67]

Still, neither Ferdinando nor Baglivi used the term *crisis* to describe the periodic dancing of the *tarentati*, probably because only a fever could have a crisis, as defined by Hippocrates and Galen. By Baglivi's time, though, the whole concept of crisis had become contested. Just before turning to tarantism in his collection of case histories, he reviewed the status of this venerable term, noting that "the greater part" of medical writers "for almost these forty Years have made it their business to deride" the ancient physicians and their theories, including the concept of crisis.[68] For this, Baglivi blames the reformists Paracelsus and Jan Baptist van Helmont, who "charg'd *Galen* and almost all of the ancients with Error and Ignorance" in order to "raise their own Sect upon the ruins of the *Galenic*." In contrast, Baglivi thought that "Crisis's and critical Days are not Chimera's but certain Motions common to every Disease by a Physical Necessity, as being the means by which they compass the due Pitch of Solution and Maturation."[69] He asserted "a Law of Nature" that "the Blood is a fermenting Liquor" that eventually reaches the crisis of the fever fixed by the character of that fermentation, so that it would be unwise to hurry that process "before the set Time of Nature; for such Violence makes 'em corrupt rather than ripen."[70] Extending metaphorically this process of fermentation to gout, Baglivi used "crisis" to describe a critical stage in gout, though that disease does not involve fever.[71] Fifty years later, we find medical references to the experience of "crisis" by *tarentati*, showing that this ancient concept was finding new life beyond its original boundaries, as we shall later consider.[72]

Whatever one calls the critical events the *tarentati* experienced, Baglivi agreed with Ferdinando that music had a real effect he ascribed to its vibrational impact. Baglivi took a purely physical approach in which "Musick is one of those Motions that smartly strike the Air, and disposes it to brisk Undulations"; unlike Ferdinando, he did not even mention *musica mundana* or Platonic notions of cosmic harmony.[73] Instead of the static ancient harmonies, Baglivi stressed the importance of motion because "whatever lives, whatever grows, and whatever undergoes the sensible Mutations of Life and Destruction, is in a perpetual Motion."[74] Through its basis in motion and vibration, music was capable of powerful therapeutic intervention because "the slightest Impressions of Motion produce admirable Effects, by communicating the *Impetus* of the Contact to the very remotest

Parts, as [Giovanni] *Borelli* has demonstrated," here referring to the recent work of a pioneer who applied physical knowledge about machines to the structures of the body in his *De vi percussionis* (*On the Force of Percussion*, 1667), to whom we will return in the next chapter.[75] These "Undulations of the Air have an influence on the solid and fluid Parts of the Body" of animals, not just humans. Baglivi instanced the "disagreeable Undulation of the Air occasion'd by the Motion of a File or a Saw" that makes the teeth of animals grate, who then feel "uneasy; and that because the Particles of the Air thus mov'd, being disproportion'd to the Pores of the Nerves inserted in the Roots of the Teeth, do so distort and twitch 'em, that they are either benumb'd or affected with Pain."[76] The painful affect comes from disproportionate vibrations. Contrariwise, he mentioned the "gentle shivering over the Skin, and a Sort of Erection of the Hair" upon hearing "an unwonted and agreeable Harmony of Musick." Such physiological responses he also noted in animals, such as angry elephants appeased by music, bees delighted by the sound of metal, swans by harps and singing.

Applying these insights specifically to the "Cure of Persons stung by a *Tarantula*," Baglivi considered that

'tis probable, that the very swift Motion impres'd upon the Air by Musical Instruments, and communicated by the Air to the Skin, and so to the Spirits and Blood, does, in some measure, dissolve and dispel their growing Coagulation; and that the Effects of the Dissolution increase as the Sound it self increase, till, at last, the Humours retrieve their primitive fluid State, by vertue of these repeated Shakings and Vibrations; upon which the Patient revives gradually, moves his Limbs, gets upon his Legs, groans, and jumps about with Violence, till the Sweat breaks and carries off the Seeds of the Poison.[77]

The vibrations through the skin then reach to the coagulations in the blood that Baglivi mentions, affecting also the humors, which he continued to consider significant, though this ancient concept was then coming under critical scrutiny.

In comparison to the liquid humors, Baglivi and other physicians were increasingly interested in the solid constituents of the body, as we shall see in the following chapter. Something of this shift can be seen also as he turned from the liquid to the fibrous parts of the body, noting that music

chiefly and immediately affects the Organ of hearing, that lies very near the Brain; and affects even the Brain it self, or the minute or fine Fibres in which the Spirits lye drooping and almost sunk, till the continual and forcible Contact of the Musick makes them march out upon the Membranes of the Brain; upon which having partly recover'd their Motion, they enter with greater Facility and Agility into the little Tubes of the Nerves and Fibres, and so recover their former Correspondence with the Humours and solid Parts.[78]

Here, the fluid parts seem to be gateways or connections to the fibers that writers before Baglivi had identified as nerves and brain tissue, fibers that will become even more important in the medicine of the succeeding century. But Baglivi wanted to maintain the significance of both these new fibers and the old fluids and humors, consistent with his general desire to reconcile ancient and modern medicine. To do so, he found that treating *tarentati* should not rely on using medicines that increase perspiration because these only affect fluids in the body, "whereas the musical Sound affects at once both the Solids and Fluids, and that very forcibly, and by vertue of its Percussion upon the small Fibres of the Brain, in which the Secretion and Distribution of the nervous Juice is immediately perform'd."[79]

In his account of the erotic aspects of tarantism, Baglivi did not mention the priapism that Ferdinando noted but commented on female behavior: "Maids and Women, otherwise chaste enough, without any Regard to Modesty, fall a sighing, howling, and into very indecent Motions, discovering [i.e., exposing] their Nakedness."[80] Though he considered the tarantula's venom dangerous and believed that music could counteract it, Baglivi thought that tarantism could be "false" as well as "true." Unlike Ferdinando, he thought that women make up a "great part" of those exhibiting tarantism and "very frequently counterfeit it under the Mask of its usual Symptoms" because they are "under the Power of Love, or have lost their Fortunes, or meet with any of those Evils that are peculiar to Women." In this gendered stereotype, such women become "mopish and melancholy" and hence "mightily delighted with Variety of Musick and Dancing; whence they feign themselves to be stung by the *Tarantula*, on purpose to enjoy the agreeable Diversion of Musick, which is only allow'd to such Persons." Baglivi cited "a Proverb with us" that called tarantism *Il' Carnevaletto delle Donne*, "the little carnival of the ladies," with its innuendo of lascivious imposture. He immediately added that "tho' Women counterfeit this Distemper sometimes, we must not therefore imagine, that all others do the same" because "some Persons, otherwise both Learned and Religious," unwisely experimented with tarantulas and would have lost their lives "unless the Musick had been ready at hand," here agreeing with Ferdinando.[81]

Baglivi used music itself to test whether the tarantism is authentic or false. He notes that "how ever much soever they [*tarentati*] vary in their particular Tunes, yet they all agree in this, To have the Notes run over with the greatest Quickness imaginable." Thus, the musicians can "easily discover the Cheat of the Women: for if they find that they presently take any Motions, and jog on indifferently, without any regard to the Swiftness, Slowness, or of other Difference of Sounds, they give to understand that the honest Woman is but in jest; and afterward Experience puts the Matter out of doubt."[82] Baglivi noted also the extraordinary sensitivity of true *tarentati* to musical nuances, for even if they are "poor Country Girls and Boys, that perhaps in all their Lives never so much as set their Eyes upon any of the better Sort of musical Instruments," if they sense that the instruments are out of tune, "one would not believe what vehement Sighings and Anguish at Heart they are seiz'd with; and in this Case they continue, till the Instrument is got into Tune again, and the Dance renew'd."[83]

Though he was aware of imposture and wondered at this extraordinary display of musical discrimination among the unlearned, Baglivi never seemed suspicious that tarantism may not be just an effect of venom. During the eighteenth century, the Neapolitan physicians who reconsidered tarantism took a much more skeptical view; the sophisticated metropolitans decided their provincial Apulian colleagues had fallen victim to a popular delusion that was a social phenomenon rather than an actual disease caused by toxic venom. The matter continued to rouse passions and public conflict. In a Neapolitan bookshop, Don Domenico Sangenito vehemently claimed the reality of tarantism and spider venom, claims hotly denied by Bernardino Clarizio, a medical student. Finally, Clarizio offered to be bitten by a tarantula and endure the consequences; if there were no ill effects, Sangenito must buy him some books to enjoy "in spite of all the tarantulas of Apulia and all the fanatics of the universe." Clarizio got his books.[84]

In *On the Tarantula, or the Phalangium of Puglia (Della tarantola, o sia falangio di Puglia*, 1742), the Neapolitan physician and academician Francisco Serao argued that this "bizarre and surprising phenomenon" had nothing to do with spiders but rather was a peculiar "melancholy delirium [*delirio malinconico*]" of Apulians, thereby returning to the general view already held by Gulielmus centuries before, though now in the context of the new natural philosophy that was critical of Kircher. For Serao as for Gulielmus, the Apulians danced the tarantella in order to cope with their endemic melancholy; Serao called tarantism "an institution," though considering it still a kind of illness.[85] Ernesto De Martino, the eminent twentieth-century sociologist, espoused a similarly expansive social and institutional view of tarantism, though he doubted the sharpness of Serao's distinction between "institution" and "disease."[86] Still, Serao preserved a kind of sympathetic understanding for the *tarentati* absent in a dismissive comment of the Neapolitan natural historian Domenico Cirillo in 1770 "that the bite of the tarantula never had any bad effects and that music never had anything to do with it."[87]

In describing the body, we have seen Ferdinando and Baglivi participate in a shift from the terminology of humors to a new language of fibers, a term Serao also used.[88] During the following century, medicine and natural philosophy increasingly considered the body as an ensemble of vibrating fibers, whose response to music and sound was essential. In terms of vibration, music offered a new kind of insight into the body and its functioning. In the process, though, a profound and ironic shift had occurred: the vibratory account of the "musical disease" tended to dismiss the music and dance whose vibrations seemed to cure its lethargic sufferers. Their condition was more a social institution than a disease in the ordinary sense; those who attempted to medicalize it came to realize only gradually the limitations of medical understandings of the dancers' "disease." Though the most celebrated and debated case of musical "cure," tarantism could not be understood in terms of medical diagnoses, nor did music simply "cure" it. As music moved away from being disembodied and divine (but static) numbers, it became the rhythmic life of vibrating bodies, reflected in the fundamental processes of physiology.

II

Sonic Turns

7

Vibrating Fibers

As physicians increasingly questioned the ancient conception of humors, they gained new insight by viewing the body in terms of its constituent fibers. Disclosed by dissection and the emergent tools of microscopy, those fibers were not static structures but could vibrate, thereby adding another dimension to a growing understanding of bodily function. In new ways, the body became a musical instrument, vibrating and resounding. Carrying forward this long development, Denis Diderot used vibrating strings and musical instruments as a crucial analogy to explain how a purely mechanical device might produce the rich complexity of living organisms. The chapters in part II consider how these vibrating fibers informed a revival of Herophilus's ideas that sought to notate the literal music of the pulse, then learned how to hear the body's actual sounds.

Baglivi's notion of bodily fibers emerged from a long history.[1] For sinew or tendon, Homer used the words *neuron* (also the word for a cord or bowstring) or *is*, the word Plato used to describe the fibrous tissue in muscles.[2] Related to the same verbal root (*histemi*, to stand upright), the word *histos* meant the beam of a loom and hence the loom or its web as well. From there, *histology* eventually came to mean the study of bodily tissues, whose ancient name suggests that they were like woven cloth. The Hippocratic writings began this comparison between cloth and bodily fabric, comparing (for instance) the finer "cloth" of a woman's bodily tissues to the coarser weave of a man's.[3] Herophilus and Erisistratus took this comparison even further. According to Galen, they noted that "the thick, hard inner tunic [now called the tunica media of the artery] with transverse fibers does not exist at all in veins," distinguishing the various different "weaves" of blood vessels.[4] Galen then developed from this his own three-fold classification of fibers into longitudinal, transverse, and oblique, in terms of which he tried (for instance) to explain the action of the stomach in digestion.[5] In the inner coat of the stomach, longitudinal fibers exist "for the purpose of exerting a pull from mouth to stomach," whereas transverse fibers "may contract upon its contents and propel them forward . . . in vomiting no less than in swallowing."[6]

In the process, Galen used various kinds of fiber not only to describe the *static* structures of bodily organs but also those organs' specific *dynamic* functions as enabled by the fibers' differing contractions. After Galen's work was transmitted by Arabic and Persian commentators, his ideas about fibers found new interest in the West. Consider the Parisian physician Jean Fernel, whom his contemporaries and immediate successors ranked alongside or even above Vesalius and Paracelsus.[7] Fernel's *Physiologia* (1567) "can be seen as the first serious attempt to present the prevailing assumptions about the nature of the human body, which underlay the Renaissance system of medical theory, in a comprehensively synoptic way."[8] Fernel considered the body to be woven of fibers from the first shaping of the fetus, which display "a kind of sketch of the similar parts, like fibers and warp [*stamina*] woven from some very fine threads like those spiders make."[9] Eventually those threads form the solid parts of the body "interwoven with fibers of much strength, in which a great deal of support resides, so that I reckon no force can ever weaken them, or only feebly and within limits, so that they can hardly yield noticeably in a lifetime."[10] To explain the "nourishment and maintenance" of organic bodies, Fernel posited four "natural faculties," of which the "attracting," the "expelling," and the "retaining" assimilate nutritive parts into the body, which the "concocting" faculty then transforms into bodily fluids.[11] Referring to Galen's threefold classification of fibers, in every organ or "natural instrument" the attracting faculty uses the movement of straight fibers, while the expelling faculty uses transverse fibers and the retaining uses oblique. For Fernel, "the variety of filaments [*villorum*] determined by dissection declares the number of faculties of any part" of the body.[12]

Among Fernel's contemporaries, Vesalius incorporated fibers throughout his foundational treatise, *On the Fabric of the Human Body* (*De humani corporis fabrica*, 1543), which often refers to "Herophilus, prince of anatomists."[13] To be sure, *fabrica* primarily meant "craft" or "art," the manner of construction or workmanship, not only in relation to the textiles often implied by the English *fabric*; for instance, Cicero described "the incredible skillfulness of nature's handiwork [*incredibilis fabrica naturae*]," meaning the amazing structure of the world.[14] Then too, *fabrica* also connoted "working," not just static structure.[15] Though Vesalius corrected more than two hundred errors he found in Galen, nevertheless he called him "the foremost teacher of anatomy" and continued to use the concept of fiber in ways that followed Galen.[16] For example, he used all three of Galen's types of fiber to illustrate the construction of venous walls (figure 7.1a). Vesalius also depicted the overall structure of the veins as a network of fibers (figure 7.1b).

Galen had already noted the fibrous structure of the muscles, which Vesalius emphasized further in figures that showed the three-dimensional structure of the various muscle layers and their respective fibers (figure 7.2). This was, in fact, one of the first illustrations that used a figurative peeling back of successive tissue layers, which (particularly in the case of the muscles) enabled a new depth of understanding of structure and function.[17] Vesalius's representations of the various systems of veins and nerves (presented separately from the surrounding tissues) invited comparisons between their functioning as well as their similarities in terms of the structure of their larger networks. Indeed,

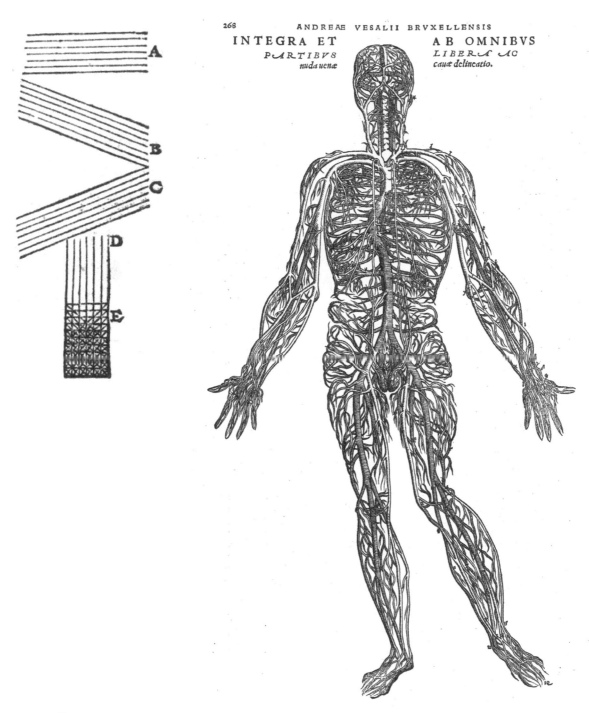

Figure 7.1
(a) Fiber structure of venous walls, showing transverse (*A*), longitudinal (*D*), and oblique (*B, C*) fibers. (b) The entire web of veins. From Vesalius, *De humani corporis fabrica* (1543).

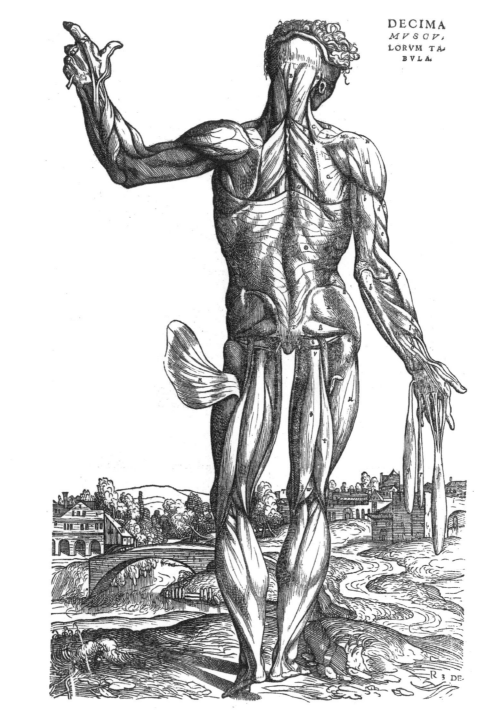

Figure 7.2
Musculature of the back, showing successive layers and their fiber structures. From Vesalius, *Fabrica*.

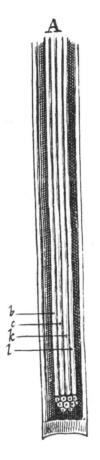

Figure 7.3
Descartes's diagram of a nerve sheath (*A*), containing four hollow nerve fibers (*b, c, k, l*). From *De homine* (1662).

his *Epitome* (1543) of anatomy began by listing fibers among the fundamental parts that characterize every part of the human body, such as the "tougher, threefold type of fibers" of the heart.[18] Likewise, Vesalius's contemporary and rival Gabriele Fallopio also emphasized fibers as an important component of tissues, responsible for movement and the conduction of fluids.[19]

In his pioneering mechanical model of the body in his *Treatise on Man* (*Traité de l'homme*, 1647–1648, first published 1664), René Descartes described "all the solid parts being made exclusively of small filaments, which stretch out and fold back, and which are sometimes also intertwining."[20] He treated nerves as hollow fibers (figure 7.3) that transmitted impulses he analogized to a church organ, so that "you can think of our machine's [the human body's] heart and arteries, which push the animal spirits

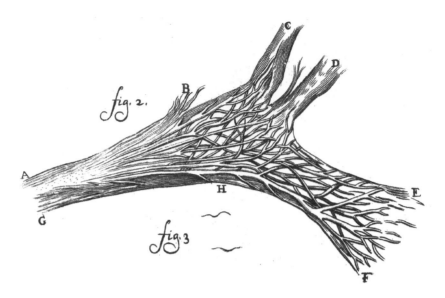

Figure 7.4
Antoni van Leeuwenhoek's drawing of the microstructure of a muscle fiber. From *Arcana naturae detecta* (1695).

into the cavities of the brain, as being like the bellows of an organ, which push air into the wind chests; and of external objects, which displace certain nerves, causing spirits from the brain cavities to enter certain pores, as being like the fingers of the organist, which press certain keys and cause the wind to pass from the wind chests into certain pipes."[21] In his model, the muscles were literally inflated by air transmitted through these taut pipes. Likewise, the brain "is composed of various tiny fibers connected together in different ways."[22]

During the following century, the development of microscopy revealed new levels of detail in fibers and fibrous structures. Antoni van Leeuwenhoek distinguished between "bundles of fibers" (*striae*) that can be seen with the naked eye and "fibers" (*fibrae*) visible only with the help of lenses. Thus, he observed the microstructure of muscles to comprise a web of interwoven fibers (figure 7.4).[23] His contemporaries described these intricate structures through the metaphors of weaving, of warp and weft; for instance, Nehemiah Grew compared microscopic images of muscles to "fine Bone-Lace."[24] However beautiful the fabric, these images intensified questions about how those intricate fibers might move or come to life.

Though Robert Hooke worked with a microscope of considerably less power than Leeuwenhoek's, he directed attention to a new aspect of natural fibers: their vibrations. In his *Micrographia* (1665), though his ultimate interest was in natural bodies, Hooke chose to begin "with the Observations of Bodies of the *most simple* nature first, and so gradually

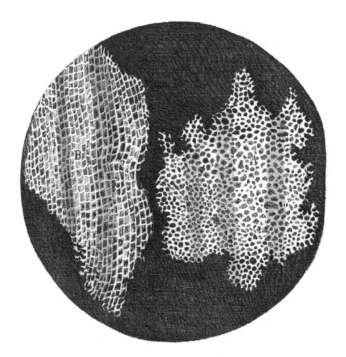
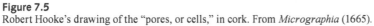

Figure 7.5
Robert Hooke's drawing of the "pores, or cells," in cork. From *Micrographia* (1665).

proceed to those of a *more compounded* one."[25] He apologized that "the Productions of art are such rude mis-shapen things, that when view'd with a *Microscope* there is little else observable, but their deformity."[26] Yet because "in *natural* forms there are some so small, and so curious, and their design'd business so far remov'd beyond the reach of our sight," we first need to probe "the imperfections of our senses" through the study of artificial objects before approaching "the Productions of a more curious Workman."[27] Following the order of geometry, he began with "a Mathematical *point*," the point of a needle. Then he studied a razor's edge (corresponding to a geometrical line) and finely woven linen threads (a surface). Natural objects begin to appear in taffeta and watered silks, though he chose to examine the silkworms' thread when artificially woven in various ways. Examining cork showed Hooke "the first *microscopical* pores I ever saw, and perhaps were ever seen," which he called "cells" (figure 7.5), that "hinted to me the true and intelligible reason of all the *Phaenomena* of Cork"—its lightness, ability to float, and springiness.[28] Further, these pores "seem to be the channels or pipes through which the *succus nutritius* or natural juices of Vegetables are convey'd, and seem to correspond to the veins, arteries and other Vessels in sensible creatures," which Hooke considered "the Vessels of nutrition to the vastest body in the World."[29] Alexander Berg noted that "from here dates the modern term *cell*, though in these images Hooke by no means had seen a structural element but rather a sap pipe system," giving a new insight into the vascular system of plants.[30] Thus, as Mirko Grmek put it, the fiber theory was the mother cell of cell theory itself.[31]

In order to understand these pores, Hooke tried "to make an Artificial pore as *small as any Natural I had yet found*" by fashioning "small glass pipes melted in the flame of a Lamp, and then very suddenly drawn out into a great length" that were "almost as small as a *Cobweb*." Observing how liquids rise up in these pipes led Hooke to inquire into the nature of fluidity, which he ascribed to heat, considered as "a very *brisk* and *vehement agitation* of the parts of a body" that makes them "so *loose* from one another, that they easily *move any way*, and become *fluid*."[32] Such a minute pipe could model biological pores and hollow fibers, as Descartes had suggested.

From this, Hooke proposed a detailed musical model of the vibratory motions in matter. Because "*all* bodies have some *degrees* of *heat* in them, and that there has not been yet found any thing *perfectly cold*," he deduced that "the *parts* of all *bodies*, though never so *solid*, do yet *vibrate*."[33] Hooke interpreted these vibrations in terms of dance and music:

I suppose the *pulse* of heat to *agitate* the small parcels of matter, and those that are of a *like bigness* and *figure*, and *matter*, will *hold*, or *dance* together, and those which are of a *differing* kind will be *thrust* or *shov'd* out from between them; for particles that are all *similar*, will, like so many *equal musical strings equally strecht*, vibrate together in a kind of *Harmony* or *unison;* whereas others that are *dissimilar*, upon what account soever, unless the disproportion be otherwise counter-ballanc'd, will, like so many *strings out of tune* to those unisons, though they have the same agitating *pulse*, yet make quite *differing* kinds of *vibrations* and *repercussions*, so that though they may be both mov'd, yet are their *vibrations* so *different*, and so *untun'd*, as 'twere to each other, that they *cross* and *jar* against each other, and consequently, *cannot agree* together, but *fly back* from each other to their similar particles.[34]

Hooke used this musical analogy to explain that alcohol and water can mix because "a *unison* may be made either by two *strings* of the same *bigness*, *length*, and *tension*, or by two strings of the same *bigness*, but of differing *length*, and a contrary differing *tension*."[35] In general, to these "*three properties in strings*, will correspond *three proprieties* also in *sand*, or the *particles* of bodies, their *Matter* or *Substance*, their *Figure* or *Shape*, and their *Body* or *Bulk*. And from the *varieties* of these *three*, may arise *infinite varieties* in fluid bodies, though all agitated by the *same pulse* or *vibrative* motion. And there may be as many ways of making Harmonies and Discords with these, as there may be with *musical strings*."[36]

Using this vibrational model of fluidity, Hooke addressed many capillary phenomena (as we now call them) such as "the rising of Water in a *Filtre*," the globules of quicksilver, or the "Rising of Oyl, melted Tallow, Spirit of Wine, &c. in the Week of a Candle or Lamp," in the end leading him to the query "whether the capillary Pipes in the bodies of small Trees, which we call their *Microscopical pores*" might be governed by the same vibrational principles as his artificial pipes.[37] Elsewhere in the *Micrographia*, Hooke inquired into the vibratory motion of a fly's wings, which "(if it be compar'd with the vibration of a musical string, tun'd unison to it), it makes many hundreds, if not some thousands in a second minute of time," using his perception of the pitch of the fly's buzzing along with knowledge about vibrating strings to infer the wings' motion.[38] Noting the still higher pitch (and faster vibrations) of a bee's wings, he speculated that "the quickest vibrating *spontaneous* motion is to be found in the wing of some creature." Yet these relatively large-scale movements seemed less fundamental than the omnipresent vibrations Hooke deduced in all matter, which offered an explanation of the nutritive functions of plants.

Indeed, vibration was at the heart of "Hooke's law" (as it came to be called) that "the Power of any Spring is in the same proportion with the Tension thereof."[39] In his published account, he announced that he had discovered this "Rule or Law of Nature" in 1660, before or near the time he began working on cork and the rising of water in small pipes.[40] He emphasized that all these projects rest on the same principles of "*Congruity* and *Incongruity*," meaning "an agreement or disagreement of Bodys according to their Magnitudes and motions."[41] He also noted that this law applies to "all other springy bodies whatever, whether Metal, Wood, Stones, baked Earth, Hair, Horns, Silk, Bones, Sinew," so that it represented a touchstone for the study of living organisms as well as inanimate mechanisms.[42] He recognized that this law does not only describe static behavior but also dynamic vibrations, as of "a spring or extended string, and the uniform sound produced by those whose Vibrations are quick enough to produce an audible sound, as likewise the reason of the sounds and all their variations in all manner of sonorous or springing Bodies."[43]

Hooke also noted the connection between his work on fluids in small pipes and his discovery that "a Spring applied to the Balance of a Watch doth make the Vibrations thereof equal." All of these phenomena depend on vibrations because "all bulky and sensible Bodies" are made up of particles whose "peculiar and appropriate motions . . . are kept together by the different or dissonant Vibrations of the ambient bodies or fluid."[44] His use here of *dissonance* is consistent with his reliance on music at other points in his argument. Thus, to explain what underlies "*Congruity* and *Incongruity*" in various bodies, Hooke argued that "as we find that musical strings will be moved by Unisons and Eighths [octaves], and other harmonious chords, though not in the same degree; so do I suppose that *the particles of matter* will be moved principally by such motions as are Unisons, as I may call them, or of equal Velocity with their motions, and by other harmonious motions in a less degree."[45]

Hooke envisioned a vibrating cosmos in which "the whole Universe and all the particles thereof [are] in constant motion" so that "the particles of all solid bodies do immediately touch each other; that is, the Vibrative motions of the bodies do every one touch each other at every Vibration [figure 7.6]"[46] He even argued that these universal microscopic vibrations are the underlying cause for the law of the spring itself, as observed in bulk matter.[47] He judged that his explanations were sufficient to account for "the Spring of

Figure 7.6
Robert Hooke's diagram from *Lectures de Potentia Restutiva* (1668) showing three vibrating bodies *A, B, C* continually mingling their vibrations by colliding at *E* and *F*.

any other Body whatever," including the air itself, as well as all the components of living organisms.[48]

Though it is not clear to what extent Hooke's radical vision was recognized by his contemporaries, in their own ways they came to understand the importance of vibration for the fibers of living organisms. Around the same time, two eminent physicians, contemporaries and associates of Hooke in the Royal Society, put forward vibrational accounts of nerves. Thomas Willis's *Anatomy of the Brain* (*Cerebri anatomi*, 1664) compared the "communication of spirits" along a nerve to the motion of waves; in the case of vision, "as often as the exterior part of the soul being struck, a sensible impression, as it were the optic species, or as an undulation or waving of waters, is carried more inward."[49] Having compared the "medullar Trunk" of the brain to "the Chest of a musical Organ," Willis applied an analogy with a vibrating string to explain how, "in sensation, the fibers receive first and most immediately the impressions of sensible things, and retain their forms by a modification of the internal particles (as musical strings do the pluckings of a pick or thumb), and represent the various approaches of the object by similar motion of the fibrils, as by a moveable and fluid stamp, whose idea the nerves convey as far as the head."[50]

To be sure, Willis did not propose that an actual sonic vibration passed along the nerves. Unlike Descartes, he did not think the nerves were literally stretched tight, poised to transmit vibrations, or that they were hollow tubes. As Wes Wallace has argued, "Descartes understood the nerve as a conduit of force, a structural entity materially involved in a transfer of force from the brain to the muscle. In contrast, Willis understood the nerve as a conduit of *information*, not involved in the transfer of force but only in its *signification*."[51] His own understanding may have been closer to the hypothesis Isaac Newton developed from it of nerves acting through chemical (and perhaps also electrical) processes.[52] Nevertheless, the metaphor of vibration remained important because it enabled the underlying transformation of information rather than of any material substance or force. Indeed, a similar comment on the "vibrations" of nerves still can be addressed to the modern view of nerve action, which has returned to Newton's bold hypothesis. On the largest level, we should bear in mind Henry David Thoreau's claim that "all perception of truth is the detection of an analogy," which may apply to the development of science no less than other realms of human thought.[53]

Appearing at the same time as Willis's work, William Croone's treatise *On the Reason of the Movement of Muscles* (*De ratio motus muscolorum*, 1664) gave another vibrational account of the nerves. Croone agreed that "all sensitive membranes in the body originate from the tunics of the brain," their fibers connected to other membranes and nerves throughout the body.[54] These membranes "are also under the greatest tension, as was said above about the muscles. They are also full of some characteristic and very subtle liquor or lively spirit, which continuously passes through all their fibers and maintains them in due tone," also apparently entertaining some chemical theory of their action.[55] Most of all,

these membranes while they maintain their tension and due tone can be considered as behaving like a bell or as the purest glass. These have the property that, if you strike any part, all the remaining [parts] are shaken by some sort of quivering vibration. In the same way, I say that all objects of the senses are carried to the brain by the intervention of either the membrane of the nerve which pertains to the true organ of that sense, or of the common membrane which involves the whole body, carried in lines as straight as possible. In the brain the different and distinct movements of the objects are perceived by the mind. . . . This is the origin of the marvelous harmony of the nervous parts.[56]

Croone also used this vibrational theory to address a difficult question: In cases of paralysis, why does sensation still persist even after movement has been lost? He hypothesized that if "the tone of the membranes is totally or partially destroyed, . . . then that vibration and oscillation of the particles which make the sensation is interrupted. In a bell or a glass in which a crack occurs, we notice that the tinkling and high-pitched sound is altered into a harsh and unpleasant clanking."[57] Crooke even thought that the vibrations of the nerves are immediately perceptible, for if "we unexpectedly chance to knock an elbow against some hard material, a tremor or painful vibration flies from the elbow down to the tips of the fingers as it were in the twinkling of an eye. Similarly, when a long rod of iron is struck at one end, it produces a quivering vibration of all the particles which you feel with your hand at the other end."[58]

During the seventeenth century, attention turned to how exactly the muscle and nerve fibers could accomplish their various functions. In the last chapter, we encountered Giovanni Borelli as the pioneer in explaining the motions of living bodies in terms of basic physical principles. This approach became known as *mechanism*, as opposed to *vitalism*, which understood living bodies in terms of a vital principle that was not reducible to material causation. Borelli was an early and important champion of the mechanist view, which he presented in terms of the body's constituent fibers. In his book *On the Movement of Animals* (*De motu animalium*, 1685), he argued against the position that flesh "fills the interstices of the fibers [and] does not really constitute the muscle." Instead, "the body is full of muscular flesh . . . constituted of many filaments or tendinous fibers . . . in some places these fibers are bound together by countless transverse fibers" (figure 7.7a).[59] Invoking a musical example, he continued, "The muscles are bundles of very thin and strong tendinous threads which contract as do the strings of a cithara [an ancient Greek lyre] or steel wires when stretched. The fibers can contract with much more force by order of the will. Such contraction cannot be understood without a machine like a spring and the contraction of all the elements of a fiber cannot be conceived without a continuous series of small machines."[60] He calculated that "twenty small machines in a row are equal to one finger breadth."[61] Borelli provided many physical models to explain the contractile power of muscles in terms of such microscopic machinery within the muscle fibers (figure 7.7b). Following Galileo's example, Borelli used Euclidean geometry to state and prove his mechanical propositions, which laid the groundwork for the "iatromechanism" (mechanical medicine) that continued through the first half of the eighteenth century. He considered his work to fall within the "combined sciences which apply geometry to natural phenomena, such as astronomy, perspective, music, mechanics, etc.," explicitly including music as an example of these "combined sciences."[62]

Though many of his "little machines" relied on the kind of smooth changes in the length of the fibers I have called "dynamic," Borelli was also interested in "forces of percussion" such as those resulting from the various kinds of impact that Galileo had begun to investigate. Thus, Borelli's treatise *On the Force of Percussion* (*De vi percussioni*, 1667) also studied the "vibration and shaking which occurs in bodies" in response to percussion, which he compared to the oscillation of a pendulum.[63] He had in mind "quivering in animals" but also sound, noting that "the weak vibration of air generated by the sound of a drum or a trumpet can shake a vast temple and induce vibration in it."[64] During an eruption of Mt. Etna, Borelli noted that buildings shook much more violently if they were directly exposed to the vibrations in the air caused by the eruption, compared to those experiencing only the ground vibrations.[65]

Borelli had a special interest in the operation of the fibers that drive the heart, whose motive force he calculated "by themselves would be able to raise a weight heavier than 3,000 pounds."[66] These fibers operate "not to contract but to swell and to fill the cavities of the heart by their volume and express and push the blood contained in the ventricles" in a process he compared "to a small weapon called the blunderbuss. The inflated internal fibers play the role of the gunpowder which is ignited and rarefied. The blood expelled from the ventricles can be compared with the shots projected by the blunderbuss," though the heart acts by means of "fibers convoluted in spirals."[67] Borelli emphasized the role of "swelling and turgescence of its fibers" in causing the all-important pulsation of the heart, surely an essential example of periodic vibration in the body.[68] He also discussed

the continuous oscillatory movement which air mixed with blood can produce. . . . This movement can never stop as long as the animal is alive because new percussions are never lacking, from breathing, from pulsing of the heart and arteries, from the movement of the muscles and from a thousand other internal and external causes which can set up again this trembling movement of the

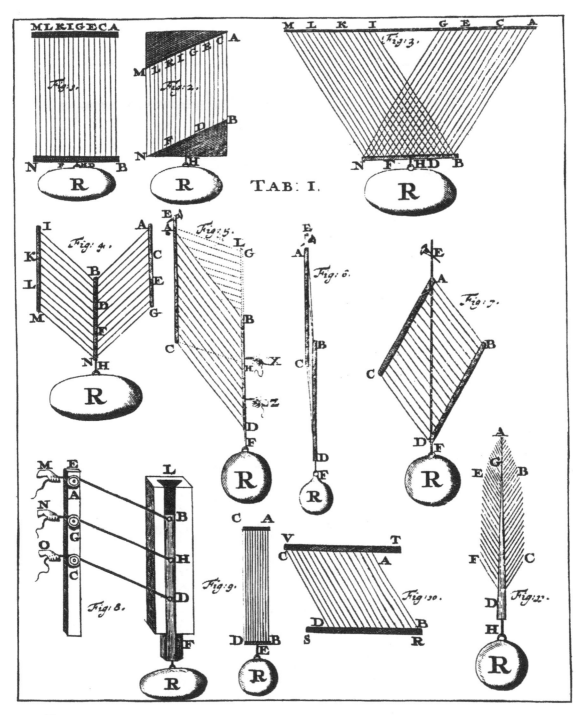

Figure 7.7
Giovanni Borelli's illustrations of (a) various fiber structures within "muscular flesh" and (b) different physical mechanisms (labeled 5–12) whereby fibers could exert contractile forces. From *De motu animalium* (1685).

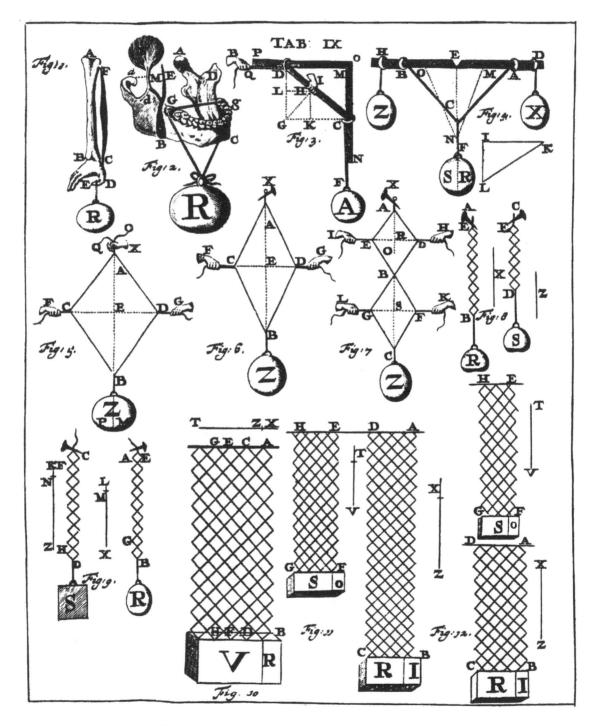

Figure 7.7 (continued)

small machines of air. Therefore, the particles of blood must always be shaken by a peculiar movement. They must be activated by the oscillatory movement of the small machines of air included
in the blood.[69]

Trying to understand the mechanism behind the "animal spirits," Borelli argued that
"it seems necessary that the spirits be activated in the brain by some motion as is required
by their mobile character. Therefore, it is possible that the cerebral juices or spirits thus
activated displace or shake the origins of the fibers of some nerve by a movement of con-
cussion or by stinging mordant (which they may be able to provoke). They thus irritate
and titillate the nerve."[70] Thus (without referencing Hooke, whose work he did not seem
to know), Borelli in his own way considered some kind of vibration or irritation necessary
to initiate nervous activity:

The structure and temper of the nerves are very delicate and sensitive, as experience shows. When
touching the inside of the nostrils or of the ears with a light straw, the nerves are shaken so violently
along their whole length that they provoke convulsive sneezing and coughing. Therefore, it is not
surprising that a slight commotion or irritation of the nerves in the brain provokes some convulsive
shaking throughout their whole length, resulting in the secretion of some droplets of this juice
which swells the canals of the nervous fibers,

arriving at "spongy structures" where the nerve meets the muscle, because "droplets
hanging from a wet sponge do not flow out. A shaking force is required to express them.
This may be the cause why the nervous juice is secreted and instilled in all the mass of the
muscle by order of the will."[71]

To his vibrational account of nerve activation, Borelli added that "the usual ability of
the animal spirits to stimulate determined nerves in the brain is not innate but acquired
by exercise and experience." He confessed "not to understand the mechanism by which
the movements of the spirits thus activated in the brain by order of the will, direct these
spirits to titillate certain well-determined muscles. If I wish to extend the hand, the spirits
are not directed to the nerves of the feet or of the chest" but only of the hand. "Attempting to provide an explanation," Borelli turned to the rhythm of these vibrational signals
because "I think that not all the acts of will are carried out by the same motions of the
spirits. In any act of will the spirits do not move to the same parts, in the same way, with
the same rhythm and velocity but in very different ways, so that there are as many movements of the spirits in the brain as there are acts of will. Hence, in any act of will, spirits
are directed and carried to a determined area of the brain where the nerves are located
which are aimed at carrying out this act of will."[72] These processes "do not result from
natural and blind necessity as the fall of heavy objects does. They may occur as a result
of practice unconsciously acquired by frequent unconscious repetition of the acts," for
which he again instanced musical practice: "The fingers of cithara players touch and pull
the different strings with incredible velocity and art without thinking but by some habit.
Conscious reflection is certainly not necessary in such action. If cithara players wish to
think and take care of moving their fingers according to the rules of art, they hesitate and
are confused rather than they emit proper harmonic sounds."[73] Borelli treated musical

practice in terms of the cithara, an ancient stringed instrument important in quadrivial accounts of mathematical music theory, but his understanding of habituation applied no less to contemporary instruments and players.

Similarly, Borelli's account of the initial quickening of life in fertilized seeds and embryos depended on periodic vibration, "oscillatory pendulums in the animal machine. Countless particles of air are included in the blood and liquids of the animal. They are oscillatory machines. They are the main and most powerful causes of the vital movement of the animals. Being successively compressed and rebounding they exert a continuous oscillation in the animal. Shaking and activating the liquids and spirits of the animal they act like the wheels of a clock and maintain the vital movement of the machine by their pendulum oscillation."[74] For Borelli, from its inception to its daily functioning, life depended on well-regulated periodic vibration throughout the organism.

Though Galen had already used the term *irritation* to describe the body's reaction to pathogenic materials or agents, the English physician Francis Glisson (whom Borelli cited) introduced a more general concept of *irritability* during the 1660s to understand the response of bodily fibers even in the absence of pathology.[75] Rather than viewing them as purely passive, Glisson described the ways in which various kinds of fibrous tissue reacted specifically to stimuli. Their irritability indicated how fibers would move in response to certain stimuli, depending on their nature. For instance, Glisson showed that the intestines and muscles immediately after death could be stimulated to movement by corrosive acids or by cold, even when removed from the body.[76] He then developed a terminology that distinguished such purely involuntary reactions from conscious activity. Glisson also used metaphors involving sympathetic vibration, such as when two notes an octave apart "tremble in agreement," to describe "examples of purely natural irritation by accord [consensus] between fibers that frequently occur."[77]

Writing several decades later than "the celebrated Borelli," Baglivi made much more of music, as discussed in the previous chapter.[78] As part of his understanding of the curative force of music in tarantism, Borelli believed that "the human Body is a Bundle of Fibres variously interwoven and corresponding to one another, which are bended this way or t'other by the Fluid that moves with in, as by a Spring: And from thence proceeds that great Sympathy and united Consent of the Parts."[79] Indeed, Baglivi considered that insufficient knowledge of bodily fibers caused the ancient physicians "many mistakes"— for instance, by not understanding that asthma "arose not from a viscid Humour crouded upon the Lungs, but from the Convulsion of the Muscles of the Breast or Midriff, or of the fleshy Fibres interlac'd with the Lungs."[80] Throughout his medical practice, from colic to blistering plasters, Baglivi considered the fibers involved and how vibrations might affect them.

Looking back at the developments in this and the preceding chapter, it seems that music, and its special role in tarantism, led Baglivi to consider fibers as vibrational elements, though in a different way from Hooke's vibrational account of heat and capillarity.

The dynamic role of fibers suggested that the body could be sounded vibrationally to reveal its inner states, which may respond to external stimulation for diagnostic or curative ends. As later chapters discuss, such a vibrationally active body can be listened to or sounded much like a musical instrument. In the remainder of this chapter, we consider early realizations of these new possibilities.

In the seventeenth century, investigations into musical sound led to a physics of vibrating bodies that would eventually affect considerations about the mechanics of bodily fibers. After Galileo Galilei and Marin Mersenne advanced the concept of pitch as frequency of vibration, Hooke's law and Christiaan Huygens's work on oscillating clocks led the way to Brook Taylor's 1713 theory of the vibrating string and Jean le Rond d'Alembert's 1743 equation describing such strings.[81] These developments were noticed by savants across Europe, including physicians who followed the development of natural philosophy in both the physical and biological sciences. We shall trace some ways in which these new understandings of vibrating bodies could address the fibers constituting the body itself.

Fibers continued to be a central theme in the mechanistic medicine of Baglivi's contemporary Herman Boerhaave and even more for Boerhaave's student Albrecht von Haller, who wrote that "the fiber is to physiology what the line is to geometry" because the human body is a "cellular fiber fabric."[82] Where Boerhaave had viewed the fibers as tubing for a hydraulic model of bodily functions, Haller developed the concept of irritability, demonstrating experimentally that (as contractility) it exclusively characterized all muscular fibers, whereas a different form of irritability he called *sensibility* characterized nervous fibers.[83] His experiments showed the physical reality and efficacy of fibers. For instance, Haller thought the heart pulsated because it was the most irritable organ in the body (in the next chapter, we consider his interest in musical understandings of the pulse).

Turning from medicine to wider biological studies, the mid-eighteenth-century Genevan natural philosopher Charles Bonnet tried to explain the amazing regenerative capacities of microorganisms in terms of their constituent fibers. He worked with *Hydra*, a genus of small, freshwater organisms about 10 millimeters long, each of which, if cut in half, regenerates into two individuals. *Hydra* remains of great interest to this day because it seems to be biologically immortal, not suffering from senescence, hence holding out prospects of understanding aging in other organisms. To explain these extraordinary abilities, Bonnet introduced what he called a "germ-fiber-unit," an automaton made of fibers able to act as mini-machines (*machinules*) that could animate the whole organism (*tout organique*) by virtue of their own capabilities: "A fiber, as simple as it might appear, is nonetheless a *tout organique* that feeds, grows, and vegetates."[84] Bonnet even described fibers as themselves being mini-machines (a concept Borelli had already used as *machinulae*).[85] The fibers can fuse into larger patterns he calls "meshes" or cells that could form an organic network (*réseau*) that Bonnet described in terms of "bended," "stretched," and "hooked" fibers forming tissues.[86] He and others described this process in terms of weaving and looms, which were undergoing increasing mechanization during his century, so that in many respects it is "both the automated production process and the product that are similar in looms and living beings."[87] Yet the sensitive interactivity of this mechanism (and controversies about the relation between body and soul) moved Bonnet to further comparisons with musical instruments:

The seat of the soul is a small machine prodigiously complicated and yet very simple in its operation, a summary of all the *kinds of nerves*, a *neurology* in miniature. One can imagine this admirable instrument of the operations of our soul as a harpsichord [*clavessin*], an organ, a clock, or some other, much more complicated machine. Here are the devices destined to move the head; there are those that make the extremities move; higher are the movements of the sense, lower those of respiration and voice, &c. And what number, what harmony, what variety in the pieces that compose those devices and these movements! The soul is the musician that executes different melodies on that machine, or that judges which are executed in the machine and that it repeats. Each fiber is a kind of a key or a hammer destined to produce a certain note. It does not matter if the keys are moved by the objects or by the moving force of the soul; it is always the same play. It can only differ in length or intensity.[88]

Bonnet's contemporary Théophile de Bordeu, an eminent physician at Montpellier who pioneered glandular studies, transformed "irritability" into "sensitivity" or "sensibility" (*sensibilité*) as a way of characterizing fibers and the organs they constitute and connect. Bordeu emphasized that this sensitivity would lead to continual movement within the sensible parts, a movement he characterized as oscillation or vibration: "Ceaselessly the body trembles, quivers, agitates itself unto the deepest of its smallest parts; these quiverings are constantly graduated and directed to maintain the regularity of the order of its functions and are operationally subject to the principle of sensibility that directs everything by laws very different than those that govern the movements of dead and inanimate bodies."[89] As with Bonnet's image of the body as a harpsichord, the vitalist Bordeu also believed that the bodily mechanism vibrates at the behest of a soul.

While taking up many of the fiber structures brought forward by Bonnet and Bordeu, Denis Diderot took the further step of eliminating a separate soul commanding the bodily keyboard. Though most famous as a leading philosophe and editor of the *Encyclopédie* (as well as the author of 7,000 of its entries), Diderot was a remarkable polymath who also had a long-standing interest in music and its relation to mathematics and the natural sciences. He was especially proud of his early *Mémoires sur différens sujets de mathematiques* (*Memoirs on Different Subjects in Mathematics*, 1748). Three of the five memoirs dealt with sound or music, often in considerable detail.[90] The first memoir gave a detailed review of the physical theory of acoustics, with special attention to Taylor's work on the vibrating string. Diderot showed close knowledge of the musical theories of the mathematicians d'Alembert and Leonhard Euler. Diderot's third memoir considered vibrating strings under varying tension; his fourth proposed a new kind of mechanical organ "on which one can play any piece without knowing music" through coding the piece's notes

Figure 7.8
An illustration of Diderot's "new organ on which one can play any piece without knowing music." From his
Memoires sur différens sujets de mathematiques (1748).

as raised marks on a cylinder (figure 7.8), whose rotation then actuates the organ, on the
model of a music-box.

In these memoirs, Diderot also showed his awareness of the theoretical work of
the celebrated composer Jean-Philippe Rameau, with whom he became personally
acquainted around this time. In his *Thoughts on the Interpretation of Nature* (*Pensées sur
l'interprétation de la nature*, 1754), Diderot began with a vibrating string, generalized its
vibration to metal bodies, then to any "sounding elastic bodies [*corps élastique sonore*],"
extending to further systems of bodies and finally the whole universe.[91] Such vibrations
could range from violent shocks that would "reduce a body to its elements" to those so
weak as to cause "only the smallest possible trembling."[92] Though Rameau initially used
a single vibrating string as his exemplar, Diderot may have had a considerable influence on Rameau's advocacy of the *corps sonore* (sounding body) as the central concept
on which to base his later accounts of harmonic theory.[93] Their conversations may also
have influenced one of Rameau's finest works, an *acte de ballet* called *Pygmalion* (1748)
in which the *corps sonore* dramatizes the crucial moment. Having fallen in love with a

Figure 7.9
The *corps sonore* (sounding body) of E brings the statue to life in Rameau's *Pygmalion* (1748). Pygmalion:
"Where do these chords come from? What are these harmonious sounds? A vivid light spreads all around"
(♪ sound example 7.1).

statue he has made of a beautiful woman, the sculptor Pygmalion prays to the goddess
Amour for help; in response, the chord of the overtones of E sounds (figure 7.9; ♪ sound
example 7.1). Pygmalion wonders, "Where do these chords come from? What are these
harmonious sounds? A vivid light spreads all around." A moment of silence falls; Amour
flies across the scene and lights a torch. Through sympathetic resonance, vibrating strings
have brought the statue to life.[94]

Diderot's concept of the all-pervasive effect of vibration could thus affect living bodies
as well as inanimate matter, perhaps enough to cross the line between them, as Rameau
evoked in music. By his time, the concept of fiber had become a commonality of ordinary speech, as when we describe someone as "high-strung." In his celebrated dialogue
Rameau's Nephew (*Le neveu de Rameau*, 1761–1762), Diderot portrayed the quirky, can-
tankerous, yet oddly touching qualities of the famous composer's nephew François, a
little-known musician languishing in the shadow of his uncle's fame. François observed
that "merely to equal the renown of your father, you must be more skilled than he; you
must have inherited his fiber . . . but that fiber is missing in me, though my wrist is limber,

my bow works, and the pot boils; it's not glory, but at least it's a living."[95] Ironies abound; the surrogate son of the musician who put the *corps sonore* on the map himself lacked the crucial fiber; though a professional violinist, he was missing a string. Indeed, François viewed all his failings in terms of faulty fibers:

Myself: How is it that with such fineness of feeling, so much sensibility for musical beauty, you are so blind to the beauties of morality, so insensible to the charm of virtue?

He: Apparently virtue requires a special sense that I lack, a fiber that has not been granted me. My fiber is loose, one can pluck it without making it vibrate.[96]

Conversely, in his "Letter on the Deaf and Dumb" ("Lettre sur les sourds et muets," 1751), Diderot noted that "in music, pleasure and sensation depend on a particular disposition not only of the ear but of the whole nervous system. If there are heads that resound [*têtes sonnantes*], there are also bodies that I gladly call harmonic: persons in whom all their fibers oscillate with such promptitude and vivacity that in their experience of the violent movements harmony causes them they feel the possibility of still more violent movements and reach the idea of a sort of music that could make them die of pleasure."[97]

In other writings, Diderot brought these musical concerns—and his interest in musical mechanisms—to bear on fundamental questions about the nature of the body and its faculties. Though not a physician, he wrote a number of works on physiological matters, including a compendious *Elements of Physiology* (*Elémens de physiologie*, 1773–1774), and was greatly interested in the constitution and operation of body and mind. In his *Elements*, Diderot regarded the simple fiber "as an animal, a worm. It is the being the animal it composes nourishes. It is the principle of the whole machine."[98] He included many ways in which the fibers influence every aspect of the life of organisms—for instance, "fibers, along with the fleshy connections, group themselves in wavy ridges one can see on the fascia and on the elementary fiber, so that the overall movement of a muscle seems only to be that of the fibers on themselves." He analyzed "love or antipathy" in terms of the effects of climate and diet on "the fibers that line the very sensitive canal of the urethra."[99] In the end, "age, which hardens the fibers, desiccates the soul."[100]

Diderot even included Bordeu as a character in *D'Alembert's Dream* (*Rêve de d'Alembert*, 1769), along with d'Alembert, Julie de Lespinasse (d'Alembert's close friend), and himself. This unique philosophical drama took the concept of fiber further still by connecting it even more audaciously to music. In it, Diderot tells d'Alembert that he

has sometimes been led to compare the fibers of the body to sensitive [*sensibles*] vibrating strings. The sensitive vibrating string oscillates, resounding long after one has plucked it. It's this vibration, this kind of necessary resonance that keeps the object present, even while the understanding occupies itself whichever quality it chooses. But vibrating strings have still another property— making other strings tremble—so that a first idea can call forth a second, those two a third, and so forth, without one being able to fix a limit to the ideas revealed and linked together to the philosopher who meditates or who listens in silence and darkness. This instrument has an astonishing range [*sauts*, literally "jumps"], so that an idea that has been revealed sometimes can make tremble a harmony that is an incomprehensible distance away. If this phenomenon is observed among vibrating strings, inert and separate, how could it not take place between living points connected together, between sensitive fibers that are continuous?[101]

D'Alembert admires this "very ingenious" idea, though he doubts whether it is true because it eliminates the difference between mind and matter. He objects that "you are making the understanding of philosophy something distinct from the instrument [the vibrating string], a kind of musician who lends an ear to vibrating strings and who judges their consonance or dissonance," presumably standing apart from those vibrations.[102]

In reply, Diderot asks him to consider "the difference between the philosopher-instrument and the harpsichord-instrument," thereby turning the abstract example of vibrating strings into a *clavecin*, the familiar mechanical instrument of contemporary music, whereby the philosopher becomes a musical instrument as well. At this point, the distinction between these two "instruments" melts away:

The philosopher-instrument is sensitive [*sensible*, has sensations]; he is at the same time the musician and the instrument. As a sensitive being, he has the momentary awareness of the sound that he plays; as an animal, he has a memory of it. This organic faculty, joining together the sounds inside himself, produces and conserves the melody. Imagine a harpsichord endowed with sensitivity and memory and tell me whether it would not repeat by itself the airs that you play on its keys. We also are instruments endowed with sensitivity and memory. Our senses also are keys that are played by nature surrounding us, keys that often play themselves—thus, in my judgment, everything happens in a harpsichord organized like you and me.[103]

D'Alembert remarks that "if this animated and sensitive harpsichord were also endowed with the ability to eat and reproduce itself, it would be alive and would beget—either by itself or with its female—little harpsichords, living and resonant." "Doubtless," replies Diderot; "what else do you think a bluefinch is, or a nightingale, or a musician, or any human being?"[104] Thus, Diderot's audacious vision rested on his understanding of vibrating strings and musical instruments. This early and influential account of life and intelligence in terms of mechanism raised itself above earlier accounts of mechanical marionettes by virtue of the qualities by which strings and instruments seem to rise above inert matter through the suggestive power of their vibrations.

After this conversation, d'Alembert falls into a strange, fitful sleep, in which he talks "a lot of nonsense about vibrating strings and sensitive fibers," as it seems to Lespinasse.[105] Concerned that he is ill, she summons Dr. Bordeu and reads him her transcript of d'Alembert's ravings, which reflect Diderot's ideas. Bordeu agrees and tells her that "the head, the feet, all the members, all the viscera, all the organs, the nose, the eyes, the ears, the heart, the lungs, the intestines, the muscles, the ears, the nerves, the membranes, strictly speaking are only the coarse developments of a network that forms itself, grows, extends, throws out a multitude of imperceptible threads."[106] Hearing this, Lespinasse interprets this network as a web, in whose center sits a spider "at the point of origin of all those threads," representing the mind.

Bordeu, however, argues that "the threads are everywhere; there is not a point on the surface of your body in which they do not connect and the spider is ensconced in a part of your head that I have called the meninges, which one can scarcely touch without inducing unconsciousness [*torpeur*] in the whole machine."[107] Mlle de l'Espinasse notes that this web vibrates, so that "if an atom made one of the threads of the web oscillate, the spider would take alarm," just as "if one lightly taps the end of a long stick, I hear the blow if I press my ear against the other end. One end of this stick could touch the Earth, the other Sirius and the same effect would be produced."[108] In this way, the vibrations of the fibers not only transmit but actually embody sensation and thought. Together, she and Bordeu arrive at the possibility that, rather than mind being an alien spider seated within the corporeal web, "intelligence unites with very energetic portions of matter," becoming an aspect of that matter's activity.[109]

Thus, the long history of bodily fibers reached a new plane of significance when those fibers were not only static structural elements or simple machines restricted to pure contraction but could also vibrate, as Hooke, Croone, and Baglivi variously considered. Those vibrations became the way the bodily mechanism could come alive, as it did for Bonnet, Bordeu, and especially Diderot, who eliminated any vital force or immaterial spirit precisely through his audacious reliance on the resonance and sympathy that mechanical instruments could evoke through their vibrating fibers. These fibers also connected with emergent ideas of evolution; in his *Zoonomia* (1794–1796), Erasmus Darwin (grandfather of Charles) explained how different species came into existence, beginning "from a single living filament; and that the difference of their forms and qualities has arisen only from the different irritabilities and sensibilities, or voluntarities, or associabilities, of this original living filament."[110] The living fiber, its characteristics and vibrations, had become fundamental to all of biology and medicine.

8

Rhythms of the Heart

During the eighteenth century, the possibility of understanding the body in terms of fibers that could vibrate and resonate converged with new versions of Herophilus's ideas about musical ratios in the pulse. These interacting streams of practice and theory led to new approaches to the heart and to physical diagnosis, particularly to an important part of the sonic turn in biomedical science. Consideration of musical and sonic aspects of the body opened the possibility of gaining access to the interior state of the body in ways that transcended the limitations of visual examination. The development of modern physical diagnosis profoundly depended on the use of sound, allowing a trained "medical ear" to transcend the limitations of "the medical gaze."[1] These developments unfolded over time in several somewhat independent areas, including Samuel Hafenreffer in Germany and Athanasius Kircher in Rome. I will concentrate on a little-known French physician, François-Nicolas Marquet, who used contemporary French musical traditions to revisit Herophilus's ideas notating the pulse in music by notating the heart's rhythms in music.

Indeed, Herophilus's ideas had remained in circulation since Pietro d'Abano's time, even becoming a commonplace referenced in literary texts.[2] Thus, when Hamlet's mother diagnoses "ecstasy" as the cause of her son's strange behavior, he replies, "Ecstasy? My pulse, as yours doth temperately keep time / and makes as beautiful music. It is not madness / that I have uttered."[3] Hamlet seems to consider the music of his pulse more an indication of his mental, rather than cardiac, condition. Yet his statement raises questions about what he really means by that "beautiful music," a concept he may have been playing with or altering to suit his larger purpose. Then too, the modern distinction between the psychic and the somatic may miss the point: If the pulse is discordant, how can the mind remain untouched? It is not clear what place such questions may have had—if any—in the prince's interrupted studies at Wittenberg.

In the decades after Hamlet, we will pause over the Tübingen physician and professor Samuel Hafenreffer and his *Monochordon symbolico-biomanticum* (1640), whose untranslatable title mingled the Pythagorean monochord (whose single string can perform the musical ratios), the symbolic, and the "bio-prophetic." His lengthy subtitle proposed to "elucidate the most abstract doctrine of pulses from the harmony of the world" in order to "resonate in the practice of medicine."[4] Though referring to Aristotle and Galen, Hafen-reffer's book was steeped in cabbalistic and mystical lore. His version of pulse music represented the experience of the four fingers taking the pulse, each finger assigned one line of a staff (figures 8.1a and 8.1b); at one point, he proposed that the different fingers can discern pulse "melodies" in the various known modes (figure 8.1c), but then wrote a different notation for "moderate" pulse (figure 8.1d), in which the higher and lower "chords" seem to denote systole and diastole. Because these examples do not have clefs, one does not know to what extent he really meant to indicate pitch, rather than a more general tactile sense of feeling the pulse through the fingers, like the tablature that notates finger positions in guitar or lute music. Because of these omissions, one cannot play or hear these as sound examples. On the other hand, his examples of "soft" and "hard" pulse (figure 8.1e) have a clef and make a notational pun by using the "soft B" (B-flat) or "hard B" (B natural), suggesting that these pulses are like "soft" or "hard" chords (minor or major, we would call them), though his example makes no sense as chords. This symbolization seems to be his invention, without any precedent in pulse music as it came down through Galen. Likewise, his notation of "undulating" pulse (figure 8.1f) is a kind of free fantasy on that concept, unless one believes that he actually felt that sequence of touches in his fingers. Though at times he gives examples on normal staff with a clef and the numerical ratios corresponding to the musical intervals, in the images of figure 8.1, Hafenreffer seems interested in conveying a symbolic representation of the experience of the fingers taking the pulse, a curious synthesis of ancient pulse music with tactile sensation. On the other hand, he was not only a speculative mystic but was practical enough to write the first German book on diseases of the skin (cited approvingly by Sir Thomas Browne).[5]

Published only ten years after Hafenreffer, Kircher's *Musurgia universalis* (1650) treated pulse music in a much more down-to-earth fashion, though both men had arcane and cabalistic interests. To be sure, Kircher also went beyond purely numerical ratios to try to notate the pulse rhythms (figure 8.2), but his imagination stayed closer to Galen's account.[6] Where Galen (following Herophilus) had distinguished the rhythms proper to each stage of life, Kircher notated them as four different voice parts from a polyphonic composition. The soprano (*cantus*) is the voice of a boy singing semiminims "in equal rhythm"; alto, a youth (*iuuenis*) whose minims are twice as slow; tenor, a mature man (*vir*) singing semibreves, four times as slow again; bass, an old man (*senex*) in breves, the longest, slowest notes of all. Together, they form an F major chord (as we would call it), a nice touch by which Kircher indicates the concord between these different stages of life; aging is merely taking another part, lower (and slower), in the chord of life (♪ sound example 8.1). The complex rhythmic relations Galen ascribed to the pulse here are reduced to much simpler ratios, each successive stage of life pulsing twice as slowly as the one before, a rhythmic proportion of 1:2:4:8.

When Kircher turned to the variety of rhythmical variants, however, his imagination took fire and presented a massive table of possibilities (figure 8.3) whose columns

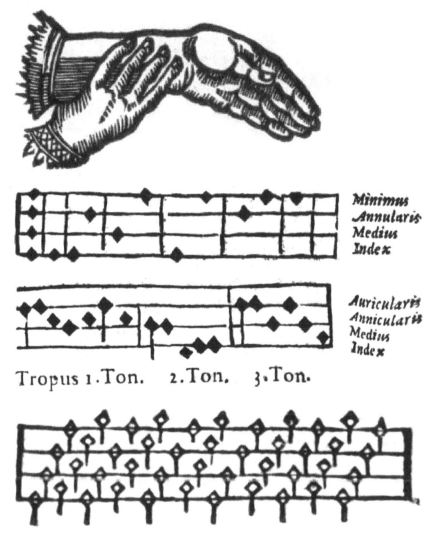

Figure 8.1
Samuel Hafenreffer's illustrations of (a) four fingers feeling the pulse; (b) the assignment of lines on a staff to each finger, labeled at right; (c) possible pulse "melodies" in three different modes; (d) "moderate" pulse; (e) "soft" (*mollis*) and "hard" (*durus*) pulse, using B-flat and B natural respectively (the clef indicates G), though the chords make no harmonic sense; (f) "undulating" pulse. From *Monchordon symbolico-biomanticum* (1640).

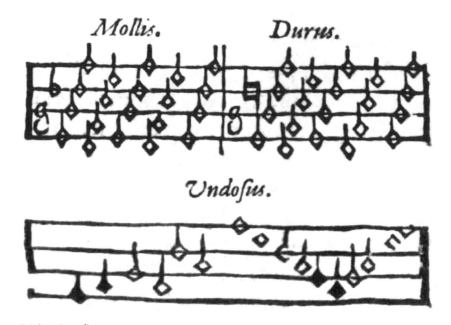

Figure 8.1 (continued)

enumerated the permutations of pulse "size" (such as "great" or "small"), speed, frequency, strength ("vehement" or "moderate"), and quality ("soft" or "hard"). As with figure 8.2, his notation reflected the musical practice of his time, its rhythmic values basically arranged by ratios of two (rather than three or five or any of the more complicated Galenic possibilities). Here, he gave no indication of pitch so that the pulse is presented as pure rhythm. In figure 8.4, he illustrated various pathological pulses: unequal, intermittent, or deficient (dropped). By using two spaces on the staff, he seems to distinguish systolic and diastolic; there is no clef or indication of any precise pitch relation between them. To indicate the "dropped" beats characteristic of deficient pulse, Kircher used the contemporary notation for rests, as if the heart were an instrument pausing for a measured silence.

A century later, François-Nicolas Marquet's musical representations of the pulse were even closer to the music of his time. Though himself a rather obscure figure, Marquet's ideas sparked interest and controversy in the larger medical and intellectual world, forming part of a more general debate about the role of mechanism in biology. Before Marquet, the examination of pulse was a purely tactile exercise; the rhythms were felt, not heard, as in Hafenreffer's notations. By notating them musically, Marquet directed attention to the larger context of those rhythms; what he considered a "normal pulse" he notated as an audible minuet, not just a series of abstract mathematical patterns. His attempt to notate various pulse rhythms drew on the conventions of pitch but fell short of considering them to be actual sounds rather than felt rhythms.

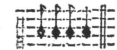

Cantus Puer.

Rhythmus æqua-
lis fiue paris pro-
por.in notis fufis

Pueriliætati refpondet, habetquecelerem
contractionê aut diftenfionê arterie, vti notæ
fufæ cantus fiuè fupremę vocis, exprimunt.

Altus Iuuenis.

Rhythm.æqualis
fiuè paris propo.
in notis femimin.

Adolefcentiæætati refpondens,contractio-
nem diftenfionemque arteriæ habet,notis fe-
miminimis Alti conuenientem.

Tenor Vir.

Rhyth.ęqualis,fi-
ue paris proport.
in notis minimis.

Virili ætati correfpondens ,contractionem
diftenfionemque arteriæ habet,notis minimis
Tenoris conuenientem.

Bafis Senex.

Rhyth.ęqualis,fi.
ue paris proport.
in not. feimbreu.

Senectuti refpondens, contractionem diftenfionemque arterię patitur , qualem femi-
breues notæ in bafi teferunt .

Figure 8.2
Athanasius Kircher's notation of the pulse rhythms appropriate to a boy, a youth, a man, and an old man, from *Musurgia universalis* (1650). Together, they form a chord of F major (reading upward, F–A–C–F); note the B-flat in the staff reserved for old age (♪ sound example 8.1).

Marquet undertook his medical studies at Montpellier (1710–1714), a famous medical center that had not yet become a center of vitalist physiology, as it did later in the century.[7] After finishing his medical training, he practiced at Nancy, where he eventually became the doyen of physicians.[8] Though he ascribed his fundamental idea to Herophilus, Marquet's *An Easy and Curious New Method of Knowing the Pulse by Musical Notes* (*Nouvelle Méthode, facile et curieuse, pour connoitre la pouls par des notes de la musique*, 1747) went much further than any of the ancients (or his contemporaries) in applying music to medicine.[9]

Marquet's cardiac history began with himself, for "having been attacked by heart palpitations, I had much leisure to examine seriously in myself the different derangements and intermissions of the pulse."[10] Marquet viewed his symptoms from the dispassionate perspective of contemporary mechanical philosophy, asserting that "the heart has the same rank and performs the same functions in man as the pendulum of a small or large clock; the veins and arteries take the place of wheels and the nerves are the cords that make the hydraulic machine go."[11] This mechanical view informed his treatment of cardiac rhythm, for

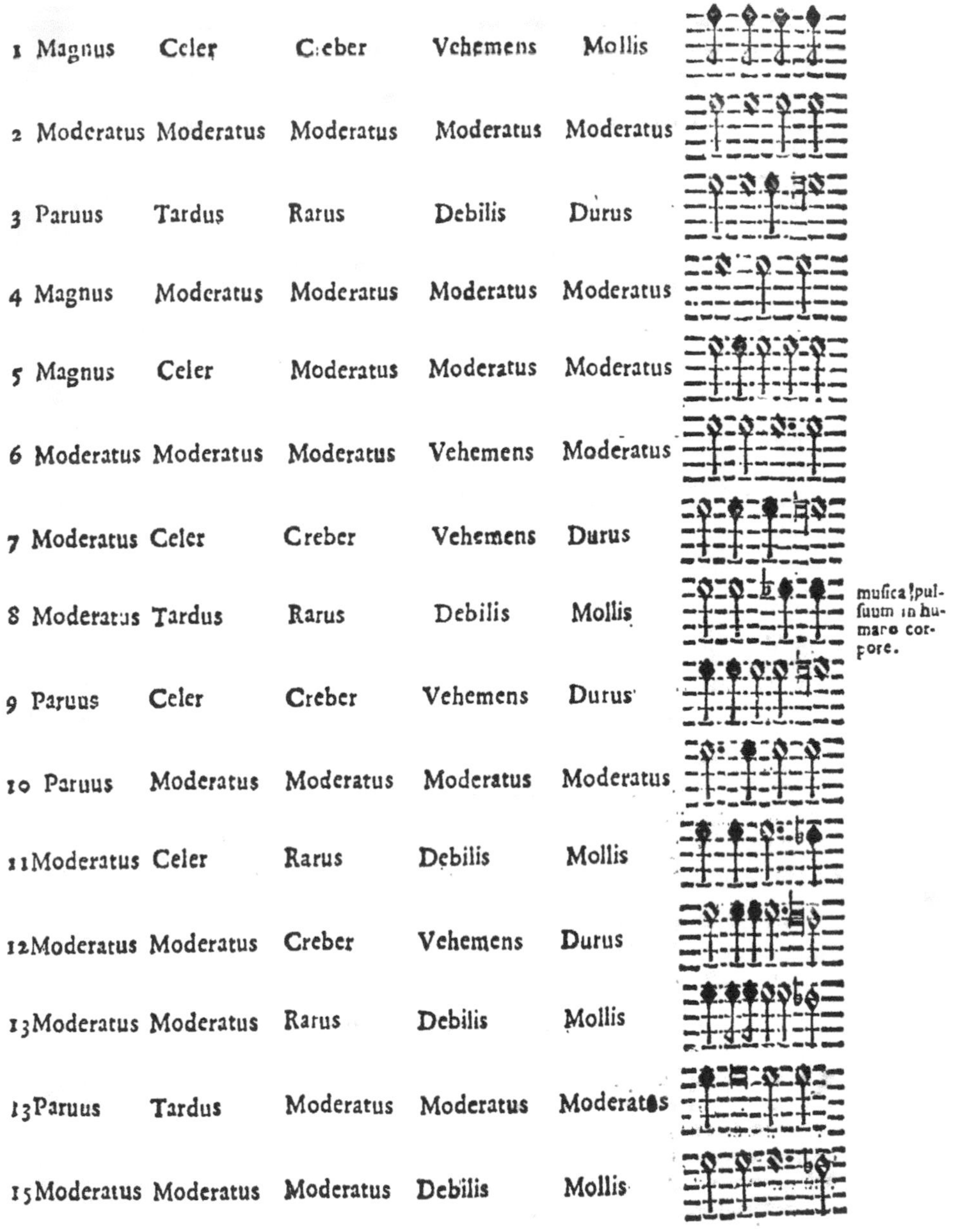

1	Magnus	Celer	Creber	Vehemens	Mollis	
2	Moderatus	Moderatus	Moderatus	Moderatus	Moderatus	
3	Paruus	Tardus	Rarus	Debilis	Durus	
4	Magnus	Moderatus	Moderatus	Moderatus	Moderatus	
5	Magnus	Celer	Moderatus	Moderatus	Moderatus	
6	Moderatus	Moderatus	Moderatus	Vehemens	Moderatus	
7	Moderatus	Celer	Creber	Vehemens	Durus	
8	Moderatus	Tardus	Rarus	Debilis	Mollis	
9	Paruus	Celer	Creber	Vehemens	Durus	
10	Paruus	Moderatus	Moderatus	Moderatus	Moderatus	
11	Moderatus	Celer	Rarus	Debilis	Mollis	
12	Moderatus	Moderatus	Creber	Vehemens	Durus	
13	Moderatus	Moderatus	Rarus	Debilis	Mollis	
13	Paruus	Tardus	Moderatus	Moderatus	Moderatus	
15	Moderatus	Moderatus	Moderatus	Debilis	Mollis	

musica pul-
suum in hu-
mano cor-
pore.

Figure 8.3
Kircher's table of pulses, with columns indicating the size, speed, frequency, strength, and quality of the pulse, each illustrated by a rhythm in contemporary notation (without showing pitch).

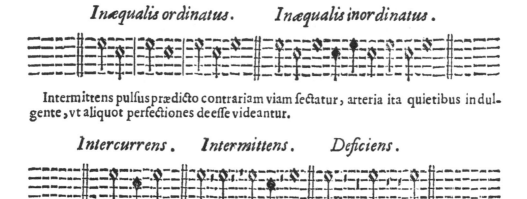

Figure 8.4
Kircher's illustration of different kinds of unequal (ordinary and extraordinary), intercurrent, intermittent, and deficient (dropped) pulses (the rests indicated by strokes between the notes).

so long and to whatever degree the movement of the heart and arteries is regular, the human body remains in perfect health, but as soon as this movement is disturbed by whatever accident, the health is found altered by an infinity of maladies. In order to detect this disturbance, the touching of the pulse has been invented, which is a science absolutely necessary for physicians and surgeons, a science which has something of the divine in that it not only informs us of that which happens in ourselves but also instructs us of the future.[12]

Marquet acknowledged that his quest for this kind of prophetic knowledge of the heart through musical notation may seem "something bizarre," but (in his view) no more so than the universally recognized practice of representing the "sounds of music itself with the same notes, or depicting numbers with numerical symbols [*chiffres*] or finally depicting words with letters of the alphabet."[13]

Marquet's work applied mechanistic physiology to the ancient theory of humors.[14] Referring to "the wonderful mechanism" Joseph Guichard Duvernay found in the human ear, Marquet thought the cochlea caused the sensation of tone by resonating to various incoming vibrations.[15] This mechanism then transmits these vibrations to the internal fluids, which thereby change their state according to the character of the music. Asserting that the sympathetic vibrations of the ear "in fact are a kind of touch" received from the vibrating air, Marquet thus understood music as a complex result of the mechanical percussion of the eardrum. Music "stimulates the auditory nerve and other sympathetic nerves, which being struck agreeably" affects the lymphatic and cardiac systems, "from which come sweet and agreeable ideas."[16]

Marquet based his theory on the "natural regular pulse" (*pouls naturel réglé*) in which the spacing between heartbeats "ordinarily equals the cadence of a minuet in movement," which he specified as about sixty beats per minute, so that each measure—and heartbeat—takes about a second (figure 8.5; ♪ sound example 8.2). His 3/4 time signature assigns three quarter notes (equivalent to six eighth notes, which he calls *tems*) to each measure. The felt pulse coincides with the downbeat, the first eighth note of each measure, followed by five eighth notes of rest between pulses. In modern terms, Marquet heard the minuet "in one," rather than "in three," feeling the dance's primary pulsation on the downbeats, which give its literal heartbeat. Marquet thus gave useful information about the performance practice of the minuet in his time, well before the establishment of metronomes (about 1800) or other mechanical tempo measurements.[17] "As if one sings or plays a minuet on some instrument, in that way one touches a tempered pulse," so that the "touch" by which the musician sensitively evokes the minuet became the model for the physician's equally sensitive—and musical—"touch."[18]

Despite the old saw that the hearts of waltz-loving Viennese beat in 3/4 time, Marquet was (to the best of my knowledge) the first to suggest that the heart beats to that rhythm, writing more than half a century before the waltz emerged as a well-known dance. His claim that cardiac rhythm should be understood in triple meter (such as 3/4), rather than duple (such as 2/4), is remarkable because the heart's "lub-dub" (to use a common American rendition of its sound) does not fall clearly into any simple musical meter. Thus, Marquet could alternatively have used another dance or musical form, such as a march (in 4/4 or another duple meter like 2/4), but seems to have chosen the minuet for musical reasons, probably because it was the primary social dance of his time and hence the most obvious rhythmic pattern for the heartbeat, or at least the most familiar to his audience. His rhetoric implied that anyone who had the least tincture of music would readily be able to call a minuet to mind, or at least feel its tempo, which he seemed to assume his readers would have danced many times.

Marquet himself indicated his awareness of different rhythmic possibilities from the one he called "natural": "If the blood is well conditioned and has a perfect equilibrium between liquids and solids, the pulse will be natural and tempered, will beat equally and have the same force and same interval of time between all the pulsations; on the contrary, if the blood lacks either quantity or quality and if the solid parts are not proportional to the liquids, the pulse will become nonnatural."[19] Those "of a lively and bilious temperament" have heartbeats sufficiently elevated that they show *six* (rather than five) *tems* between beats; the blood of those who are "pituitous or melancholic" circulates so slowly that they require six *tems* between beats, thereby demonstrating their notably lower vital tempo. He advised physicians to be aware of their patients' temperaments so that that can adjust to their respectively "natural" cardiac function, "for they should be counted as natural if they continue in the same *mouvement* [tempo]." He also noted that some physicians consider four heartbeats between regular respirations as "natural," three being "too slow, but this rule is vague and unsure."[20] Here too he seems influenced by musical practice, which does not observe breath (controllable by a skillful singer or player) so much as more deeply ingrained rhythms. His judgment of "natural" cardiac function finally rested

Figure 8.5
Marquet's notation of the "natural regular pulse," overlaid against a minuet (♪ sound example 8.2).

on a musical judgment that "all kinds of pulse that approach most closely" the pulse of his minuet (figure 8.5) "are counted the best." Never before had a dance served as a standard of cardiac rhythm; on the contrary, until that time, musicians used resting pulse as a standard of tempo.[21]

Marquet correlated every degree of physical condition and activity with the respective pulse rhythms. Following the traditional Galenic terminology, he distinguished between "strong," "small," or "worm-like" pulses, which he notated with whole notes, half notes, or tied half notes, respectively (figure 8.6a). Further musical notations on various lines designated still other kinds of pulse, "elevated" or "superficial." In terms of *tems* between each heart pulsation, "if one counts more or less of these spaces between each beat, the pulse will be irregular or unequal [*inégal*] in movement; if the note is not placed between two parallel lines, it will be unnatural in its force, as well as if it is a whole note, half note, or quarter note."[22] Marquet deployed a combination of the traditional mensural notation of music (written left to right, following a certain tempo), along with a temporal microstructure using the pitch-indicating capabilities of the staff, indicating the fine structure of "nonnatural" pulsations in the upper register of the output. In this way, he notated information about the varying "force" of the pulse. Marquet's notation is comparable to contemporary notation for drums, such as Rameau used in his opera *Hippolyte et Aricie* (1733, figure 8.6b; ♪ sound example 8.3). Like a racing heartbeat, Rameau's accelerating drumbeats underline the climax of his orchestral fanfare.

Marquet asserted that his notations provided "the rules by which one can very easily acquire a knowledge that has been for a long time imperfect, so regulated by the notes of music, which should not be disdained, for one has not yet found surer methods to imprint strongly the ideas of the pulsations of which one wanted to give knowledge than those that can enter the memory by the most evident signs that one can show the eyes."[23] He felt that "there does not remain much to do to perfect this method, by which one will show clearly to the finger and to the eye all the differences between natural and unnatural pulses." Marquet required only that "those who wish to be instructed in these procedures have at least some little tincture of music, so that in beating a regular measure they accustom themselves to know the correct cadence of the pulse by comparing it with that of music."[24] Marquet seems unaware of anatomical diseases of the heart, such as had been discussed in the previous century by Raymond Vieussens or by Jean-Baptiste Sénac in his own.[25] Thus, Marquet followed the ancient tradition in which pulse was a measure of fever; his musical notation therefore was a teaching device within that tradition, not a method of cardiac diagnosis.

Figure 8.6
(a) Marquet's notation of various kinds of pulse. (b) The notation for drums (*tymbales*) from Jean-Philippe Rameau's *Hippolyte et Aricie* (1733), act I, scene 4 (♪ sound example 8.3).

2 Exemple du pouls grand ou plein.

3 Exemple d'un pouls petit.

4 Exemple d'un pouls profond.

5 Exemple d'un pouls Superficiel.

6 Exemple d'un pouls dure tendu ou éleve.

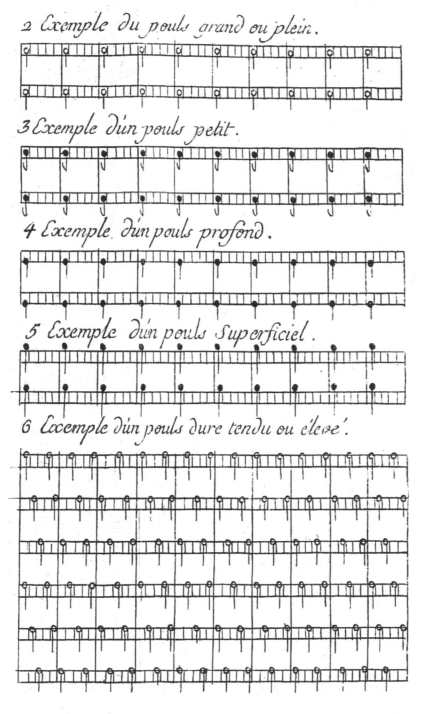

Tymballe

Though he drew on the conventions of musical staff notation (in which pitch is notated vertically), Marquet indicated more nuances for the pulse than did contemporary musical notation for drums (such as figure 8.6b). Nonetheless, there is no evidence that he considered the pulse actually should be heard, as opposed to felt in a purely tactile way.[26] Marquet's ensuing topics also looked to musical archetypes. The more rapid pulse of children he considered a fourth higher than that of adults because the ratio between them, 80:60 beats per minute, respectively, equals the ratio 4:3 (the Pythagorean perfect fourth). Marking the gradual declination into mortality, "from age sixteen, [the pulse] slows more and more as the blood thickens and becomes [less] rapid; one becomes aware from time to time of its inequality and of some intermissions," whose awareness had spurred his own studies of the pulse.[27] These regular pulsations, musical and cardiac, form the background against which Marquet then posed his discussions of irregular pulses.

The contemporary compositional technique of *notes inégales*—the dotted rhythms so characteristic of contemporary French music—was both notationally and theoretically the obvious correlate of unequal pulse, *pouls inégales* (figure 8.7a). For instance, the passage in figure 8.7b (♪ sound example 8.4) shows a typical example of *notes inégales*, familiar in Marquet's period and illustrated by an example from Rameau's *Hippolyte et Aricie*. Though ostensibly using musical notation as a neutral means to represent the heartbeat, Marquet's musical experience shaped his hearing. Rather than taking unquestioningly the notion that the heart beats *notes inégales*, we may more plausibly conclude that the innumerable *notes inégales* he had heard in musical works prepared him to identify that pattern in heartbeats.

When he turned to even more irregular cardiac patterns, we can discern the shapes of familiar rhythms. His notation of "irregular and intercadent" pulses (figure 8.8a) had the same pattern of rapid upbeats that characterizes the French *ouverture* (figure 8.8b; ♪ sound example 8.5), again illustrated by Rameau's opera. Marquet's notation allowed medical practitioners much clearer knowledge of the heartbeats' complex time structures by referring them to the shared sonic world of French music. Marquet took music as the means by which he gained and conveyed information about the pulse.

The French reaction to Marquet's work was dominated by the polemic between vitalists and mechanists. An early critical response came from Théophile de Bordeu, the eminent Montpellier graduate, advocate of vitalism, and pioneer of glandular studies we considered in the preceding chapter.[28] In his *Investigations on the Pulse* (*Recherches sur la pouls*, 1756), Bordeu wrote that because "the perfect [*parfait*] pulse of adults is disposed to take all kinds of modifications," depending on various glandular secretions, "it is only in that sense that one could say with Herophilus that the movements of the pulse have some relation to the laws of music; but if one wanted to apply to the pulse the laws of music, as a modern has tried to do, one will not avoid entering into painful details [*détails pénibles*] that will not be either useful or better founded."[29] Here the anonymous "modern" is clearly Marquet, his approach ascribed to its ancient protagonist Herophilus, as

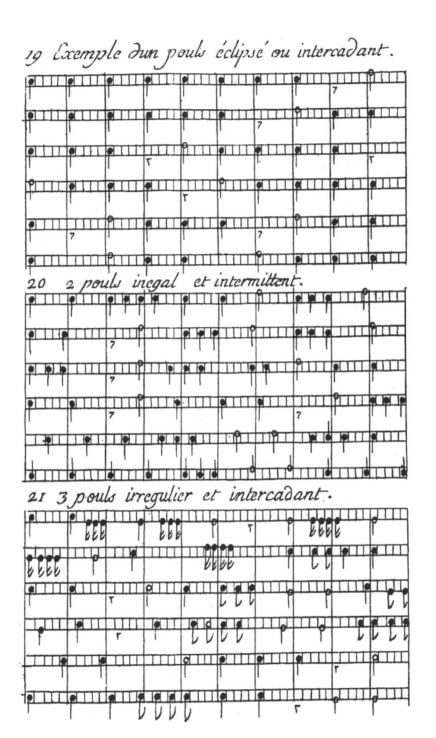

Figure 8.7
Marquet's notation of (a) "intercadent," "unequal and intermittent," and "irregular and intercadent" pulses.
(b) A passage in *notes inégales* from the Prologue to Rameau, *Hippolyte et Aricie* (1733; ♪ sound example 8.4).

Figure 8.7 (continued)

if Marquet himself were merely the misguided epigone of a much older view. Bordeu himself classified further types of pulse whose distinctions required much sensitivity he thought went beyond the possibilities of musical notation, but gave no evidence to confirm later claims that, for him, "there was no division between pulse sound and medical illness. When a physician observed illness, he also perceived its *sound*."[30] Bordeu's exquisite distinctions between pulses remained purely tactile.

A second edition (1769) of Marquet's *Nouvelle Méthode* indicated that his views were widely noticed by his peers.[31] It included several critiques, such as satirical verse that made fun of the idea that the heart might be reduced to musical rhythms.[32] This new edition presented an excerpted version of the section dealing with "the doctrine of the pulse following music" in the article "Pouls" from Diderot's *Encyclopédie* (1765), written by Joseph-Jacques Ménuret de Chambaud, who wrote many of the medical essays in the later volumes of the *Encyclopédie*.[33] This sizable treatment of Marquet's ideas, part of this important (and widely read) compilation, placed Marquet's work before a wide audience.[34]

Like several other medical contributors to Diderot's project, Ménuret was trained at Montpellier and advocated the vitalism prevalent there.[35] As had Bordeu, Ménuret's review of Marquet's theory begins by ascribing the origin of this theory to Herophilus. Ménuret critiques Marquet's work as "only an absurd and bizarre [*singuliere*] mélange of several dogmas of the Galenists, the mechanists, and the chemists." Still, Ménuret devoted a considerable amount of space to the details of Marquet's ideas, even giving fourteen examples in musical notation. After presenting these examples, Ménuret conceded that

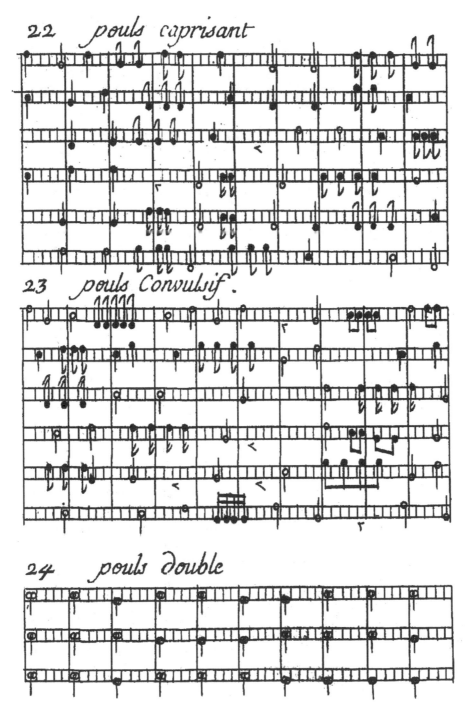

Figure 8.8
(a) Marquet's notation of "capering," "convulsive," and "double" pulses. (b) A French *ouverture*, to Rameau, *Hippolyte et Aricie* (1733; ♪ sound example 8.5).

Figure 8.8 (continued)

one cannot deny that there is a quite sensible connection between the laws of music and the pulse; it is nevertheless no less true that the painful details [*détails pénibles*] into which this author has descended are almost without foundation and utility; at most, this comparison and these figures, if they were indeed really accurate [*bien justes*], could serve to allow conception of that which it is necessary to express, [namely] to give a more palpable idea of the modifications of the pulse by depicting them for the eyes; and if the author had only this object in view, he was not carried far from his goal and his work surely would have been very advantageous, if the system on which it were based conformed less to that of the mechanicians, were less reasoned, and (in a word) were closer to observation.[36]

Thus, Ménuret's objection was not so much to the basic connection between music and pulse but to what he considered the mechanical reductiveness of Marquet's approach. Ménuret, writing a decade after Bordeu, seems to have picked up his phrase *détails pénibles*, with its implications of futility and worthlessness (perhaps caught by the English idiom "gruesome details").[37] The transmission and repetition of this barbed comment show a clear line of polemic unity connecting these fellow vitalists as they joined forces to reject Marquet.

Ménuret, among others, devoted a great deal of attention to Chinese pulse theory, which evidently struck many of the *poulsistes* as even more detailed and subtle than Western accounts. Indeed, in his *Encyclopédie* article "Pouls," Ménuret turned to the Chinese theory immediately after his discussion of Marquet.[38] As Ménuret described it, both theories involve music underlying the pulse because

man, following the Chinese, through his nerves, muscles, veins, and arteries, is a kind of lute or harmonic instrument whose parts give various sounds, or rather have a certain kind of temperament proper to themselves because of their shape, location, and various usages. The different kinds of pulse are like the different sounds and different touches of these instruments, through which one can judge infallibly their disposition, just as a string that is more or less tight, touched in one place or in another, more or less strongly, gives different sounds and allows us to know whether it is too tight or too loose.[39]

Arguably, Ménuret treated this musical approach at far greater length than Marquet's because the Chinese seem far closer to his own vitalist presuppositions. Nevertheless, their exquisitely sensitive classification of pulse did not introduce any quantitative measures, as Marquet did by measuring its tempo.

Though he commented critically on what he called the "chaos" of the Chinese description of various pulses, Ménuret expressed respect for the antiquity of their doctrine, only recently received in the West: "I desire that one suspend judgment on things one does not know, that one should not condemn them only after a ripe examination based on repeated observations." His commentary brought forward but did not make explicit the powerful suggestion that the human body "is a kind of lute or harmonic instrument whose parts give various sounds." Ménuret remained at the level of analogy: "the different kinds of pulse are *like* the different sounds and different touches of these instruments," implicitly raising but not exploring the possibility that the different kinds of pulse *are* the different sounds of the body as musical instrument.

Ménuret indicated other ways in which he wanted to carry further the connection between music and medicine. Beside the eighty medical articles he contributed to the *Encyclopédie*, he also wrote for it an extensive entry, "Musique (effets de la)," in which he discussed the medical uses of music alongside many instances of its mythic powers over soul and body. Regarding these, he quoted the mechanist Boerhaave that "there is reason to presume that the prodigies due to enchantment and the verses used in the cure of maladies ought to be credited to *music*, in which ancient physicians excelled."[40] Thus, Ménuret instanced Orpheus's use of music in rescuing Eurydice from hell as signifying "the cure of the wound a serpent had made, for which *music* is extremely efficacious."[41] Ménuret cited the use of music against tarantism and melancholia, as had Marquet, though dismissing Giovanni Battista Porta's "bizarre idea" that music could provide a panacea against all illnesses.[42]

To understand the basis of music's medical effects, Ménuret began by noting that "only considering the human body as an assemblage of more or less stretched fibers and fluids of different natures, an abstraction made from their sensitivity, life, and movement, one will conceive without difficulty that *music* ought to have the same effect on these fibers as it does on the strings of a nearby instrument," through sympathetic vibration.[43] Despite his vitalist training, he shared with Marquet this mechanistic account of the body, to which we shortly return. Further, Ménuret mentioned that he himself had published findings that music can have curative effect against what he calls *passion hystérique*, a malady then considered specific to women.[44]

Ménuret thus was a rival of Marquet in the realm of musical medicine. Though both used mechanistic language to describe the general effect of music on the body, they parted company over the "painful details," the specific quantification that musical notation applies to pulse. Ménuret's critical attitude reflected his adherence to vitalism. Yet close reading of Ménuret's article shows that he shared many of Marquet's ideas.

Marquet's name sank from sight in subsequent French sources, reflecting the wide distribution of Bordeu and Ménuret's deprecatory remarks and the power of Montpellier anti-mechanist physicians.[45] Bordeu's work inspired three major works imitating him (including Ménuret's own treatise), the coterie that came to be called *poulsistes* and advanced a vitalist philosophy that relied on "natural" curative measures. In accord with their rejection of mechanism and its correlate quantitative bias, Bordeu and his followers considered accurate measurement of pulse rate to be futile and useless.

Mechanist physicians were far more receptive to Marquet than were vitalists. His findings found a place in a larger movement to apply quantitative measurements to the pulse. For instance, the English physician John Floyer had commissioned a special watch for the timing of the pulse that included the innovation of a second hand, which he described in *The Physician's Pulse-Watch* (1707), along with extensive treatment of the Chinese approach to the pulse. Marquet showed no awareness of these developments.[46] Mechanists who continued Floyer's work accumulated data on pulse rates, including Marquet's pulse tempo. In so doing, they were following the lead of Albrecht von Haller, who advanced his teacher Boerhaave's view of living tissue as composed of fibers, whether "irritable" (contractile muscles) or "sensible" (nerves).

Where Ménuret had treated them analogically, Haller's detailed laboratory demonstrations gave those fibers physical reality, as noted in the preceding chapter. The pulsation of the heart, the most "irritable" organ in the body, was a paramount manifestation of the behavior of its constituent fibers.[47] In his comprehensive *Elements of the Physiology of the Human Body* (*Elementa physiologiae corporis humana*, 1757–1766), Haller approvingly mentioned Marquet's "reduction of the pulse to musical notes," as well as Floyer, who "above all reduced [the pulse] to numbers."[48] Haller also included Marquet's pulse tempos in his collection of comparative pulse data for various ages.[49] Citing Haller in this connection, later writers on the pulse continued to refer to Marquet.[50] In the process, Marquet's specific approach was generally forgotten, subsumed in the larger movement to understand the body in mechanistic terms. Yet Haller acknowledged the closeness between numerical and musical understandings of pulse, which seemed natural for fibers whose physical constitution necessarily involves vibration.[51] By Haller's time, the well-studied physics of vibrating bodies had opened the way to all kinds of studies of the sound and resonance of living bodies.

9

Songs of the Blood

The work of Leopold von Auenbrugger and René Laennec stood at later points in the sonic turn that unfolded in the decades after Marquet. In the process of using Marquet's musical data in their larger mechanistic project, Albrecht von Haller and Gerard van Swieten directed attention to the physical behavior of fibers and fluids. Here, the physics of vibrating bodies converged with the sonic awareness schooled by musical sensitivity. Both approaches paid close attention to the sounds bodies could emit and therefore were ready to consider their diagnostic significance. Educated in both this mechanistic physiology and music, Auenbrugger and Laennec thus were well situated to exploit the overlap between them through which they set forth a new sonic diagnostics using percussion and auscultation, which treated the body as a musical instrument whose sounds conveyed much significance.

The son of a hotelkeeper in Graz, Auenbrugger completed his medical training in Vienna under Boerhaave's student Gerard van Swieten, who became personal physician to the Empress Maria Theresa and was a prominent advocate of reforms in the practice and teaching of medicine. Van Swieten played the bass viol, which he recommended along with attendance at the theater.[1] In addition, Gerard's son Gottfried went on to a distinguished career as a diplomat and librarian in Vienna, where he was an important patron of Haydn, Mozart, and Beethoven, especially noted for his advocacy of the music of J. S. Bach, who at that time was less well known.

Like Haller, Gerard van Swieten's encyclopedic *Commentaries on Hermann Boerhaave's Aphorisms* (*Commentaria in Hermanni Boerhaave Aphorismos de cognoscendis et curandis morbis*, 1742–1772) continued their teacher's mechanistic approach to medicine based on the twofold constitution of the body in terms of fluid and fiber, for "of the solid parts of the body, we can conceive none more simple than a fiber," which van Swieten compared to "a mathematical line" in its fundamentality and geometric simplicity.[2] Commenting on Boerhaave, van Swieten began his encyclopedic treatment with the successive diseases that could be found in "a simple solid fiber," "a weak and lax fiber," or "a stiff and elastick fiber."[3]

Van Swieten's awareness of the properties of fibers and fluids led to his attention to their possible vibrations and the concomitant sonic symptoms. Discussing empyema in the thorax (a collection of pus between the lung and the space around it), van Swieten related that "Diemerbroeck had the care of a merchant of Nimmeguen, in whom he could plainly hear the fluctuation of the matter contained within his breast, upon bending his body backward and forward," matter which the man passed out through his ureters.[4] In his discussion of pneumonia, van Swieten described the mechanics by which compression of the air vessels of the lungs can lead to coughing. Because of this, "Generally there is also at the same time a disagreeable rattling in the breast, which arises from the collision of the air against the mucus here collected, or else from the dried vesicles of the lungs rattling like dry parchment, when they are expanded by inspiration."[5] This description of pulmonary crepitation connected its characteristic sound (sometimes audible to the unaided ear) with the underlying physiological mechanics of the afflicted lungs. These examples may be closer to the Hippocratic practice of succussion (the intentional "shaking" of a patient); as Jacalyn Duffin and others have noted, the fundamental techniques of percussion may have long predated Auenbrugger.[6] Indeed, van Swieten himself regularly used abdominal percussion to judge whether abdominal swelling came from fluid or from gas.[7] In all these cases, attention to the mechanical behavior of fibers and fluids underlay van Swieten's attention to sonic signs.

This context underlies his student Auenbrugger's *New Discovery to Detect Diseases within the Chest from Percussion of the Human Thorax* (*Inventum novum ex percussione thoracis humani interni pectoris morbos detegendi*, 1761), in which the only authority he cited was "the commentaries of the very illustrious Baron [Gerard] van Swieten because there the true observer finds resolved in every respect everything one could ever desire."[8] A much-repeated story connects Auenbrugger's early experience with testing the levels in wine casks by knocking on them with his discovery of the diagnostic value of percussion, as if the thorax were akin to a cask. Yet Max Neuburger, the historian who put forward this story, also noted that it lacked any documentary evidence.[9] Auenbrugger's training in the behavior of bodily fibers and his musical abilities may have been far more germane, for in his book, he initially described the sonic qualities of the thorax in terms of drums and their variable timbral possibilities, only later mentioning an analogy with a cask.[10]

Auenbrugger's musical activities went far beyond the rather high standards of familiarity and education expected of a *Kulturträger* in his milieu.[11] He wrote the libretto for an opera by Antonio Salieri, *Der Rauchfangkehrer* (*The Chimney-Sweep*), performed repeatedly in Vienna and elsewhere between 1781 and 1788, whose text Wolfgang Mozart described as "a wretched work."[12] Auenbrugger knew Salieri well enough to be the witness for his wedding in 1775. Further, Auenbrugger maintained his connections with the van Swietens in Vienna, including Gottfried. Auenbrugger was also a friend of the Mozart family, to whom they turned when the Archbishop Colloredo (Wolfgang's irascible employer in Salzburg) traveled in 1773 to Vienna, where the Mozarts sought a new life.[13]

Figure 9.1
The title page of Marianna Auenbrugger's *Sonata* (ca. 1781) (♪ sound example 9.1).

Auenbrugger's daughters Marianna and Katharina were both distinguished keyboard players, to whom Joseph Haydn dedicated six piano sonatas (Hob. XVI:3–59, 20), writing in 1780 that "the approval of the *Demoiselles* von Auenbrugger . . . is most important to me, for their way of playing and genuine insight into music equal those of the greatest masters. Both deserve to be known throughout Europe through the public newspapers." Leopold Mozart also noted that "the daughter of Dr. Auenbrugger . . . play[s] extraordinarily well and [is] thoroughly musical."[14] Marianna also studied composition with Salieri; her *Sonata per il clavicembalo o forte piano* (ca. 1781), her "first and last work" (as its title page proclaimed), was published along with an ode by Salieri, who signed himself "a friend and admirer of her rare virtues" (figure 9.1, ♪ sound example 9.1).[15]

All of these connections, both Auenbrugger's own and those of his daughters, embedded him strongly in Viennese musical culture. His *Inventum novum* used precise musical terminology to describe various thoracic sounds. After noting that "the healthy human thorax sounds, if it is struck," he added that "the sound the thorax gives is observed to be such as in drums [*tympanis*] when they are covered by a cloth or by another fabric made of thick wool."[16] Auenbrugger's description showed close knowledge of contemporary drum technique, including the use of muting cloths called *coperti* to muffle the sound of the timpani. This practice was long familiar by the 1750s, when he began his clinical studies of percussion; such muffled drums were customary in military music to achieve a funereal effect. Mozart used them in *Idomeneo* (1781), as did Haydn in his Symphony no. 102 (1794).[17] For instance, in *Die Zauberflöte* (1791), Mozart used these muted timpani to accompany Prince Tamino's arrival at Sarastro's mysterious temple (figure 9.2, ♪ sound example 9.2). Auenbrugger justified his comparison with a drum by emphasizing its utility for clinical description, in which "we lack the specific notions that express the character of the thing conceived": in this unfamiliar realm of sonic phenomena evoked by thoracic percussion, he thought that music was the best (if not the only) way to give a precise description.

Auenbrugger's work remained largely unknown until translated and published in 1808 with extensive commentary by Jean-Nicolas Corvisart, who had been Napoleon's personal physician since 1804 and whose advocacy of percussion brought that technique to the forefront of medical attention.[18] Corvisart emphasized the significance of Auenbrugger's musical terminology, especially its nuances of timbre, remarking that this comparison with a drum was "quite ingenious and above all quite accurate to express the sound one perceives from a struck human thorax," whereas "a metaphysical definition would have nothing to offer to fix the understanding."[19] Though admitting that all such efforts to describe these nuances "have always been vain or enormously insufficient," nevertheless, "in order to give birth to the idea of the things sought, it is necessary to make a comparison offered to the competent organ. . . . How to depict musical sounds, the song of birds, other than by ear?" By justifying the competence of the ear, Corvisart affirmed its legitimacy for cardiac diagnosis in general and also in the specific cases for which Auenbrugger provided even more detailed sonic descriptions.[20]

To establish the range of variability of the slightly muffled drum note sounded by the percussion of a "completely healthy" person's chest, Auenbrugger noted the need for careful sonic training to adjust for differences in the fat, muscle, and general constitution of each individual, for which he advocated first of all self-percussion, then examination of other healthy subjects. Even within the same subject, percussion of different locations will elicit various sounds, to which the practitioner needs careful attunement. Auenbrugger classified those sounds as "higher, or deeper, or clearer, or more obscure, and sometimes almost completely suffocated"; Corvisart glossed Auenbrugger's Latin in terms of his own French auditory vocabulary, in which timbre seems even more important than absolute pitch.[21] This shift from relying on pitch to including timbre implied new dimensions of sonic awareness that could be correlated with more refined cardiac findings.

Corvisart thereby continued the direction of Auenbrugger's own investigations, which went from the range of "normal" muffled sounds heard on percussing a healthy chest toward a spectrum of tones correlated with underlying pathology. In general, Auenbrugger

Figure 9.2
Mozart's use of muted timpani (*coperti*, sixth line down) at the beginning of the Finale to Act I of *Die Zauber-flöte* (1791) (♪ sound example 9.2).

asserted that "if a sonorous part of the chest, struck with the same intensity, yields a sound more superficial [*altior*] than natural, disease exists in that part" and likewise for a sound that is "duller [*obscurior*]" than normal. Corvisart agreed that "in fact, this result is invariable and never deceives."[22] He noted further that in such cases, percussion goes far beyond other clinical signs because often such patients show no other symptoms of disease. Through percussion, the physician can diagnose illnesses that may have no other clinical manifestation but nonetheless be very grave.

Corvisart instanced the case history of a forty-five-year-old Parisian laborer being treated for lead poisoning, whom he percussed and found "an absence of natural sounds under the sternum, in the region occupied by the heart," which he diagnosed as an aneurysm. The workman had experienced no palpitations or pains but died of his aneurysm eighteen months later, just as Corvisart's percussion had diagnosed. Thus, sonic cardiology was not merely an adjunct to preexisting modes of diagnosis; it could discern illness far beyond the abilities of any other medical tool then known. Going far beyond pulse observations or scrutiny of such manifest signs as stool, urine, or the visual appearance of the skin, percussion treated the body as a musical instrument capable of producing deeply revealing sounds, for the first time allowing the prediction of disease in the absence of any other symptoms.

As befits a diagnostic tool of such unprecedented scope, Auenbrugger detailed and refined the differential signs evoked by percussion toward more specific sonic signals. He used the change in percussed sound when the patient took a deep breath to judge how deeply the disease extended into the chest cavity. He graded the various qualities and timbres of struck sound: "The duller the sound and the more nearly approaching that of a fleshy limb struck, the more severe is the disease."[23] This and other signs (as well as studies of autopsies) convinced him that these sounds emanated from greater or lesser masses of fluids effused within the body as part of the disease process. Therefore, "the more extensive the space over which the morbid sound is perceived, the more certain is the danger from the disease."[24] Because they indicated wide-ranging or especially dangerous masses of fluid, among the most alarming sonic indicators was the total "destitution of sound" in regions of the lungs or heart, which Auenbrugger diagnosed as fatal. So great was the power of sonic diagnosis that the *absence* of percussive sound was, in some cases, the truly fatal sign.[25]

Auenbrugger's percussive methods also revealed the morbid effects of "affections of the mind," especially "*nostalgia* (commonly called *Heimweh*)," the intense homesickness that particularly affects young people. When they are

forcibly impressed into the military service, and thereby at once lose all hope of returning safe and sound to their beloved home and country, they become sad, silent, listless, solitary, musing and full of sighs and moans, and finally quite regardless of, and indifferent to, all the cares and duties of life. From this state of mental disorder nothing can rouse them, neither argument, nor promises, nor the dread of punishment; and the body gradually pines and wastes away, under the pressure of unsatisfied desires, and with the preternatural sound of one side of the chest. This is the disease *nostalgia*. I have examined the bodies of many youths who have fallen victim to it, and have uniformly found the lungs firmly united to the pleura, and the lobes on that side where the obscure sound had existed, callous, indurated, and more or less purulent.[26]

To the sensitive and trained ear, the abnormal sounds of their nostalgic bodies give voice to the mortal melancholy of these sufferers.

By the time of Corvisart, the problem of clinical nostalgia had grown in importance as the huge conscript armies of the Revolutionary and Napoleonic periods included more men suffering from debilitating homesickness, at times rising to epidemics affecting 5 percent of the army.[27] Corvisart noted that this dangerous nostalgia could even affect those who voluntarily "go to colleges, the cloister, commercial houses, etc."[28] He observed that the "prostration of their forces grows rapidly and becomes total in a very short time; the face and extremities are pale and a freezing cold spreads over the whole body; the pulse is small, weak, contracted; sometimes slow, at other times frequent and unequal; the difficulty of breathing, the oppression in the pericardial region is very marked." In particular, "the Swiss, mountain-men of the Alps, transported away from their country for whatever cause, are the most often attacked, without having experienced bad treatment or subject to too harsh labor."[29] Even the slightest mention of their families or native customs can

suddenly plunge them into a state that quickly becomes desperate, unless the wise precaution is taken of allowing them a trip to their country. One might know the rustic air called the *Ranz des vaches*, which Swiss soldiers cannot hear without the strongest emotion and which can precipitate them into a melancholy so deep that it could suffice to lead them quickly to their grave. In order to stop the progress of this frightful illness, it was necessary to forbid the playing of this air on pain of capital punishment.[30]

The celebrated Swiss song (figure 9.3, ♪ sound example 9.3) caused, exacerbated, but also relieved the terrible effects of the *mal du Suisse*.

Nostalgia was supposed to affect Bretons particularly, to whom their commanding officers learned to give periodic home leave in order to mitigate otherwise crippling attacks of *mal du pays*. Perhaps, then, it was appropriate that the next step in the physiological study of nostalgia and a crucial advance in sonic medicine came from a Breton, René Theophile Hyacinthe Laennec. As with Auenbrugger, Laennec's clinical innovations took place in a notably musical context, as well as in the mechanistic physiology in which he was trained.[31] From youth, he sang, played the flute, and took private dance lessons.[32] The eminent literary critic C. A. Sainte-Beuve, who studied medicine under Laennec, recalled "exciting encounters" in which Laennec played the flute while the historian Claude Charles Fauriel sang Breton songs: "Fauriel knew the words, but Laennec knew all the tunes, tunes learned in his childhood never to be forgotten. He brought his flute (and you have to have seen Laennec to really imagine him cast as Lycidas) and while his friend recalled the words, he tried to jot them down."[33] Laennec studied the Breton language, made contact with Celtic scholars in Britain, and advocated the merits of Breton-speaking clergy.

Figure 9.3
The Swiss *Ranz des Vaches*, as transcribed by J. J. Rousseau, *Dictionary of Music* (♪ sound example 9.3).

During his medical studies under Corvisart, Laennec learned the diagnostic value of percussion, which in his later work became a model of the kind of objective evidence that could lead to his chosen goal of a new "pathological anatomy," an explanation of disease in terms of anatomically locatable physical lesions. After he formed this goal (by 1804), the sonic modalities of percussion seemed directly connected to his subsequent discovery of mediate auscultation. Though Corvisart described heart sounds so loud that they could be heard "very close" to the patient's chest, Laennec asserted that his teacher had "never" put his ear directly to the chest; after his close friend and colleague Gaspard-Laurent Bayle actually did so, Laennec tried this technique.[34]

Bayle's own illness deeply influenced his and Laennec's clinical studies. Bayle conducted nine hundred autopsies to clarify the criteria for the diagnosis of pulmonary phthisis (consumption), from which he died in 1816. Laennec himself showed marked symptoms of this disease, which he and others called the "nostalgia of Bayle," whose wasting symptoms abated during rare visits to Bayle's native village "in one of the most dismal parts of the Alps."[35] Bayle's resort to direct listening at the chest, as an adjunct to percussion, represented an intensification of his efforts to solve his own mortal enigma.

He attributed his own fatal relapse to his despair at the return of the "usurper" Napoleon from Elba in 1815. At that point, Laennec still agreed with Bayle about the importance of such a "vital lesion" that might be caused by profound despair and clinical nostalgia, but after Bayle's death, Laennec reconsidered the nature of consumption as he elaborated the techniques of intensive listening he had seen Bayle use. Much later, this work gradually led to the revised view of tuberculosis as an organic disease, of which clinical nostalgia was more an effect than the cause.

As he pushed forward the clinical practice of diagnostic listening, Laennec drew on the terminology introduced by his friend Matthieu-François-Régis Buisson's 1802 thesis, which distinguished passive hearing ("audition") from active ("auscultation").[36] In those terms, Corvisart practiced audition, not auscultation, which Laennec believed Bayle and he had pioneered. Both approaches sought sonic evidence of underlying physiological and anatomical conditions, which auscultation probed actively and more deeply. Though he may have begun with "immediate" auscultation as pioneered by Bayle by placing his ear directly on the patient's chest, Laennec began practicing "mediate" auscultation about 1817, using a stick to transmit the vibrations from the chest to his ear.

As Laennec himself noted (and has often been repeated), he was moved to do so by factors of modesty (in the case of female patients) but also because many of his patients' bodies were "disgusting": sweaty, ill smelling, lice-ridden, or too fat. Though undeniable, these factors may have been less significant than the need to hear the internal sounds more clearly than immediate auscultation would allow. In any case, the heightened activity of auscultation bespeaks the kind of acute, engaged, critical hearing that was already familiar to Laennec from his musical interests.

His "stick" metamorphosed into what he at first called a *cylindre*, a simple roll of paper that developed into a funnel-shaped wooden tube he turned on his home lathe (figure 9.4). Devoted to studies of ancient medicine and skilled in classical Greek, Laennec came to prefer the name *stethoscope* (literally "looking into the chest") to other contemporary appellations he found "improper" or "barbarous," such as *sonomètre* or *pectoriloque*, but also *cornet médicale*.[37] This last term ("medical trumpet") reflected the cylinder's likeness to a musical instrument, whereas the more dignified "stethoscope" subsumed these musical, auditory references under the aegis of visual perception, as when Corvisart had urged physicians to develop diagnostic means "to see with a better eye."[38] Such appeals to what seemed a higher court of visuality bespoke a certain unease with the status of purely auditory evidence, as had already plagued the reception of Auenbrugger's percussion: the epistemic claims of seeing versus hearing remained at issue.

No amount of visual imagery, though, could conceal the essentially auditory character of mediate auscultation and stethoscopic technique, of which Laennac was both discoverer and first virtuoso. His remarkable abilities were clearly enabled by his acute sense of hearing, which he tried to impart to his readers in the detailed auditory imagery he provided in his *Treatise on Mediate Auscultation* (*Traité de l'auscultation médiate*, 1826), in

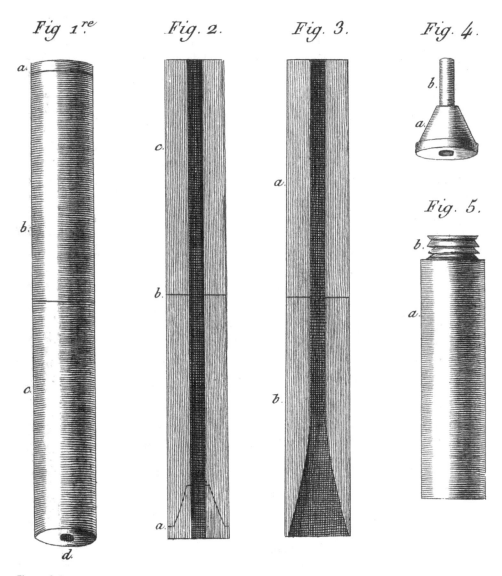

Figure 9.4
Laennec's original stethoscope, from his *Traité de l'auscultation médiate* (1826).

which "one is impressed by the frequency of musical references," as Saul Jarcho noted.[39] Laennec developed a whole new sonic vocabulary to describe and differentiate the vast variety of crackles, rumbles, whistles, and wheezes (*râles*) needed to make adequate differential diagnoses from stethoscopic evidence. His *Treatise* is an extended guide to this new world of sound and of acoustic medicine, every page presenting some new facet of sonic diagnosis, sensitizing its practitioners to ever more subtle gradations of timbre, pitch, and volume as well as to startlingly new sounds.[40]

In some ways, this new sonic world was distinctly apart from contemporary musical practice because it involved subtle gradations of sound and novel sonic objects; even a trained musical ear would struggle to discern stethoscopic crackles and fizzes that (to use anachronistic terms) sound more like twentieth-century *musique concrète* or artificially synthesized sounds than anything remotely considered "music" during the nineteenth century. The ordinary study of melody or harmony did not prepare the physician to cope with the noises of the body, even if amplified and made more accessible for investigation. Laennec's *Treatise* is a fascinating exercise in awakening and expressing whole new realms of sonic awareness and timbre. Though he developed a new vocabulary to describe "unmusical" sounds with greater precision, at many points the shaping force of previous musical and sonic experience seemed to guide his choices.

Laennec referred to familiar sounds as he made his first steps into the bizarre world of thoracic sounds, which otherwise could have been dismissed as a mélange of insignificant noises. Because of the amplifying effect of the stethoscope, its user enters into a world of microsound, suddenly rendered equal in volume to normal environmental sound.[41] A meaningful diagnostic "signal" is embedded in the chaos of other, unrelated sounds; to this day, medical students struggle to distinguish subtle heart or lung sounds. For decades, they relied on recorded examples, but until the availability of such recordings (and even after then), the primary source was direct instruction by someone versed in stethoscopic listening. No less than learning the subtleties of the "medical gaze" by working with an experienced practitioner, learning to hear the body was, since Laennec's time, very much like learning a musical instrument, requiring practice guided by a master artist.[42] The difficulties of this practice evoked some resistance and doubt about Laennec's claims for the stethoscope, but on the whole, "the difficulty of auscultation contributed to its value, made it a mark of initiation, a form of virtuosity."[43]

Laennec had to teach himself this art through prolonged trials and interactions with many patients, colleagues, and students (figure 9.5). Though he began with cardiac sounds, his first breakthrough to realizing the full potential of stethoscopy seems to have come with his discovery of what he called "pectoriloquy"—literally, "the chest speaks" (as "ventriloquy" means "the stomach speaks").[44] In such cases, Laennec noticed that the patient's voice was markedly intensified when heard through a stethoscope positioned at some well-defined site on the chest. He associated this striking acoustic finding with a cavity in the underlying lung. His insight depended on the recognition of voice as a primal

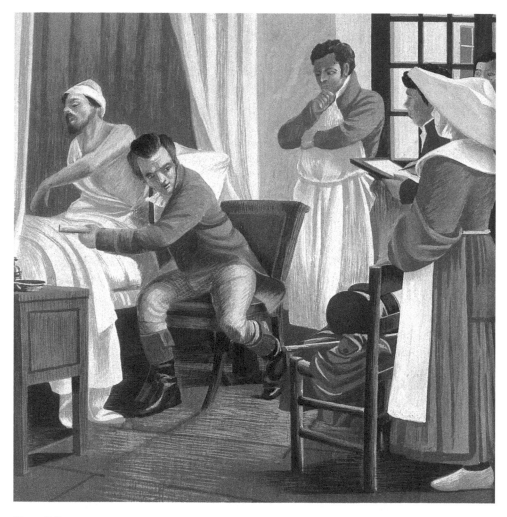

Figure 9.5
Laennec Examines a Consumptive Patient with a Stethoscope in Front of His Students at the Necker Hospital
(1816). Note the stethoscope in Laennec's left hand. Gouache after painting by Théobald Chartran. Courtesy
of the Wellcome Collection.

sonic phenomenon, here remarkably magnified but still recognizable by comparison with
the ordinary experience of the voice or its more muffled rendition as heard stethoscopically elsewhere on the chest. The vocal localization of the site guided Laennec to the
underlying lesion in the lung. But the startling intensity of "chest speaking" seemed to
be a revelation of the nature of the primal voice itself, heard now even more directly than
when ordinarily emitted through the mouth.

In petoriloquy, the primordial sound of the human voice seemed to guide Laennec's initial forays into the new territory of stethoscopic hearing. In general, Laennec connected
an acoustically vivid sign (here, the intensified voice sounds) with deductions about the
likely underlying physical causes (the underlying cavity due to tubercles). In so doing,
he carried forward his mechanistic quest for physical lesions or conditions that would
reasonably explain their attendant sonic manifestations through considering the behavior
of fluids in various configurations and spaces. In the case of what he called "egophony,"
Laennec encountered a distinctively nasalized quality of the human voice heard stethoscopically through the chest wall overlying a collection of fluid in the thoracic space, a
sound he named after the bleating voice of a she-goat (in Greek, *aigos*). But much of what
he heard was alien. About the nuances of *râles*, he confessed that "often I lack words to
express them, or at least it will be difficult for me to describe them in way sufficiently
exact to give a fair idea to someone who had never heard them."[45]

Laennec tended to use the French word *bruits*, noises, rather than *sons*, in the more
neutral sense of "sounds." Searching for descriptions that would make these noises recognizable to his students and colleagues, Laennec often drew on terms from craftwork
and the soundscape of his Parisian milieu. He began with the sounds of the craft shop
in which he himself worked, a hospital ward; thus, in describing the stethoscopic sound
of a *râle*, he noted as its most common meaning the death rattle, the sterterous, labored
breathing of those in extremis.[46] This ominous sound would have been all too familiar in Parisian hospital wards, where the mortality rate was commonly around 20 percent. By connecting stethoscopic rattles to the sounds made by those approaching death,
Laennec noted a kind of sonic prefigurement or thematic recurrence. Those with pulmonary complaints were emitting microacoustic death rattles and for precisely analogous reasons: as dying breaths rattled through fluid-filled lungs, those in earlier stages of
pulmonary disease might emit microrattles from their fluid-compromised lungs. Based
on this basic picture of the fluid mechanics of the lungs, Laennec went on to distinguish a considerable variety of *râles*, encompassing more or less dry rattling, gargling,
whistling, or snoring.

In each case, he drew the specific name and significance of the stethoscopic sign from
the sounds of various kinds of agonized and suffocated catarrhs heard audibly from
his more advanced patients. In this variety of alarming sounds, Laennec recognized a
"trembling very much analogous to what the voice itself produces on the thoracic wall."
Extreme *râles* can be felt by the hand on the chest, not just the ear, "and in some cases

is even much more sensible" than the voice itself. In such cases, the diseased body itself begins to "speak" directly and more loudly than its vocal tract. In one case of particularly advanced consumption, Laennec heard "a tinkling like a little bell that had just finished ringing or a fly buzzing in a porcelain vase."[47] In his ward, he must often have witnessed priests coming to the ward bearing the last sacraments for the dying, according to custom accompanied by the sound of "a little bell."

Then too, Laennec's hospital practice was populated with lower-class craftpersons and workers (among which metalworkers were outnumbered only by dressmakers and female domestics).[48] Thus, he described a certain characteristic as "metallic tinkling [*tintement métallique*]" he heard stethoscopically by evoking an eerily diminutive craft scene, hearing "a noise perfectly like a blow struck on metal, glass, or porcelain, which one strikes lightly with a nail or on which one drops a grain of sand."[49] Sounds on this scale tend to be deceptive. At first, he thought this tinkling, with its artificial character, was somehow produced by the stethoscope itself, only gradually concluding that it was independent of the material of the instrument. Connecting this curious noise with its physiological source requires further effort, absent ordinary links between environmental sounds and the visible behavior of objects. He began by noting its association with the patient's breathing, talking, or coughing, then gradually connected the metallic tinkling with what he deduced as its physical cause, a kind of echo in the chest due to pneumothorax with a small quantity of fluid or alternatively a vast cavity in the lung.

At times, Laennec's investigations went beyond metaphor to a direct encounter with stethoscopic music emerging from the body. On March 13, 1824, he examined a woman "in whom I found a few signs of pulmonary phthisis" and heard stethoscopically a "moderately loud bellows-sound [*bruit de soufflet*]" near the subclavian artery. This kind of sound, with its associations of the blacksmith's forge or even the domestic hearth, was common enough in his experience, but on this occasion

I wanted to see if this sound also was above the carotid [artery]. I was strangely surprised to hear, in place of the bellows-sound, the sound of a musical instrument playing a rather monotonous song, but quite distinct and capable of being notated. I thought at first that someone was making music in the apartment below. I prepared my ear attentively; I put my stethoscope at other points, but heard nothing. After assuring myself that the sound came from the artery, I studied the song: it rolled through three notes forming nearly a major third; the highest note was *out of tune* and a little low, though not sufficiently to be able to mark it as a [semitone] flat. With respect to the *length* or duration, these notes were quite equal between them. Only the *tonic* was prolonged, from time to time, and formed a *hold* [*tenue*], whose length varied. Consequently, I notated the song as follows:

The sound was weak, as if coming from far away, a little shrill [*aigre*] and much like the sound of a *guimbarde* [mouth harp], with the difference that that rustic instrument can only produce staccato notes [*notes pointées*] and that here all the notes were slurred. The passage from one note to another was evidently determined by the arterial diastole, which in the holds themselves, perfectly rendered the light jerks that musicians call *coulé-pointé* [dotted rhythm in a flowing style, a variant of *notes inégales*]. The weakness of the sound made me believe at first that it was passing away into the distance, but by listening attentively and touching a finger to the artery I recognized that the sound was linked to a slight quivering of the artery that, in diastole, seemed in vibrating to come to rub the end of the stethoscope. From time to time the *melody* would suddenly cease and give way to the very loud noise of a rasp. At the risk of using an odd comparison, I can only give an idea of the effect of this alternation by comparing it to a military march in which the sounds of the martial instruments are interrupted from time to time by the rustic noise of a drum.[50]

Laennec's "ear" and pitch sense were manifestly acute; musical terminology sprang immediately to his mind. Most of all, he grasped the auditory passage as a melody, not just a series of pitches. His reference to the rustic *guimbarde* and *tambour* brings to mind his keen knowledge of Breton tunes, as noted by Sainte-Beuve.

Consider, in comparison, the Breton song "Alan al louarn" ("Alan the Fox," figure 9.6; ♪ sound example 9.4).[51] Though this Breton melody is not identical to what Laennec notated from his patient's carotid (♪ sound example 9.5), both begin with repeating permutations of C, D, E and have the recognizable *melos* of folk song, supported by his suggested "orchestration" with a *guimbarde*. Laennec's rendition of the arterial melody strongly suggests that he assimilated it to folk song, perhaps even the Breton songs he knew so well.

Certainly Laennec was startled to hear this music emitted by his patient. After listening for more than five minutes, he asked a medical colleague at the bedside to join him. When they listened again, they only heard the bellows sound; the mysterious melody had fallen silent. Laennec noted his patient's pulse, a regular 84 beats per minute; she had coughed for months, bringing up blood, and was subject to marked attacks of nervous agitation. Laennec gave the musical scores of two subsequent patients whose arteries respectively whistled (*sifflaient*) the melodies (♪ sound examples 9.6, 9.7):

and

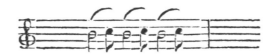

Figure 9.6
The Breton song "Alan al louarn" ("Alan the Fox"), from Théodore Hersart La Villemarqué, *Barzaz-Breiz, Chants populaires de la Bretagne* (1867). Text: "The bearded fox yaps, yaps, yaps in the woods; woe to foreign rabbits! His eyes are two sharp blades! Sharp are his teeth and rapid his feet; his nails red with blood. Alan the fox yaps, yaps, yaps: Beware!" (♪ sound example 9.4).

Another woman, one with a particularly "nervous" constitution, at times emitted a sibilant bellows sound "analogous to the sound of an octave." Incongruously, this most primal consonance emerged from a patient in a notably dissonant physical state.

Laennec's investigations into the causes of these sounds led him to consider experiments by Paul Erman and William Wollaston that provided possible physical explanations for the bellows sound based on the noises made by the heart valves.[52] According to them, the sound might be the overall impression of a rapid series of intermittent valve noises, on the model of rapid pulsations of a thumb striking a wooden rod. Laennec analyzed this model using musical data, estimating the most rapid notes a musician could execute in arpeggios or ornaments (*agréments*) like the *port de voix* (a rapidly rising scale fragment familiar in vocal music from the Baroque period). He calculated the speed of such rhythms and concluded that Erman and Wollaston's suggestion seemed finally void of justification; other physical arguments led him to prefer the hypothesis that the bellows sound "is due to a true spasmodic contraction, whether of the heart or of the arteries."[53] Such rapid vibrations, he noted, "seem to announce a phenomenon that is dependent on an anomaly of nervous influx."

Laennec thus implied a link connecting the extreme nervousness of some patients, their cardiac disease, and the rapid vibrations heard from their circulatory systems. "When the bellows-sound exists in the aorta, the carotids, or in the arterial trunks of the members, the patient is in a state of anguish and extreme anxiety. If the heart and the greater part of the arteries present the same phenomenon, life is in peril; however, it is very rare for the patient to succumb if there is not at the same time organic affections of the heart."[54] Over all this, the question of the mysterious arterial melodies remained suspended, with the implication that they might be heightened or even more rapid states of vibration than the bellows sound. If so, the strange melodies Laennec heard from within his cardiac patients were indeed their swan songs.[55] Yielding their final vibrations, these patients' fibers showed their bodies to be musical instruments sounding their last strains.

III

Sounding Minds

10

Music, Melancholia, and Mania

We now turn from sonic modes of physical diagnosis (using percussion and mediate aus-cultation) to ways in which sound opened new approaches to mental illness. This chapter recounts how a symphony described a new form of "mania" and tried to cure it. The remaining two chapters in part III concern changing concepts of hypnotic suggestion as they first emerged in the work of Franz Anton Mesmer and his glass harmonica, then decades later in the different approach of Jean-Martin Charcot and his tam-tam. In all of these cases, different styles of music and different instruments were associated with dif-ferent modalities of treatment and therapeutic understanding.

During the eighteenth century, the long-standing use of music to treat melancholia expanded to include other mental maladies, naming and bringing into focus some that previously had not been understood as distinct conditions. Among these, obsession had particular significance as the besetting disorder of modern consciousness, if not its hall-mark. During the nineteenth century, monomania became the most frequently diagnosed form of mental illness. For instance, during the period 1826 to 1833, 45 percent of the inmates at the Charenton asylum in Paris were diagnosed with various forms of mono-mania, according to Jean-Étienne Dominque Esquirol, the French physician who coined this term and brought it to wide awareness. He initially defined it as a single pathological preoccupation in an otherwise sound mind.[1] Later known as idée fixe, monomania gener-ally corresponded to what in the following century came to be called obsessional neurosis (Freud), then obsessive-compulsive disorder (in late twentieth-century psychiatry).[2]

Yet decades before Esquirol, a symphony entitled "Il Maniático" ("The Maniac") writ-ten in 1781 by Gaetano Brunetti, an Italian working at the Spanish royal court, gave a detailed musical depiction of *manía*, the word he applied to the obsessive motives of his symphony's solo cello.[3] Brunetti's symphony gives a somewhat subversive take on the musical style of his time in the context of the disturbed conditions familiar among his royal Spanish patrons, beset by royal insanity since the beginning of the century.[4] Bru-netti's symphony has a special place in the unfolding sonic turn in medicine. In contrast to the political dangers of discussing the varieties of royal insanity verbally, Brunetti's musical depiction of obsession could be direct and detailed with less risk of incurring royal ire. He not only evoked the static outward signs of mania through the mannerisms of a solo cello but also the dynamic interaction between the solo cello as "maniac" and the surrounding "normal" world, portrayed by the rest of the orchestra. He presented the different stages of their interaction with wit and sympathy in ways that might well have had special significance for his royal patron, soon to become Carlos IV, who had grown up with the mental maladies of his elders. Brunetti ultimately turned the tables on the "normalcy" of contemporary music to reveal obsessional elements that lie beneath the elegant surface of classical style. In a dénouement that recalls similar reversals of perspective in classic Spanish dramas, Brunetti disclosed unsettling resemblances between "sanity" and "madness." In so doing, he gave musical treatment in a new way to themes that had become famous in works like Cervantes' *Don Quixote.*

From the traditional meanings of various terms for insanity emerged the new developments we discuss concerning "mania" and "melancholy."[5] According to one standard medical text, Philip Barrough's *The methode of phisicke* (London, 1583), "Mania in Greeke is a disease which the Latines do call *Insania* and *furor,*" characterized by "furiousnes" and behavior that is "unruly like wild beastes. It differeth from the frenesie, because in that there is a fever. But *Mania* commeth without a feaver. It is caused of much bloud, blowing up to the braine."[6] Short of the "furious madnes" of mania, melancholy "is an alienation of the mind troubling reason, and waxing foolish, so that one is almost beside him self. . . . The most common signes be fearfulnes, sadnes, hatred, and also they that be melancholious, have straunge imaginations, for some think them selves brute beasts. . . . Moreover, they desire death, do verie often behight and determine to kill them selves."[7] The standard remedies were sleep, exercise, "moderate carnall copulation," and "delectations of the mind," especially to "heare musical instruments and singing."[8] Other accounts, however, noted that the wrong kind of music could cause rather than cure such maladies.[9]

Barrough's text was widely reprinted until 1652. After that, the picture became more complicated. Thomas Willis, who coined the word *Neurologie,* gave one of the most extensive accounts of mental illness in *Pathologiae cerebri* (1667) and *De anima brutorum* (1672). Willis turned from explanations of mental conditions in terms of humors (or of the uterus, in the case of hysteria) instead to the influence of "the Brain and Nervous Stock."[10] He attributed melancholy to "the passion of the heart," mania to "vice or fault of the Brain."[11] Further, Willis considered these to be related, for "Melancholy being a long time protracted, passes oftentimes into Stupidity, or Foolishness, and sometimes also into Madness [*mania*]."[12] Indeed, melancholy and mania "are so much akin, that these Distempers often change, and pass from one into the other, for the *Melancholick* disposition growing worse, brings on *Fury*; and *Fury* or *Madness* [*mania*] growing less hot, oftentimes ends in a *Melancholick* disposition. These two, like smoke and flame, mutually receive and give place one to another."[13] We will shortly return to this connection as it was theorized in Spanish medicine.

In contrast, what later came to be called obsession, monomania, or idée fixe was first described in religious contexts, then more broadly, though without clear identification using what later became the standard medical terminology of mania or melancholy. Religious texts from the 1500s address "scruples," repetitive thoughts that are "a great trouble of minde proceeding from a little motive," as Jeremy Taylor put it in *Ductor dubitantium, or the Rule of Conscience* (London, 1660), referring to the meaning of *scruple* as a tiny weight that nevertheless can lead to "a sad plight": "Some persons dare not eat for fear of gluttony, they fear that they shall sleep too much, and that keeps them waking."[14] Indeed, since the sixteenth century, *obsessio* specifically meant being besieged by the devil, an exacerbated form of *fascinatio*, the diabolic "fastening" of the mind.[15] The term *superstitio* also was used to describe various forms of obsession or compulsion, such as what James Boswell later called Samuel Johnson's "superstitious habit" of counting his steps.[16] By the eighteenth century, obsession had become the province of physicians rather than clerics; unitary explanations in terms of diseased organs or ill-mixed humors gave way to a notion of "nervousness" that could manifest itself in more complex mixtures of rationality and obsession.[17] There was no general clinical term for such involuntarily recurring thoughts or actions until the descriptions of idée fixe and monomania by Esquirol and others in the nineteenth century. Decades before that, though, Brunetti applied the term *mania* to such behavior via a musical evocation of obsessional mental states and their changing relation to the surrounding social milieu.

After the seventeenth century, the terminology of the Romance languages regarding madness changed significantly. Though *manie* is found in French dictionaries as early as 1606 (glossed as "*Fureur*, Mania, Furor"), *El Ingenioso Hidalgo Don Quijote de La Mancha* (whose two parts first appeared in 1605 and 1615) described himself as *loco*, "crazy"; the term *mania* never appears in the book, only *locura*, "craziness."[18] Thereafter, Don Quixote loomed large as the most famous "madman"—obsessed with chivalry—in European literature, only Prince Hamlet having comparable significance and fame.[19] To be sure, Cervantes left much ambiguity in his descriptions of Quixote as "a man whose power of reasoning is weak [*hombre de flacos discursos*]" or having "no judgment at all [*de ningún juicio*]."[20] Much turns on the understanding of mental conditions and its changing vocabulary.[21] In this case, *mania* as obsessional fixation (as opposed to a social craze like tulipomania) appeared in Spanish usage decades before the 1780 symphony we will consider, which arguably was reflecting this new use of this word.

Brunetti's symphony addressed a Spanish court with a long history of mental maladies. Felipe V, grandson of Louis XIV, ascended the Spanish throne in 1700 as its first Bourbon monarch. Contemporary observers reported that he was prone to "black fits," terror, and bouts of melancholy.[22] In constant mental crisis throughout his forty-six-year reign, he would bite his arms and legs, stay in bed for days, and refuse to cut his hair; for a time, he believed that he was a frog, among numerous other symptoms.[23] To address the king's alarming mental instability, in 1737 the queen summoned the famous castrato Farinelli (born Carlo Broschi) in the hope that his singing might calm the monarch. According to a contemporary report, the queen had "arranged a concert in an apartment adjoining to that where the king was in bed, where he had lain for a considerable time; and from which no persuasion could induce him to rise. Felipe was struck with the first air sung by Farinelli, and at the conclusion of the second, sent for him, loaded him with praises, and promised to grant him whatever he should demand."[24]

Farinelli thereafter sang nightly for the king, whose mental stability seemed to depend on hearing this music and often howled for hours on end in imitation of his chosen singer.[25] Farinelli ascended to incomparable influence at court, which continued during the succeeding reign of Fernando VI (r. 1746–1759). Though far more stable mentally than his father, Fernando was also given to bouts of melancholy, which he too assuaged with Farinelli's singing; music and shooting, he said, were his only pleasures. His wife, Maria Barbara, was the patron of Domenico Scarlatti, who had come to the Spanish court in 1731 as her music master and remained until his death in 1757. Maria Barbara's long and close association with Scarlatti and his sonatas added depth of musical involvement to the Spanish court.[26] Her death in 1758 left Fernando prostrate; during the remaining year of his life, he wandered about in his nightgown, unwashed and unshaven.

More than a half century of royal madness led to much perplexity. In 1759, the royal physician, Andrés Piquer y Arrufat, gave a detailed case history of Fernando's condition, which he described by the novel term "melancholic-manic illness [*affectio melancholico-manaíca*]. Melancholia and mania, although treated in many medical books separately, are the same disease."[27] Here, he seems to have gone beyond Willis's position that melancholy and mania were merely "related."[28] Thus, some modern scholars credit Piquer with describing the disorder later known as manic-depressive insanity (now called bipolar disorder) almost a century before Jean-Pierre Falret and Jules Baillarger described what they respectively called *folie circulaire* or *folie à double-forme* (1854).[29]

The childless Fernando was succeeded by his half-brother Carlos III (r. 1759–1788), a man of some intelligence who became "the perfect type of the benevolent despot of the eighteenth century . . . thoroughly imbued with the ideas of the French Encyclopaedists . . . , one of the best, greatest, and most patriotic monarchs that Spain has ever known."[30] In contrast to his predecessors, Carlos III was pointedly uninterested in music, though he built one of the largest contemporary opera houses (in Naples). His punctilious adherence to schedule led him to carry out his public functions always at the same hour with great precision, invariably eating the same things at the same times of day. To mitigate his tendency toward melancholy, he pursued hunting with a comparable kind of obsessiveness; not long before his death, he told a foreign ambassador that he personally had killed exactly 539 wolves and 5,232 foxes.[31] Carlos III's own melancholy and obsessive hunting did not keep him from exercising his duties, but he barred his eldest son, Felipe, from the succession on grounds of "imbecility": Felipe had to be restrained from his violent erotic tendencies, not to speak of his penchant for putting up to sixteen gloves (each larger than the one before) on one hand.[32] As Piquer's work showed, contemporary Spanish medical discourse struggled to find new concepts and vocabulary for these perplexing (and ominous) royal maladies.[33]

Though himself unassuming, hard working, and competent, Carlos III could hardly conceal his contempt for his younger son and designated successor, Carlos (IV), "a good-natured man with a fair memory, and he was nowise deficient in judgment once his interest had been aroused, but his development was what is known as somewhat 'arrested.'"[34] His father told him he was a "complete fool" for naively believing that his wife would never become amorously involved with men of "inferior rank," as in fact she did.[35] Exceedingly fond of hunting and a great collector of clocks, Carlos IV preferred the company of grooms to courtiers, practiced carpentry, advanced Goya to painter-in-ordinary at the court, and was extremely fond of music, playing the violin especially in string quartets, his favorite musical genre.[36]

In 1767, when the nineteen-year-old Carlos was prince of Asturias and heir designate, Gaetano Brunetti (then twenty-three) joined the court as his violin teacher. Born in Naples, Prince Carlos never mastered Spanish; his court remained basically Italian, the prince surrounding himself with Italians such as Brunetti.[37] By the time he came to Spain in 1759 on his father's accession to the throne, Prince Carlos already had "an excellent knowledge and appreciation of musical technique. He was perhaps not exempt from a certain boastful arrogance, for he was a violinist himself."[38] Brunetti had studied with the eminent violinist and composer Pietro Nardini; for his royal pupil, Brunetti wrote pieces of considerable difficulty, indicating the prince's level of skill.[39] In subsequent years, Brunetti continued to teach Prince Carlos while steadily advancing in rank among the royal musicians, in 1770 becoming director of music for various court festivals, thereby "reaching the highest office and position that any musician at the court could hope to obtain."[40]

The years after 1776 brought considerable political stress and controversy as Spain considered how to respond to the American Revolution, facing "the dilemma of taking an imperial power into an anti-colonial war by pursuing exclusively Spanish interests without allying directly with the United States and without recognizing American independence."[41] Allied with France, Spain then engaged in a war with Britain (1779–1783). During this period, Prince Carlos became embroiled in various court intrigues that pitted him against his father, "who had kept him poorly educated, confined to childish amusements, trusted with nothing, and debarred from even the appearance of doing business."[42] He became so outspoken in cabinet meetings advocating views that opposed the king's that Carlos III directly warned him that "if it is thought that division exists now between father and son, then there will certainly be people in the future who will suggest to your family to do exactly the same to you."[43] The underlying tension between the obsessive king and his rebellious son formed one of the subtexts of the music we now consider.

Such was the tense political and dynastic situation in 1780 when Brunetti introduced his Symphony no. 33, subtitled in the manuscript score "Il Maniático" ("The Maniac").

Like all his other works composed at court, this symphony was designated solely for Prince Carlos's use and implicitly dedicated to him, indicating that its peculiarities should be read within this context; Brunetti never traveled abroad, and the circulation of any of his works outside the court would have required special permission lest it fall afoul of the royal administration or even of the Inquisition. Thus, this symphony remained unpublished until the twentieth century.[44] What, then, was the significance of Brunetti's depiction of a *maniático*, that is, one suffering from *manía*? Though Brunetti wrote his own explicit program in the score presented to the prince, we will examine it against the primary evidence of the music itself and contemporary understandings of *manía*.

The symphony begins with an extended introduction in C minor, marked "Largo," in a serious, almost tragic style. Peremptory forte chords of the full orchestra alternate with softer statements from the strings, beginning with a sighing semitonal motif shown in figure 10.1 (♪ sound example 10.1), seemingly a gesture of lamentation. The introduction continues in that vein, adding arresting unison statements fortissimo, abrupt silences, and other hallmarks of the "sensitive style" (*empfindsamer Stil*).[45] Indeed, such introductions also were found in the symphonies of Joseph Haydn, generally taken to have been the exemplar whom Brunetti followed.[46] Normally a slow introduction would lead to an allegro; instead, after a fermata on the dominant, the strings put on their mutes and begin another slow movement, marked andantino, with a rather dreamy pianissimo phrase, still in C minor and in the prevalent mood of sadness, ending with a sighing semitone. At this point, a solo cello enters, indicated in the score just above the other celli and basses, his line marked *Manía*. Here, then, is the *maniático*, who merely repeats a semitone in pensive thirty-second notes on the dominant (G F#; figure 10.2, ♪ sound example 10.2).[47] It seems likely that Brunetti would have known the Spanish court poet Tomás de Iriarte y Oropesa, whose didactic poem *La Música* had appeared in 1779, the year before, and who described the chromatic semitone as "primordial": "Let him o'er whom or tears or fears prevail, / Study the tones of the chromatic scale," noting that semitones belong to sad emotion and anguish, "Moanings and groans that speak unnumbered woes / And death's last agonies."[48] In Brunetti's symphony, the orchestra's melancholy semitones seem to be the trigger for the cello's *manía*. In the following passage, though the *maniático* repeats this semitone gesture in its initial form, he transposes it quite appropriately within the larger harmonic context of the orchestra.

Let us pause over Brunetti's terminology. As noted above, earlier texts did not use *manía* to describe such repetitive ideas, for which there seemed to have been no established term. As a servant of the court, it seems likely that Brunetti knew Piquer as the long-serving royal physician, a fixture of the court; Piquer's 1764 medical textbook made public his hybrid coinage of "melancholic-manic" illness.[49] By 1780, Brunetti (as well as other courtiers) might well have become aware of this new term because it specifically described the malady besetting the preceding Spanish kings according to their official physician.[50] Yet Brunetti does not use this new term; indeed, he takes *manía* in a different direction from Piquer's melancholia-mania.

Figure 10.1
Gaetano Brunetti's Symphony no. 33, the beginning of the first movement. Note the "sighing" semitone in the first and second violins (measure 1–2). (♪ sound example 10.1)

Contrary to its usual furious connotation, Brunetti's *manía* is not at all a frenzied outburst but rather a melancholy obsession characterized through a repeated semitone. His own written description specifies that this symphony "describes (as far as possible, using only instruments and without the help of words) the fixation of a madman [*maniático*] on one sole purpose or idea. This role is given to the solo cello. Other instruments assume the roles of friends pledged to free him from his delusion, offering him an infinity of other ideas, in the form of various musical motifs."[51] The actual unfolding of this plan, however, includes quite a bit of dramatic interaction between the various instruments. Evidencing his awareness of contemporary symphonic style, the madman shapes his obsessive

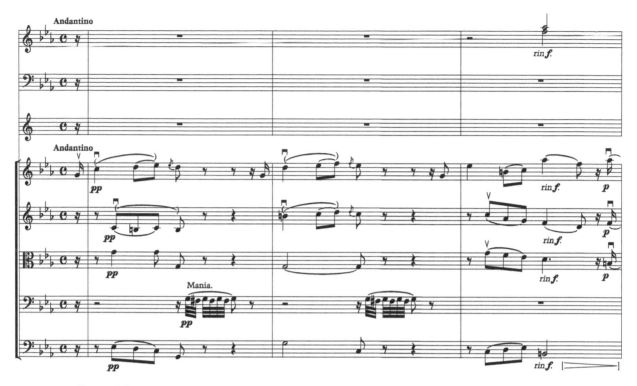

Figure 10.2
The first three measures of the Andantino, showing the first statement of *manía* in the solo cello (whose line is second from the bottom, just above the line for the other celli), a repeated semitone G F# (♪ sound example 10.2).

semitones to answer anything the orchestra can offer him. For its part, the orchestra seems to keep up the rather lyrical mood of this Andantino, as if good-naturedly trying to ignore the solo cello's obsessive repetitions. Still, at several moments, the orchestra's patience seems to run out, leading to a fortissimo outburst of frustration, shown in figure 10.3 (♪ sound example 10.3). As the Andantino unfolds, the orchestra alternates between such outbursts but more frequently reverts to "sensitive" passages that seem to humor the *maniático*. In the final measures of the Andantino, the strings take off their mutes while the winds accompany the solo cello's continued obsessing, his mutterings dying out (*più piano e morendo*).

After a fermata general pause, the orchestra launches fortissimo into an allegro in E-flat major, as if to try the effect of this upbeat tonality and tempo. The *maniático* listens silently for seventy-five measures, evincing no reaction to this cheery, mostly tri-adic material. Well into the second group of its sonata form, a phrase with descending semitones moves him to respond with his obsession, expressed in the prevailing tempo

Figure 10.3
The orchestra's outburst at mm. 43–45 of the first movement (♪ sound example 10.3).

and tonality (the dominant). The orchestra responds with a (seemingly exasperated) fortissimo, which does not deter the cello from repeating his phrase more forte but pizzicato; the orchestra redoubles its outburst, but (not backing down) the *maniático* likewise repeats his fortissimo as well.

To this point, the orchestra and solo cello have generally alternated, but now both begin to overlap their heretofore separate utterances as the *maniático*'s obsessive semitones accompany the orchestra's lyrical second theme. This denotes a changed relation between them: the solo cello alone begins the development section by taking its obsession through a rising scale of semitones, as though self-consciously showing his awareness of the conventions of sonata form, using his obsession to begin the modulatory "tour of keys," a standard procedure for this part of the movement. In such a reading of the cello's agency, he even shows his lucidity by always conforming to the changing harmonies around him.

On one hand, he still repeats the same semitonal obsession; on the other, he moves that gesture up and down the scale in such a way as to fit the developmental agenda. The orchestra even seems to appreciate this contribution, at one point even adopting a similar semitonal gesture.

Whether simply to humor the *maniático* or out of a more sincere appreciation for his efforts, the progression to the moment of recapitulation utilizes versions and inversions of his semitonal obsession (figure 10.4, ♪ sound example 10.4). During the development section, the relation between *maniático* and orchestra alters, though it never really breaks the hold of his obsession. Afterward, a kind of standoff ensues; the recapitulation fastidiously repeats all the byplay of the exposition, but now in the tonic major. The movement ends on this inconclusive note after having briefly opened the possibility of some deeper rapport between the solitary obsessive and his orchestral milieu. Throughout, we can read this thematic byplay in terms of widely used procedures in contemporary composition alongside the programmatic and dramatic interpretation offered here, invited by the composer's own program. As we shall see, these two readings converge at the end of the work, in which Brunetti will emphasize the ironic similiarities between these musical conventions and the *maniático*'s obsession.

The middle movement ("Quintetto: Allegretto") begins with winds alone, still in E-flat major; the "Minore" section uses only strings, in which the solo cello makes brief appearances relying on his semitone, though he allows it to move upward at the very last minute, at the end of figure 10.5 (♪ sound example 10.5). By comparison, though he often used a single theme as a unifying principle throughout a movement, Haydn never used it past the first movement. Though generally following his example, here Brunetti seems to take Haydn's practice even further in order to underline the extent of the *maniático*'s obsession.

As figure 10.6 (♪ sound example 10.6) shows, the final allegro spiritoso seems to mark a decisive break that evidently corresponds to the next phase of Brunetti's program: "The *maniático* for some time clings to his original fancy, until he meets an allegro motif which attracts him, and he joins the others."[52] For the first time, the *maniático* joins with the other cellos in playing the same material, seemingly totally integrated with his peers. Though the movement's theme seems an unexceptionable example of the "brilliant" style, it retains aspects of obsession—attractive to the *maniático*—by taking the initial semitone trill and transforming it into a diatonic descending motif, thrice repeated (figure 10.7, ♪ sound example 10.7).[53] He continues to play this highly repetitive material with the other cellos throughout most of the movement, occasionally stepping forward for brief solo appearances. But as figure 10.7 (♪ sound example 10.7) shows, at the very end of this movement, one of these solos merges back into his initial obsession, the orchestra falls silent, and even the metric pulse fails, marked *ad libitum, senza rigore di tempo*. As Brunetti's program specified, "Soon he falls back into his previous manner." The *maniático* has relapsed, his initial obsession reasserting itself above his participation in the orchestral whole.

Figure 10.4
The passage leading up to the recapitulation, the first violins presenting upwardly "sighing" semitones, an inversion of the *maniático*'s downward semitones (♪ sound example 10.4).

Figure 10.5
Measures 6–19 of the Minore section of the second movement, showing the *maniático*'s slightly varied semitones (m. 18) (♪ sound example 10.5).

Figure 10.6
The first six measures of the Allegro spiritoso (♪ sound example 10.6). Note that here, the solo cello plays along with the others.

Figure 10.7
Measures 73–76 of the Allegro spiritoso, the "relapse" just before the return of the Andantino (♪ sound example 10.7).

The orchestra now resumes the muted Andantino in C minor that preceded the *maniático*'s obsession in the opening movement. Aside from its status as a quasi-cyclical form, one could read this as a mournful recognition of his relapse but also as a way the orchestra joins him in that relapse by recalling its part in its causation.[54] The concluding Allegro spiritoso is in C major, corresponding to the final phase of Brunetti's program: "At last, swept along by the general feeling, he ends with the others quite happily."[55] Indeed, the solo cello now joins in the exuberant texture, at times soaring above the other cellos to join the violins, at times playing along with their tutti. Though these key relations fall within conventions widely shared by contemporary composers, we should reconsider the status of such conventions in light of the satirical context of this symphony.

Though Brunetti's program does not mention it, in the midst of this general rejoicing, the strings launch into an extended unison repetition of the *maniático*'s obsession, which the solo cello and winds accompany with jaunty eighth notes (figure 10.8, ♪ sound example 10.8). The obsessive gesture, it seems, is here exalted, practically canonized. One might read this as the orchestra mocking the obsession after it has finally been dispelled.

Figure 10.8
Measures 9–16 of the final Allegro spiritoso, showing the orchestra playing the *manía* theme (♪ sound example 10.8).

Yet given the maniac's history of relapse and the orchestra's repeated experience of having triggered those relapses, I think it more likely that the orchestra itself has been overtaken by the obsession, now not in a melancholic but a jubilant form, while the solo cello (seconded by the winds) stands back and enjoys it. Indeed, the orchestra had indulged in various forms of the "obsessive" semitone trill many previous times but less consciously, using such repeated motifs as part of the conventional style or perhaps as a stratagem to allure the *maniático*. In its musical intercourse with him, the orchestra has seemingly awakened and acceded with good humor to the obsessional quality of its own common practice, to the classical style's notable dependence on repetition of simple motives.

Within Prince Carlos's circle, the *maniático* may have alluded to a well-known figure, the cellist and composer Luigi Boccherini. There is some evidence that Boccherini stood apart from the prince's court and from Brunetti.[56] Boccherini (who arrived a few years after Brunetti) served in a rival aristocratic circle: that surrounding Prince Carlos's uncle, Don Infante Luis, whose strong claims to the throne held by his brother, Carlos III, led the king to treat him with marked care and respect. Further, an anecdote recounted by Alfredo Boccherini (the composer's great-grandson) indicates Prince Carlos's irritation with Boccherini precisely on the issue of obsessional musical motifs. One day, Prince Carlos expressed the desire to hear Boccherini's latest quartets:

The prince took up his bow, with the intention of playing the part allotted to the first violin. This part consisted of a series of supremely monotonous bars, the notes *do, si, do, si* [the solfège syllables spelling a repeated semitone such as C B C B or G F# G F#].

Utterly exasperated, he rose to his feet and said angrily: "This is abominable! Any beginner could write stuff like this! *Do, si, do, si*!"

"Sire," replied Boccherini, "will Your Highness graciously pay attention to the modulations that the second violin and the viola are executing, and to the pizzicato that may be heard in the part of the cello while the first violin repeats itself. The uniformity of the first violin ceases to be monotonous the moment the other instruments enter and take part in the dialogue."

"*Do, si, do, si*! And it goes on like that for a half an hour. A delicious dialogue indeed! It is the music of a beginner and of a bad beginner!"

"Sire, before pronouncing such a judgment it is necessary to understand music."[57]

According to one of Boccherini's biographers, this cutting retort led to his being excluded from Prince Carlos's palace, where his name was never to be mentioned henceforth.[58] Nor was this exceptional: other stories relate Boccherini's prickly behavior in defense of his compositional practices.[59]

Prince Carlos's ill-tempered comment probably touched a sore point: Boccherini's penchant for repetition. Indeed, Rudolf Rasch judged that among classical composers, Boccherini's "works make the most abundant use of repetition procedures."[60] Though the exact work that irritated the prince eludes identification, in one candidate (the third movement of the quintet op. 28, no. 2, G. 308, 1779), the second violin and viola remain stuck on *do* for almost the whole movement, occasionally interspersed with *si*, while the cello is allowed a virtuosic flight at the movement's close.[61] A celebrated virtuoso on that instrument, Boccherini was closely identified with the cello, which he often used soloistically in his chamber music. Regardless of whatever may have been the composition in question, Brunetti's symphony can plausibly be read as a satire on Boccherini by making his chosen instrument the voice of the *maniático*, whose obsessional *do si do si* fits the tenor of the story and may well have amused the prince.

In addition, within the context of Prince Carlos's circle, Brunetti's symphony also offered a mirror to the various forms of royal insanity that had plagued the preceding reigns. Though far less subject to "mania-melancholia" than his predecessors Felipe V and Fernando VI, Carlos III was somewhat melancholic, to which his obsessive regularity seemed to have offered a kind of counterbalance. Though there was considerable tension between them, any open mockery of the king before Prince Carlos would have been out of the question. Still, Prince Carlos may have appreciated a veiled satire on his father's obsessiveness, especially expressed in music, an art to which he was devoted, despite (or perhaps because of) his father's aversion. Having served as his violin teacher for thirteen years, Brunetti was a familiar and well-accepted figure; every known detail of Prince Carlos's relation to him suggests their sympathy. In the stormy days of 1780, this symphony's gentle but pointed picture of obsessiveness may well have lightened the prince's mood. At the Spanish court, instrumental music was allowed a degree and kind of dramatic mimicry that would scarcely have been permitted in a texted work.

As this analysis has pointed out, Brunetti's symphony goes far beyond a parody of obsession to give a more complex picture of the *maniático* as deeply embedded in his social context (not merely an isolated figure), while also reflexively critiquing aspects of the prevailing symphonic style. Brunetti presents obsession as both emerging from yet coping with melancholy. Though there is a certain pathos in the *maniático*'s mournful repetitions, Brunetti also underlines the ways in which the *maniático* is able to accommodate to his surroundings, both maintaining his obsession while reshaping it to fit its surrounding harmonic context. By alternately humoring him and trying to shake him into normalcy, the orchestra undertakes a certain effort of compassion that (in my reading) ultimately becomes a kind of self-recognition: the *maniático*'s private obsession turns out to be a part of a larger tissue of obsession that marks the stylistic world all the instruments inhabit together.

Thus, straightforward versions of the *maniático*'s "cure" are overshadowed by a far more ironic interpretation.[62] Prince Carlos may have savored this irony: the singular *maniático* at the center of the composition (perhaps standing for the king or even the prince himself) is finally reconciled to his world through a universal realization of the obsessionality—even the madness—of their shared world. Though Pedro Caldéron de la Barca's reputation by then had faded somewhat in Spain, one wonders whether the prince identified himself with the crown prince Segismondo, protagonist of Caldéron's play *Life Is a Dream* (*La vida es sueño*, 1635), imprisoned and tormented by his father through illusory scenes of reconciliation. Brunetti's symphony could likewise have held up a mirror to the tormented mental states and relationships known to father and son. Even more, it seems likely that Spanish listeners to this tale of obsession and melancholia would have thought of Don Quixote, the famous "knight of the melancholy countenance," obsessed by chivalry.

Apart from such topical references, Brunetti's symphony presents a careful musical rendition of the phenomenology of obsession, an unusual (perhaps even unique) example of a mental condition being anatomized musically before being described clinically. Brunetti does not present musically as simple and triumphant a narrative of cure as his program might suggest (and as many of its interpreters have assumed). His premise is deeply social: the *maniático* can only be understood in terms of his environing musical world, in which he participates even through his obsession. Even in the score, his line stands next to the other celli, not apart from them, as later became the convention for concertos. The *maniático*'s obsession ultimately reflects the stylistic practices of his time, which his story brings to consciousness, first to the orchestra's awareness and thereby to the listeners as well. The larger irony is that the whole orchestra (and the world it represents) is obsessed no less than the *maniático* himself. Brunetti shows his obsessions as congruent with (and responsive to) the orchestra around him. At first, the orchestra treats the *maniático* as irrelevantly repetitious but in the course of the development gradually involves itself in his material in service of what increasingly seems a shared endeavor. Such an irony would have been familiar to its courtly listeners, reared on *Don Quixote*, in which successively more and more characters (including a duke and duchess) come to be obsessed with the melancholy knight, in the process becoming (as Cervantes noted) madder than Quixote himself.[63]

Indeed, during the eighteenth century Don Quixote's obsessions became the premier exemplar of idée fixe in the Western tradition.[64] In defining monomania and setting forth his case histories, Esquirol noted that "one finds in Don Quixote an admirable description of monomania, which reigned over almost all of Europe after the Crusades: a mixture of amorous extravagance with chivalrous bravura, which in many individuals was a real insanity [*folie*]."[65] This brings us back to our starting point and to Spain, where the first musical portrayal of obsession appeared after decades of royal insanity, in a court ruled by an obsessive monarch in the land of Quixote. To be sure, literary depictions of obsession preceded and prepared its clinical description, which (as with Esquirol himself) sometimes indicated their indebtedness to the artistic originals that illuminated so many maladies.[66] Yet Brunetti's symphony, so long forgotten, added significantly to the description and understanding of *manía* precisely because it could give an entirely new—musical—form to what had previously only been expressed in words.

11

Composing the Crisis

Through the eighteenth century, physicians used the "animal spirits" infused by music to treat ailments like melancholia or "spleen."[1] Increasingly, though, music seemed no longer a vital part of medical therapy, compared to more materially oriented treatments, such as bloodletting, or long-practiced regimes of diet and exercise.[2] In an era increasingly devoted to the mechanistic view of the human body, the immaterial powers of music seemed increasingly suspect and marginal. Yet by the century's end, music played a significant new role in a controversial form of medical therapy, the "animal magnetism" of Franz Anton Mesmer.

Sound can affect the psyche in ways that bypass reason. Mesmer used his musical skills to structure his therapeutic practice and theoretical self-understanding; he used his own playing as an important element of his therapeutic procedures. Above all, music allowed him to control the timing and duration of his patients' crises, which medicine previously considered unalterable. Mesmer's practices deeply changed the concept of crisis by extending it to a much larger class of maladies than allowed by ancient medicine and by showing how it could be artificially induced and modulated. Mesmer controlled his patients' crises in ways that depended on his preferred musical instruments (especially the recently invented glass harmonica) and (even more significantly) drew on the dramatic structure of classical music itself.

From their time to the present, controversy surrounded the work of Mesmer and his successors, including accusations that they relied on suggestion, self-deception, and even outright fraud. Nonetheless, investigating what they themselves thought they were doing requires attending closely to their express terminology and self-understanding, while remaining cognizant of the respects in which their contemporaries and later readers may have differed or even disagreed radically. Accordingly, our emphasis on the actors' own words should not be mistaken for uncritical acceptance of their claims. Our goal is understanding how sonic and musical considerations informed Mesmer's and (in the following chapter) Jean-Martin Charcot's therapeutic innovations, especially the new directions in medicine they tried to pioneer. Yet even if one judges their work as relying on suggestion or deception, their strategies relied on musical structures and instruments in ways that reflect prior developments in classical and Romantic music.

Mesmer began his career as a rather orthodox medical practitioner, according to the standards of his time.[3] His doctoral thesis, *Physico-Medical Dissertation on the Influence of the Planets* (*Dissertatio physico-medica de planetarum influxu*, 1766), investigated the possibility of tidal effects of the moon and planets on the human body, within the confines of the natural philosophy and gravitational theory he ascribed to "the great Newton."[4] This maiden work demonstrated Mesmer's interest in applying the new natural philosophy and its doctrine of forces to medicine, drawing heavily on work by Richard Mead, a friend of Newton.[5] As Jessica Riskin observed, "Mesmer did not depart from the standard wisdom by relating health to the regulation of imponderable fluids in the body; he merely stated a commonly held belief among natural philosophers."[6] By requesting observational trial and approbation of his later work from the leading scientific bodies and authorities, Mesmer represented himself as an obedient child of Enlightenment natural science, not a rebel against it.

In 1768, Mesmer established himself in Vienna, where he played the cello, the clavichord, and, above all, the glass harmonica. The history of musical glasses goes back, according to Boethius, as far as the legendary Pythagoras (figure 11.1); Diderot referred to their use in ancient Persia. First described by Francis Bacon, musicians began using the excitation of sustained vibrations in musical glasses; the composer Christoph Willibald Gluck played several recitals on them between 1746 and 1751.[7] In 1761, Benjamin Franklin devised a mechanized improvement, fitting the glasses concentrically around a revolving crank worked by a pedal (figure 11.2).[8] By crafting the glasses to the exact size and thickness needed to sound each pitch, Franklin eliminated the use of water for their tuning. At first he called the improved instrument a "glassy-chord" but later chose "armonica," alluding to the Italian word for harmony, *armonia* (though "harmonica" became the more frequent spelling). Franklin thought that the tones of the harmonica "are incomparably sweet beyond those of any other," providing "an instrument that seems peculiarly adapted to Italian music, especially that of the soft and plaintive kind."[9] This new instrument became associated with him, his discoveries, and his political significance as exemplar of the new American nation. As Rebecca Wolf observed, Thomas Jefferson as well as Franklin hoped that the harmonica, "a new instrument, easy to copy and to play," would become "an American counterpart to the European pianoforte"; she also notes the harmonica's "very close connection to the human voice . . . often described as having a direct connection to the nerves and soul of the listener."[10] Franklin's contemporaries noted its "pure transcendent sound," suitable for the "celestial ravishment" of the "sacred choir."[11] Thus, some used musical glasses to accompany religious services.[12] The eerie ringing of the glasses seems to open a sonic portal on another world. As François-René de Chateaubriand put it, "Sometimes the ear of a mortal seems to hear the plaintive tones the echoes of the divine harmonica, those vibrations that have nothing of the terrestrial."[13]

Figure 11.1
Pythagoras and the musical glasses. From Franchinus Gaffurius, *Theoria musicae* (1492).

Franklin also pioneered the use of this instrument to alleviate mental distress. In London during spring 1772, he met Prince Adam Czartoryski, heir apparent to the Polish throne, who brought his twenty-six-year-old wife to see the sixty-six-year-old savant. She later recalled that

I was ill, in a state of melancholia, and writing my testament and farewell letters. Wishing to distract me, my husband explained to me who Franklin was and to what he owed his fame, since I barely knew that a second hemisphere existed. Franklin had a noble face with an expression of engaging kindness. Surprised by my immobility, he took my hands and gazed at me saying: *pauvre jeune femme*. He then opened an armonica, sat down and played long. The music made a strong impression on me and tears began flowing from my eyes. Then Franklin sat by my side and looking

Figure 11.2
The Swiss painter Angelica Kauffmann playing the glass harmonica. Engraving by Carl Rahl published as a frontispiece for Helfrich Peter Sturz, *Schriften* (Vienna, 1819). Pedals below the instrument drive the rotation of a spindle on which the glasses are mounted horizontally. The easel and objects d'art seem to be part of her studio, in which the harmonica forms part of her artistic life.

with compassion said, "Madam, you are cured." Indeed that moment was a reaction in my state of melancholia. Franklin offered to teach me how to play the armonica—I accepted without hesitation, hence he gave me twelve lessons. I have retained memory of him for my whole life.[14]

Princess Czartoryski's stormy life probably influenced her state. She was married at fifteen, and at least three of her six children were born out of wedlock from liaisons with various powerful men (including the last king of Poland). Though healed by Franklin's musical intervention, later she suffered relapses and attempted suicide. The charged musical interaction between this troubled young woman and the aging Franklin was repeated five years later when he formed an intimate friendship with Madame Brillon, who also was unhappily married, subject to melancholia, and sensitive to the harmonica. Franklin's protégée Marianne Davies became the first person to give public performances on the harmonica and toured Europe starting in 1768, where she moved in the highest society. Sometime before 1773 she met Mesmer, who acquired a harmonica from her.[15]

In 1768, Mesmer became a close friend of the Mozart family at the point when the twelve-year-old Wolfgang returned to Vienna after a triumphant tour of Europe. Returning to his native country, though, he experienced setbacks, as so often in his later Viennese career. Though the Emperor Joseph had commissioned the boy wonder's first opera, *La finta semplice* (K51), court intrigue forced the cancellation of the performance. At this point, according to Georg Nikolaus von Nissen (one of Mozart's earliest biographers and the second husband of Constanze Mozart), Mesmer stepped in and offered his garden theater as an alternative venue, thus earning the special glory of having produced Mozart's first staged opera, *Bastien und Bastienne* (K50), whose libretto parodied Jean-Jacques Rousseau's bucolic music-drama, *Le devin du village*.[16]

As with Rousseau's *devin*, who was a fortune-teller or wizard, Mozart's miniature opera involves a benign magus, Colas, who intervenes in a romantic misunderstanding between young Bastien and Bastienne. Mesmer sponsored Mozart's spectacle of a kindly healer whose mysterious words somehow patch up the quarrel between the lovers (figure 11.3, ♪ sound example 11.1). Colas's words, "Diggi, daggi," make no sense in any language but operate as a kind of purely musical charm or enchantment, beyond semantics or meaning. His magical persona and unorthodox cure form a strange mirror reflecting the controversial magus Mesmer was shortly to become.

In subsequent years, Mesmer became active as a harmonica player. In 1773, the ever-critical Leopold Mozart wrote to his wife, "Did you know that Herr von Mesmer plays the harmonica remarkably well? He is the only one in Vienna who has learnt to do so and has a much finer glass instrument than Miss Davies's. Wolfgang too has learned to play it. How we should like to have one!"[17] These connections stimulated Wolfgang to compose several important works for the harmonica, including the Adagio and Rondo in C minor (K617, with flute, oboe, viola, and cello) and the Adagio in C major (K356/617a), both written in May and June 1791, a few months before his death, for Davies's successor, Marianne Kirchgessner, who revived the interest in the harmonica that Mesmer had first aroused.[18]

The comparison between the Adagio (K617a) and the famous motet *Ave verum corpus* (K618), probably written almost simultaneously, may illuminate what Mozart (and his contemporaries) found compelling about the harmonica. *Ave verum corpus* was written to accompany the Benediction of the Blessed Sacrament, a Roman Catholic service in which the consecrated host, made visible at the center of a cruciform monstrance, is adored by the faithful, who seek its blessing (figure 11.4a, ♪ sound examples 11.2). Mozart's famous music for this liturgy reflects its sacred sublimity, the sacramental mystery of transubstantiation refracted through the subdued intensity of his surprising harmonic shifts, as when the flatted sixth accompanies the mention of the piercing of Christ's side during his crucifixion (measures 22–29). Though Mozart's harmonica solo is neither a transcription nor paraphrase of the sacred motet, there are many points of similarity in their texture, melody, and harmonic layout, most of all a common evocation of ethereal sounds, whether by a chorus of rapt adoration or the slow strains of the unearthly instrument (figure 11.4b,

Figure 11.3
Colas' aria, "Diggi, daggi." From Mozart, *Bastien und Bastienne* (K50, 1768) (♪ sound example 11.1).

Figure 11.4
Mozart: The opening measures of his (a) *Ave verum corpus* (K618, ♪ sound example 11.2) and (b) Adagio for glass harmonica (K356/617a, ♪ sound example 11.3), both composed in 1791.

♪ sound example 11.3).[19] A similar impulse to evoke transcendent sound seems to have moved certain British evangelicals to use the musical glasses in their churches during the nineteenth century.

Like much of Mozart's ecclesiastical music (and contemporary Catholic spirituality), this motet blends the sensual and the sacred. Similarly, he evoked a similarly contemplative ecstasy through the glass harmonica. The eerie ringing of the glasses seems to open a sonic portal on another world, which echoed but altered the more familiar strains of sacred hymnody, without any linkage to conventional theology yet sharing a common mood of hushed awe.[20]

More ominously, though, some regarded the glasses' unearthly ringing as capable of inducing insanity, their high, piercing overtones exciting the nerves very differently than did ordinary music. Davies herself experienced "nervous complaints," which some blamed on the effect of the harmonica. In 1798, though he disagreed with this view, the music critic Friedrich Rochlitz wrote that

there may be various reasons for the scarcity of harmonica players . . . , most of all the almost universally shared opinion that its playing is damaging to the health, that it excessively stimulates the nerves, plunges the player into a nagging depression and hence into a dark and melancholy mood, that it is an apt method for slow wasting away. . . . Others say the sharp penetrating tone runs like a spark through the entire nervous system, forcibly shaking it up and causing nervous disorders.[21]

Davies also advertised the harmonica as "the instrument of electrical music."[22] As Wolf notes, contemporaries often described the harmonica's "etherial, sometimes uncanny sound in a manner similar to the very popular experiments with electricity and the context of mesmerism."[23] For Rochlitz, those suffering from any nervous disorder should not play the harmonica; those who are ill should not play it excessively; those who are melancholy should not play it or should restrict themselves to "uplifting" pieces; those who are tired should avoid playing it late at night. These suspicions apparently led the police in some German towns to ban the musical glasses outright "on account of injury to one's health and for the sake of public order."[24]

In his 1766 dissertation, Mesmer had already interwoven musical considerations along with new ways physical forces might affect the human body. He considered the human body to be "a musical instrument furnished with several strings, the exact tone resonates which is in unison with a given tone. Likewise, human bodies react to stellar [planetary] configurations with which they are joined by a given harmony," depending on the time of year and the sex and temperament of the patient, among other factors.[25] Mesmer was a seasoned physician, aged forty when he took his first steps toward a new medicine that might take advantage of these hidden forces. In 1774, the year after Leopold Mozart praised his harmonica playing, Mesmer began to treat a young woman who was a mutual friend of the Mozarts. "Fräulein Franzl" suffered from "continual vomiting, inflammation of the bowels, stoppage of urine, excruciating toothache, earache, melancholy depression, delirium, fits of frenzy, catalepsy, fainting-fits, blindness, breathlessness, lameness, lasting some days, and other horrible symptoms," as Mesmer recorded in his case study. After drinking an iron preparation and having magnets applied, she reported feeling streams of a mysterious fluid running through her body, which Mesmer described as "animal magnetism."[26]

Mesmer went on to apply these same methods to several more cases, including Maria Theresia Paradis, who suddenly became blind at age three yet went on to be a well-known pianist. She too was part of the circle of friends Mesmer shared with the Mozarts; Wolfgang later wrote his Piano Concerto no. 18 in B-flat major (K456, 1784) for her. Paradis's case was especially spectacular and stormy. Moved by her condition, the empress had given her a pension to further Paradis's musical education and had her chief physician treat her.[27] This eminent practitioner tried leeches and even cauterization; following the latest scientific advances, he also experimented with electrical therapy, applying three thousand shocks to her eyes that led to agonizing pain and "hysterical fits" yet did not restore her eyesight.

Using only his hands and a wand, Mesmer was able to restore her eyesight through "animal magnetism." But Paradis's new-found sight disoriented her at the keyboard, where she now made mistakes and lost her place even in simple pieces. She became depressed and yearned for her former blindness. The whole episode brought forward both the seeming power of Mesmer's methods and their strange overlap, even interference, with musical activity, for in an earlier case, he had permanently restored the sight of a girl who did not have Paradis's musical abilities. The ensuing controversy led Mesmer to move to Paris in 1778 in order to bring his new therapy to a more sympathetic capital. The following year, he played for Gluck, who was enchanted by Mesmer's playing and urged him to keep on extemporizing rather than playing printed music.[28] Mesmer followed this advice and left no written compositions. To his improvisations on the piano or harmonica, he sometimes sang along when he was particularly moved.[29]

During this period, Mesmer became active in practicing his animal magnetism on the ever-growing crowds of patients who came to his home at the Hôtel Bouillon. There, Mesmer perfected what he considered the ideal ambiance for his magnetic cures. In those elegant surroundings, Mesmer used music as an integral part of his therapy, which he now practiced on whole groups, not just single patients.[30] His rooms were thickly curtained and somberly lit, shutting out the outside world so as to intensify the subtle sensations within. Several patients would sit in a *baquet*, a large tub containing bottles of water that could communicate the magnetic influence, as did ropes that joined the patients in a magnetic chain.[31] Deep silence prevailed, with only whispering allowed, so that soft music provided by wind instruments, piano, and especially the harmonica could immerse the patients in all-pervasive musical magnetism. Mesmer used this music both to calm patients when needed, but also to excite and disturb them when he felt it necessary to bring on the critical state he called a *crisis*. According to his follower Caullet de Vaumorel, Mesmer's patients responded sensitively to these changes in the therapeutic music (see figure 11.5).[32]

Consider, for instance, the integration of music with other modes of transmission of animal magnetism in the case of an army surgeon afflicted with gout, recounted by his friend Dr. Le Roux, who brought him to Mesmer for treatment:

After several turns around the room, Mr. Mesmer unbuttoned the patient's shirt and, moving back somewhat, placed his finger against the part affected. My friend felt a tickling pain. Mr. Mesmer then moved his finger perpendicularly across his abdomen and chest, and the pain followed the finger exactly. He then asked the patient to extend his index finger and pointed his own finger toward it at a distance of three or four steps, whereupon my friend felt an electric tingling at the tip of his finger, which penetrated the whole finger toward the palm. Mr. Mesmer then seated him near the harmonica; he had hardly begun to play when my friend was affected emotionally, trembled, lost his breath, changed color, and felt pulled toward the floor.[33]

Figure 11.5
Le magnetisme animal, colored etching after Claude Louis Desrais (1746–1816), showing Mesmer's tub (*baquet*) and violinists in an alcove (upper right, under the painting of a house). Courtesy of the Wellcome Collection.

In this account, the effects of Mesmer's harmonica playing were even more spectacular than the tingling magnetism he seemed to project from his finger. When Mesmer treated his associate Charles d'Eslon by playing the harmonica or the piano to convey animal magnetism, d'Eslon begged for mercy because of the extreme intensity of the sensations he experienced.[34] Because of its intense effect, music seems to have had a more central agency in Mesmer's practice than merely being used "to stimulate patients'" receptiveness to mesmeric and "magnetic" therapies."[35]

Alongside his increasingly flamboyant theatrics, Mesmer still thought of himself as a sober scientific investigator who sought to incorporate his practice of animal magnetism into a theoretical framework consistent with Newtonian science. In Paris, he consistently pressed for verification of his theories and clinical results from established scientific bodies; finally, in 1784 a royal commission headed by Franklin himself (and including Antoine Lavoisier) judged the results obtained by d'Eslon (acting as an exponent of Mesmer's methods) to be the product of suggestion. This did not, however, appreciably diminish the tremendous demand for mesmeric treatments or the cachet of Mesmer himself, for whom

Queen Marie-Antoinette arranged a sizable royal pension. As Robert Darnton noted, especially during the 1779–1784 period but also thereafter, mesmerism occupied more space by far than any other topic in French newspapers and journals. As the *Mémoires secrets* put it in 1784, "Men, women, children, everyone is involved, everyone mesmerizes."[36]

Not content with this extraordinary public éclat, Mesmer continued to press for scientific vindication and produced detailed case studies and theoretical writings that would prove he was no magician or mere purveyor of suggestion. For instance, his "Dissertation on the Discovery of Animal Magnetism" (1779) ended with a list of twenty-seven propositions summarizing his findings, particularly that "a universal and continuous fluid" propagates "to other animate and inanimate bodies . . . the action and properties of Animal Magnetism," which can be "intensified and reflected by mirrors, just like light," and which "is communicated, propagated, and intensified by sound."[37] In his official report on Mesmer's methods, Franklin noted the common belief that "to communicate the [magnetic] fluid to the piano-forté, nothing more is required than to approach to it the iron rod; that the person who plays upon the instrument furnishes also a portion of the fluid, and that the magnetism is transmitted by the sounds to the surrounding patients."[38] If so, the transmission of "magnetism" was really dependent on sound. Thus, Mesmer's use of the harmonica and other musical instruments informed his conclusions about the nature and propagation of animal magnetism. Conversely, the harmonica became strongly associated with the practice of mesmerism, so that when Théophile Gautier used a harmonica in his story "Avatar" (1874) to allow a character to see the trembling rays of his heroine's soul, this music implicitly had mesmeric force.[39]

Mesmer used improvised music to prepare and bring on the crisis he considered necessary for each particular patient and case. In the case of the army surgeon quoted above, Mesmer used first his finger and then the harmonica to bring on the crisis, which here involved the patient's physical collapse under the effect of the musical mesmerism. Not all mesmeric crises were violent; some could involve falling into a deep sleep, the somnambulistic state Mesmer called "critical sleep," a restorative period from which the patient could reemerge into health after processing the intense mesmeric drama.[40] In his 1799 "Dissertation on His Discoveries," Mesmer described crisis in terms of the irritability of muscular fibers, expressing "a general law, that the activity of movement always requires an effort against resistance, and that this effort must be proportional to the existing state in order to overcome it. This effort is called *crisis*, and all the effects resulting directly from this effort are called the 'critical symptoms.'"[41] Complementing his dictum that "no disease can be cured without a *crisis*," next to his regular treatment room Mesmer had a mattress-lined "crisis room" (*chambre des crises*) into which patients could be carried when they had reached the critical state and could become so violent or convulsive that they might do themselves harm (figure 11.6). Naturally, the skeptical and satirical press made much fun of the ribald possibilities that might lie behind—or be licensed by—the crisis room.

THE MAGNETISM.

Figure 11.6
Different stages of a mesmeric séance. At the left, a woman is shown composedly conversing with a mesmeric practitioner. On the right, she is shown going into crisis. The background depicts a later stage in her story: seized by convulsions, she is carried into the crisis room. Laurent Guyot (1756–1806), *The Magnetism*, engraving after a color aquatint by Antoine Louis François Sergent. Courtesy Bibliothèque nationale de France.

Mesmer's use of "crisis" clearly depended on the Hippocratic tradition discussed in chapter 1, even though he took that concept in a new direction. For the ancient physicians, the crisis was an innate, fixed part of the nature of the disease itself as it unfolded in specific cases; the physician's task was to observe attentively this timing and do whatever might be appropriate when the critical moment came in order to bring about recovery if possible. Galen and other ancient authorities limited the concept of crisis to diseases with fever, which they measured only by touch and less well-defined measures of pulse rate and physical status.[42] Mesmer, in contrast, was engaged in artificially inducing and regulating crises even in chronic conditions that, left to themselves, might not for long or ever at all lead to such a critical state.[43] To do so, he also greatly extended the meaning to crisis to include maladies that did not necessarily exhibit fever. Describing Mesmer's practices, d'Eslon observed in 1780 that "if he undertakes to cure a madman, he will only cure him by occasioning fits of madness. The vaporous will have vaporous fits, epileptics fits of epilepsy, &c. The great advantage of animal magnetism thus consists in accelerating crises with no danger. For example, one may suppose that a crisis happening in nine days through Nature, left to its own forces, will be obtained in nine hours with the help of animal magnetism."[44]

Mesmer's therapeutic practice used music in ways that follow the contemporary classical style regarding general structure as well as dramatic pacing. Baroque music had treated affects rather statically, as states to be contemplated rather than dramatized; its tonal procedures emphasized the "solar" relation of keys to the tonic. In contrast, classical key relations tended to be "polar" and dramatic.[45] As Charles Rosen noted, a Baroque aria lessens tension toward its center, whereas a classical sonata intensifies it.[46] Similarly, Franklin used the harmonica not to sooth his melancholic listener but to intensify her sadness and bring her to cathartic tears. Mesmer deployed "magnetic" forces dynamically so as to bring about the needed crisis that a "static magnetism" could not provide. To be sure, Mesmer did not always use music in any given therapeutic session; sometimes merely his glance, touch, or even outstretched finger sufficed. In the judgment of the royal commission, suggestion, rather than any putative fluid or magnetism, was the essential element.

Nevertheless, even when he used his hands or eyes without playing a musical instrument, Mesmer's therapy closely followed the same dramatic procedures he would have been familiar with in contemporary music and probably used at the harmonica. The musical art of his time provided him with the means and the methods by which psychic "magnetism"—even if viewed as suggestion or deception—could be deployed therapeutically. Thus, Anton Reicha (1814) described the highest point of dramatic and musical tension in sonata allegro form as *intrigue, ou le noeud* [knot] (which later accounts called "development"), the moment of "catharsis" according to neoclassical dramatic theory.[47] Although none of Mesmer's (primarily improvised) compositions survive, the extant accounts of his playing emphasize its power without noting any stylistic peculiarities, hence imply its general conformity with common musical practices. Accordingly, he

Figure 11.7
The mesmerism scene from act 1 of Mozart's *Così fan tutte*: [Despina] "This is the magnet, that mesmeric stone which originated in Germany and then became so famous in France." *She shakes the magnet over the heads of the supposed invalids.* [Fiordiligi, Dorabella, Alfonso] "How they writhe about, twisting and turning! They're almost banging their heads on the ground" (♪ sound example 11.4).

Figure 11.7 (continued)

would have shaped the intrigue of his musical (and therapeutic) catharses according to the familiar rising dramatic arcs of sonata form or aria.

Opera may provide the clearest and strongest example of this kind of drama. Mesmer was steeped in it as far back as his involvement with Mozart's first opera. Indeed, Mozart's late opera *Così fan tutte* (K588, 1790) looks back on the astonishing career of his friend Mesmer, reaching back to his own first opera. At the end of act 1 of *Così*, after the scheming lovers Ferrando and Guglielmo fake suicide to trick their fiancées, the cynical Despina pretends to be a mesmerist in order to "cure" them (figure 11.7, ♪ sound example 11.4). She shakes over their heads the "mesmeric stone . . . which originated in Germany and then became so famous in France," whereupon the supposed invalids "writhe about, twisting and turning, almost banging their heads on the ground," seemingly in "mesmeric crisis." Despina's vocal trill mimes the waving of the magnet, which then causes sympathetic trills throughout the orchestra and in the bodies of the "suicidal" lovers.

Appropriately enough, at the height of the imbroglio, Mozart, the musical arch-mesmerist, ironically appropriated mesmeric trappings in order to precipitate *his* operatic "crisis." Part of the joke is that mesmerism and opera so nicely reflect—and feed—on each other. Best of all, the mesmeric and operatic treatments work in concert: the "dying" lovers miraculously return to life, and their ladies, far from seeing through the deception, are all the more gratefully deluded as they proclaim to their disguised mesmeric "healer" that "this doctor deserves all the gold in Peru!"

Mozart could afford to phrase serious feelings in terms of comedy or opera buffa. In contrast, Mesmer needed to maintain a certain awesome solemnity in order to conduct his séances. Mesmer's darkened, mirrored rooms, his mysterious mise-en-scène, even his magnetic touch and glance, all depended on and used musical sounds and structures to create an all-encompassing, irresistible therapeutic drama, in which his patients were both participants and audience. As his social movement unfolded, Mesmer enlarged the dramatis personae of this music-drama into the larger international manifestation he called the Society of Universal Harmony. The political harmony thus sought was not a bland pacification but envisaged the same harmonic process of induced crisis by which Mesmer healed his patients. For this mesmeric cure to act on the sickened world, as Charles Fourier put it, "it is necessary to throw all political, moral, and economic theories into the fire and to prepare for the most astonishing event. . . . FOR THE SUDDEN TRANSITION FROM SOCIAL CHAOS TO UNIVERSAL HARMONY."[48] As it had induced a mesmeric healing crisis in so many individual sufferers, music might in similar ways bring about the apocalyptic birth of a new world. In the end, Mesmer drew from music the means and structure by which he sought to compose and control the therapeutic crisis, a strategy of intervention that characterized a new approach to medicine both on the individual and global scales.

12

Catalepsy and Catharsis

Mesmer's successors moved toward different domains of therapeutic experience and, in the process, recruited other methods for inducing "animal magnetism," a term still in use in the 1880s, though the alternative term, *hypnotism*, introduced by James Braid in 1843, gained traction during the nineteenth century.[1] These new methods—and terminology—reflected changing conceptions of the meaning of "crisis." They also bore the clear influence of ways in which Romantic music used new instruments and dramatic procedures to produce ever more shattering dramatic effects. As Mesmer's successors turned their attention from evoking spectacular crises to inducing sleep-like states of somnambulism and catalepsy, they turned from the harmonica to the tam-tam, another new instrument in Western musical practice. The contrast between these instruments' different vibrational characteristics affected their hypnotic effects, the harmonica inducing violent crises, the tam-tam overwhelming normal consciousness and facilitating the operation of suggestion. In so doing, hypnosis relied on instrumental techniques and dramatic strategies Romantic music used to intensify its crises to levels past what classical usage allowed. The salient changes from classical to Romantic musical styles directly informed the nature and structure of hypnotic therapy from Mesmer to Charcot.

Though he considered himself one of Mesmer's most faithful disciples, Amand-Marie-Jacques de Castenet, marquis de Puységur, was troubled by the dangers of convulsive crises.[2] Writing in 1786, he thought the crisis room, "which one should rather call a convulsionary hell [*un enfer à convulsions*], ought never to have existed" and was only forced on Mesmer as a makeshift to deal with crowds of patients.[3] In contrast, Puységur stated, "I do not mean by crisis a convulsive or disordered state: on the contrary, I mean a state *of physical sleep*, which has to be seen to give any idea: I dread the state of *convulsions* as much as anyone and believe that the true goal of a magnetizer should be to make them cease when they occur."[4] In 1784, Puységur had first produced in his mesmeric patients a sleeplike state he called *crise magnétique complette* characterized by "all the most marked characteristics of somnambulism."[5] As noted in the previous chapter, Mesmer considered "critical sleep" just one among many of different sorts of crises he could manipulate. In contrast, Puységur and those who followed him preferred to use somnambulism, which allowed the use of post-trance suggestion without the volatile and possibly dangerous aspects of what he called *crise magnétique* or *crise ordinaire*, which he considered "incomplete" by comparison.

Puységur himself tended to use the simplest means to magnetize, such as his voice or touch, even trees he had previously magnetized; he emphasized that the magnetizer's will was the real force at work. Still, in his commentaries, he noted the particular efficacy of music and especially of the harmonica. He and Mesmer were always in accord "regarding the effect that music could produce on men," though

this effect is more or less great because of their sensitivity, but all are susceptible of feeling it. There are those who assert they never felt such emotion; I can almost affirm that is more the fault of the musicians they have heard than a defect in their own organization for, in the end, any being whatever is sensitive in its own way and music, above all sung music, is only an emanation of sensitivity. *Love, tenderness, gaiety, sadness*, all the sentiments are expressed through words and song, and these two when combined must necessarily please all the world. It is beyond doubt that our nerves are the organs of our sensations. *Music* works therefore immediately on the nerves; and united with the agent of nature can give it a reinforcement that cannot be anything but favorable in the beneficial effect one wishes to obtain.[6]

In one particular case history, Puységur noted that music "probably contributed to *divide* the nervous crises into a good number of periods, greater than had previously presented, and lessened their force in sustaining their duration" from four seizures a day to only one. "This example well sustains the proceedings of M. Mesmer. The instruments that he played proved to effect the help that he felt could be drawn from music; and the choice of his instrument proves also his profound reflections. In fact, the harmonica can be considered as the reassembly of small electrified glass plates [*petits plateaux éléctriques*] whose accumulated movement is manifested by sound, which, combined with animal movement, should produce very efficacious magnetism."[7]

Indeed, for later nineteenth-century practitioners such as Charles Richet (writing in 1884), "Mesmer was merely the initiator of magnetism, but not its true founder," whom he identified as Puységur.[8] After attending Jean-Martin Charcot's famous clinic at the Salpêtrière, Richet went on to pioneering work in immunology that won him the 1913 Nobel Prize in Physiology or Medicine. For our purposes, though, Richet provides insight into the practices and attitudes of Charcot's circle, especially their interest in the extreme form of somnambulism called catalepsy. Richet recalled that "if, at a course at the Salpêtrière, one gives a strong stroke to a tam-tam [a type of gong], immediately three or four patients suddenly will stop, raising their arms in the air (for example) in a pose of fear, their eyes wide open; they will remain thus immobilized, frozen (so to speak) in that pose until one modifies the central innervation in one way or another. Their muscles thus are in catalepsy and one can immobilize them in such a position for an indeterminate time."[9] Such patients formed an important part of Charcot's investigation of what he and his disciple Paul Richer called *la grande hystérie* using *la grand hypnotisme*, part of what Anne Harrington calls "neo-mesmerism."[10]

Figure 12.1
The beginning of the mad scene from Donizetti's *Lucia di Lammermoor* (1836; act III, scene 5, 10 bars before rehearsal number 24): "The sweet sound of his voice strikes me." In the manuscript version, the "dolce suono" was performed by an "armonico" (as Donizetti spelled it), though the published score (shown here) uses a flute (♪ sound example 12.1).

While studying convulsive patients, beginning in 1870 Charcot and Richer sought to differentiate true epileptic crises from imitative simulations of those crises presented by some of their "hysterical" patients. Indeed, Jules Claretie (director of the Théâtre Français) thought that "hysteria is the sickness of our century. One finds it everywhere."[11] For instance, consider Gaetano Donizetti's original version of the mad scene from *Lucia di Lammermoor* (1836), in which the harmonica accompanies Lucia's flight into madness (figure 12.1, ♪ sound example 12.1). (For practical reasons, Donizetti later transferred this part to the flute, but some performances have restored his original version.)[12] Lucia identifies this instrument as "the sweet sound"—literally the blow or shock (*il colpo*)—of the voice of Edgardo, her lover, who, she imagines, has come to wed her, though in her madness she had in fact killed her bridegroom, Arturo.[13] As Heather Hadlock noted, this scene "recalls earlier scenes of Mesmeric 'crisis.'"[14] Donizetti used the harmonica to lead Lucia to her own crisis, a transformation even her act of murder had not induced. What then was veiled in delusion emerges through the mediation of the harmonica, whose eerie sounds seem to act directly through or inside her rather than as the effect of any external mesmerist.

Though presumably unfamiliar with Mesmer's theoretical writings, Donizetti's dramatic and musical instincts reflected mesmeric procedures that had become widely known by 1836. Our understanding of this scene grows through every detail by which the harmonica induces and inflames Lucia's increasingly close absorption in Edgardo's "dolce suono": for her, the harmonica *is* his voice. Not only does the harmonica introduce the melody Lucia recognizes as Edgardo's "sweet song," but it first presents the melody she then sings at the climactic moment: "Alfin son tua, alfin sei mio" ("At last I am yours, at last you are mine"). Her complex recognition and transformation have received different interpretations. For instance, in the subsequent cadenza, Diana Damrau sang in dialogue with the harmonica, whereas Natalie Dessay sang the same material completely unaccompanied, as if the harmonica had been completely interiorized into Lucia.[15] Here, the circle of influence closes: what began as a novel instrument generated an eerie and sensitive new genre of compositions that Mesmer adopted for his practice and eventually circled back to give mesmeric overtones to Donizetti's dramaturgy.

Lucia served as a potent example. Romana Margherita Pugliese noted that "if it seems likely that Donizetti's opera and other staged representations of female madness informed Charcot's view of hysteria, it is almost a certainty that Charcot's displays in turn had an important impact on the culture of the Parisian operatic stage in the last decades of the century, sparking a resurgence of operas centering on the theme of madness and the creation of new hysterical heroines."[16] In 1878, partially under Richet's influence, Charcot began a systematic study of hypnotic techniques. Using what he called *la grande hypnotisme*, some of his hysterical patients manifested what he thought was an invariable series of stages, successively lethargy, somnambulism, and catalepsy.[17] In 1882, Charcot wrote that catalepsy "can manifest itself in the first instance under the influence of an intense sound, of a bright light placed before the gaze, in consequence of the prolonged fixation of the eyes on whatever object."[18]

The tam-tam was a perfect source of such an intense sound and seems to have been widely used in Charcot's clinics to induce hypnosis (figure 12.2a).[19] In addition, Carmel Raz has insightfully discussed Charcot's use of gigantic tuning forks to induce catalepsy (figure 12.2b). Though Charcot and his associates may have used (unpitched) tam-tams more commonly because they were more convenient, the tuning fork showed that even a simple octave by itself could induce catalepsy.[20] These manifestations characterized highly suggestible hysterical patients, rather than others who were less suggestible and not marked by what Charcot called *névroses*. His associates Désiré-Magloire Bourneville and Paul Régnard noted that these manifestations "are not so well seen as in subjects already habituated to hypnotism: in fact, habit makes the *névrosis* much easier to develop. One day, one of our patients, G . . . , in playing with a tam-tam found in the laboratory let it drop and became cataleptic: no longer hearing it vibrate, one of the assistants found her immobile, fixed, and sleeping."[21] Other kinds of music could have similar effects: "Another time, R . . . , hearing military music fell into catalepsy at a moment when the brass suddenly began a reprise in the middle of a very soft passage."[22] On another occasion at the Salpêtrière, a dance was given to distract the patients, but "a stroke on the bass drum had the effect of a gong and the dancers fell into catalepsy, frozen in sudden tableaux vivants."[23]

These uses of the tam-tam took place in a larger musical context. Though some French sources used the general term *gong*, most of those we have been considering used the

Figure 12.2
(a) The "sudden and unexpected sound of a tam-tam" used to induce catalepsy. (b) "Catalepsy provoked by the sound of an octave. . . . The stopping of the octave instantaneously stops the catalepsy and ends the hypnotic sleep." From Bourneville and Régnard, *Iconographie photographique de la Salpêtrière*, 1876–1880.

more specific term *tam-tam*, which denotes a broad circular metal percussion instrument that produces no definite pitch, whereas other gongs (often having a raised central boss) can produce a definite pitch (figure 12.3, ♪ sound example 12.2).[24] First found in China during the Western Han period (second century BCE), gongs of various sorts are also found in Japan and Southeast Asia. The tam-tam was introduced to France via the Jesuit missionaries in China. In 1784, Père Jean Joseph Marie Amiot described this instrument in a letter to Henri Bertin, *ministre sécretaire d'état*, to whom he had sent a *lo* (gong); he had sent another "that makes even more sound" to the duc de Chaulnes. Amiot added that "I think that such an instrument could be marvelous in your operas when one wishes to deafen or frighten the audience. It could also serve in studying the theory of sound and in demonstrating that each isolated sound has more or less perceptible harmonics, according to the nature of the instrument and the acuity of the listeners."[25] As Gundula Kreuzer noted, "Amiot acted as a mediator between Eastern customs and Western science

Figure 12.3
Tam-tams of different sizes (♪ sound example 12.2).

and art. His hope that the gong would be helpful in both operatic and academic contexts proved prophetic."[26]

The tam-tam entered the Western orchestra during the throes of the French Revolution in François-Joseph Gossec's *Marche lugubre*, first performed on September 20, 1790, at the Champs de Mars during funeral ceremonies for Revolutionary "soldier brothers who died for the maintaining of the law."[27] As Raymond Monelle notes, there had been no "funeral marches" as such in eighteenth-century military collections; this work by Gossec "set the standard for future funeral marches."[28] According to Kreuzer, "The instrument could hardly have received a more prominent aural and visual introduction to Parisian society."[29] Scored for winds, brass, and percussion, his march begins with muffled drum rolls, followed by disjointed brass and wind phrases. Often broken by dramatic silences, these plaintive semitone moans have scarcely any melody in the ordinary sense. Suddenly a crash from the tam-tam initiates a fortissimo strain from the band (figure 12.4, ♪ sound example 12.3). The sheer power and shock of these unfamiliar musical techniques created an enormous effect on the first audiences.

Figure 12.4
François-Joseph Gossec's *March lugubre* (1790, ♪ sound example 12.3). The tam-tam part is the lowest line, just underneath the bass drum.

As contemporary reports noted, the tam-tam played a large part in the unprecedented effect of this music, which was played at a whole succession of grand Revolutionary funerals, including those for Mirabeau and Marat.[30] For Mirabeau's obsequies (1791), the nineteenth-century historian Henri Martin wrote that "an innumerable throng of people followed and pressed around the procession, which filed through the streets until midnight, amid funereal music composed by the musician Gossec and to the sound of strange and terrible instruments heard for the first time in France—the trombone and the tam-tam. Modern history has no record of such funeral rites."[31] At the time, the official *Gazette Nationale, ou le Moniteur Universel* wrote that "the lacerating harmonies, broken up by silences and marked by veiled beats of the tam-tam, truly chilled the public and 'spread a religious terror in the soul.'" Another pamphlet, the *Révolutions de Paris*, wrote that "the notes, detached from one another, crushed the heart, dragged out the guts."[32] When Gossec's *Marche lugubre* accompanied the ceremonies in which the remains of Voltaire were transferred to the Panthéon, the poet André Chénier wrote that "one could hear from afar, in the horror of darkness, / The prolonged chords of funeral trombones, / The somber rolls of the muffled drum, / And the sad howls of the Chinese cymbal [*timbre chinois*]," the tam-tam.[33] Through the double meaning of its "Chinese *timbre*," Chénier made the oriental strangeness of the tam-tam, its lamenting howl (*hurlemens*), the crux of this unprecedented evocation of mourning. Paradoxically, this "strange and terrible" Chinese instrument unleashed the deepest patriotic emotions of the French.[34]

As Père Amiot had anticipated, the tam-tam went on to become a familiar orchestral instrument; indeed, one of the tam-tams he sent home found its way into the orchestra of the Opéra.[35] In France, the tam-tam would have been well known through its operatic use in Spontini's *La vestale* (1807), Bellini's *Norma* and Meyerbeer's *Robert le diable* (both in 1831), and Halévy's *La juive* (1835). Hector Berlioz often used tam-tams (no fewer than four of them in his 1837 *Grand Messe des morts*). His *Treatise on Orchestration* (first published in 1844) specified that "the tam-tam, or gong, is used only for scenes of mourning or for the dramatic depiction of extreme horror. Played forte along with strident brass chords on trumpets and trombones, its tremor can be terrifying and exposed pianissimo strokes on the tam-tam, with their gloomy reverberations, are no less alarming."[36] Thus, a tam-tam and bass drum add a pianissimo shudder to the moment in *La damnation de Faust* (1846) when Faust signs away his soul (figure 12.5, ♪ sound example 12.4), while a tam-tam hidden in Aeneas's shield makes a "long lugubrious sound" when struck by Mercury's caduceus in *Les Troyens* (1863).[37]

In light of the many descriptions of the hypnotic effect of such tam-tam strokes, one might wonder how many of the members of Berlioz's audience (or for the other works just instanced) were really rendered cataleptic then and there. Indeed, theatrical representations of somnambulism and other dream states abounded in Paris during the 1820s, including Eugène Scribe and Ferdinand Hérold's ballet-pantomime *La somnambule* (1827), the source for Vincenzo Bellini's *La somnambula* (1831).[38] As Sarah Hibberd has noted, *La somnambule* "evokes the atmosphere of a magnetic séance," with the larger implication that "magnetism came to be perceived as infiltrating musical practices because both were believed to share an 'inexplicable' power and a spiritual dimension."[39] Then too, Eugène Scribe's influential theatrical structure for "well-made plays" (*pièces bien faites*) emphasized the use of cathartic crisis and neoclassical Aristotelian *peripetia* (reversal), which became standard for grand opera.[40] The crash of a tam-tam, like such rhetorically charged harmonic devices as half-diminished chords, became staples of Romantic dramaturgy.[41] Yet even when no tam-tam was used, the ultimate musical response to ever-heightened Romantic music was a collapse into unconsciousness, as when the young composer

Figure 12.5
Berlioz's use of the tam-tam and bass drum to underline Faust signing away his soul in *La damnation de Faust* (part I V, scene 17, bars 66–70) (♪ sound example 12.4).

Guillaume Lekeu fainted during an 1889 performance of the *Tristan* prelude and had to be carried from the theater, overcome by the intensity of the musical experience.[42]

Despite the dramatic demonstrations of hypnotic somnambulism that convinced Charcot and many others, his work was surrounded by controversy. Charcot was well aware of what he called "intentional and deliberate simulation, in which the patient exaggerates real symptoms or creates an imaginary symptomology from scratch," especially among hysterics, where "we encounter it with every step we take in the clinic of this neurosis and there is no denying that this accounts for the low opinion that is at times attached to the study of hysteria."[43] Though he himself seemed unaware of any deliberate imposture, Charcot became pessimistic about the fate of his work toward the end of his life. Later it became known that many of his patients were in fact performing to his expectations, sometimes even coached by other patients or his staff.[44] Full judgment, however, requires weighing complex factors. As Anne Harrington observed, "French asylum hypnosis and hysteria researchers exploited and manipulated their patients, but at least some of them were not a little awed by the results they were producing, and not a little susceptible to being manipulated themselves."[45] Within a decade after his death in 1893, Charcot's teachings about hysteria were disowned by most of his disciples, among whom Joseph Babinski argued that hysteria was a result of suggestion and could be cured by "persuasion."[46]

Whether "animal magnetism," "hysteria," or "suggestion" is used to describe the phenomena we have been considering, their stages and processes have thoroughgoing analogies with the dramatic arc of musico-dramatic tension and crisis. These analogies were not lost on their contemporaries. Already in 1814, E. T. A. Hoffmann had asked: "May not the musician then behave towards the natural world surrounding him like the mesmerist towards his patient, since his active will is the question which nature never leaves unanswered?"[47] Indeed, in his 1814 story "The Magnetizer" Hoffmann's mesmeric protagonist Alban initially preferred using violent crises but then adopted Puységur's approach, which he used to exercise his dark mastery over the other characters.[48] Building on Hoffmann's connection between the musician and the mesmerist, let us look back on the different modalities of hypnotism that span Mesmer and Charcot in terms of contrasting instrumentations and musical techniques.

In Franklin and Mesmer's hands, the harmonica could evoke moods and modulate between them in precisely controllable ways, its penetrating, high-pitched timbre seeming to play directly on the nerves, bypassing and even eclipsing the ability of reason to deal with the instrument's peculiar intensity. Yet even when he was using his hands or gaze without any sound, Mesmer's actions followed the ground plan of musical dramaturgy, specifically that of the classical style around him. The accounts of his practice emphasize his gesture and music but do not usually include any verbal instruction. The dramatic power of his actions was sufficient for his therapeutic ends; he used the harmonica directly to intensify or delay the impending crises of his patients without the need of verbal commands. In so doing, he acted so as to bring to a critical point factors already at work in each patient. To that end, the harmonica acted as a kind of psychic amplifier, capable of bringing forward certain themes in each patient, almost as a conductor might bring forward one or another voice within the orchestra. Here, the dramatic arc of musical form was not merely a metaphor but the immediate source on which Mesmer drew.

The tam-tam, in contrast, produces a diffuse, very wide spectrum of sound. The accounts cited above emphasized the surprising quality of its sound, its sudden and unexpected eruption into the subject's awareness, thereby precipitating an abrupt break with the previous state of mind. The sense of horror often associated with the tam-tam's sound in many of the accounts seems connected with its peculiar timbre, blending so many overtones in such a sudden, complex, and unfamiliar way. As Père Amiot had noted, the tam-tam demonstrates the physics of overtones through having a far broader spectrum (in terms of both pitches and their harmonics) than any other known instrument, blended into a single sonic experience. Perhaps the tam-tam might be best compared to a bass drum (with which it was so often paired, as in the Berlioz example in figure 12.5, ♪ sound example 12.4), which also is pitchless, producing a plethora of many low-pitched sounds blended together. Yet whereas the bass drum's boom dies away quickly, the tam-tam can sustain its complex crash for a long time so that the listener's attention cannot escape its surprise (♪ sound example 12.2). By causing the whole range of hearing to fire simultaneously and sustainedly, the tam-tam evokes a strong generalized reaction within the nervous system—what the mathematician and astronomer Rodolphe Radau described in 1869 as "an explosion of enormous tones."[49] This prolonged stimulation that can thereby interrupt any preceding state—thus the characteristic gesture of cataleptic arrest—yet renders the hearer intensely aware and suggestible. This may be the underlying cause of the tam-tam's extraordinary effect, which is permanently strange not just because of its exotic origins but because of the instrument's innately arresting, disruptive effect. Even the original Chinese context relied on this effect when using a tam-tam as a solemn signal, a warning alerting officials of the exact rank of an approaching visitor.

Yet the contrast between these two hypnotic practices has even larger musical dimensions. Mesmer improvised music calculated to elicit the desired sequence of emotional responses leading to therapeutic crisis, following familiar classical compositional procedures; Charcot sounded a tam-tam whose sonic effect seemed almost the opposite of musical composition, as understood in his time, disrupting mental states without developing a formal sequence of themes. The sound of the tam-tam rendered its subjects so far from their usual state, so suggestible that one understands anew why Puységur thought catalepsy a "complete crisis" more far-reaching and powerful than purely emotional manipulation. If music was effective, sufficiently intense sound could be even more powerful.

The sounds and sights of his days in the Salpêtrière remained "indelible memories" for Sigmund Freud, who learned from Charcot "the genuineness of hypnotic phenomena and their conformity to laws."[50] Freud probably witnessed for himself the dramatic effects in Charcot's clinic aroused by the tam-tam. The concept of therapeutic crisis remained important in Freud and Josef Breuer's "cathartic method," which used hypnotic techniques to give access to hidden traumatic memories (eschewing direct treatment by suggestion).[51] Their method reflected widespread Viennese interest in the Aristotelian concept of catharsis (stimulated by an 1880 book by Jacob Bernays, the uncle of Freud's wife), as Juan Dalma has shown.[52] Even when Freud abandoned hypnosis and changed the name of his therapy to psychoanalysis, the essential structure of catharsis through crisis remained intact, though now mediated not by musical instruments, penetrating gazes, or direct suggestion. Still, Freud noted that he retained from hypnotism "my practice of requiring the patient to lie upon a sofa while I sat behind him, seeing him, but not seen myself," a new form of therapy mediated mainly through sound: the "talking cure."[53] Here, the sounds were those of human speech, not instruments; the therapeutic rapport and its suggestive power were constantly under analysis. To be sure, Freud confessed that he was "almost incapable of obtaining any pleasure" from music because "some rationalistic or perhaps analytic turn of mind in me rebels" when he could not "explain to myself what [music's] effect is due to."[54] Nevertheless, his therapy retained the dramatic musical structure of crisis and catharsis he ultimately inherited from Mesmer and Charcot. If, as E. T. A. Hoffmann suggested, composers act as mesmerists, so also did mesmerists—and their successors in psychotherapy—act as composers, giving form to the crises and catharsis of their patients.

IV

Sounding Bodies

13

Flying in the Dark

The final part of this book concerns ways living bodies could be investigated through new modes of stimulation, at first purely sonic, then electrical but often activated or mediated by sound. These led to ultrasonic imaging and to techniques of listening to neurons that remain widely used and continue growing in their applications. This chapter traces how the changing understanding of bat navigation led to the discovery of ultrasound, which the following chapter considers as a powerful new way to sound the body. The following three chapters present the use of sound and sonically controlled electrical signals first to stimulate and "tune" the nerves, then listen to them telephonically, finally to amplify them through loudspeakers and thereby reach important insights about the nature and functioning of the nervous system.

Thus far in this book, all the sounds involved have been audible to human hearing, beginning with musical pitches and intervals. After 1800, a new sonic realm emerged through investigating how bats navigate in the dark using ultrasound, whose frequencies lie above those audible to the human ear. This sonic turn in the biology of animal communication awaited the development of a new technology capable of producing and detecting those previously unheard sounds. Ultimately those developments (especially the heterodyne principle) depended on the analysis of the "Tartini tones" first noticed by musicians and their generalization as "beat frequencies." The story of bat echolocation intertwines biological investigations with naval warfare and the songs of insects.

Already in the seventeenth century, speculations considered the possibility of sounds that went beyond ordinary hearing. Marin Mersenne's *Harmonie Universelle* (1636) argued that "all the movements that are in the air, in the water, or elsewhere can be called Sounds, inasmuch as they lack only a sufficiently delicate and subtle ear to hear them."[1] This opened the possibility of finding such an ear. Written a year after his *Micrographia*, Hooke's "Method of Improving Natural Philosophy" (1666) judged that though "the Sense of Hearing does not so much instruct as to the Nature of things as the Eye, though there are many Helps that this Sense would afford by a greater Improvement."[2] Hooke imagined the possibility that "Otocousticons" might make audible "Sounds very far distant," so that "the Variations and Changes of the Weather might be predicted much longer before-hand than now they are, and Ships at Sea might perhaps discover an Enemy of Weather coming by the Hearing, as well they can now discern an Enemy's Ship by the Sight." As telescopes show "the smaller Parts and Rocks of the Moon," perhaps a similar instrument might allow us to hear "Noises made as far off as the Planets."

Turning to biological matters, Hooke envisaged "a Possibility of discovering the Internal Motions and Actions of Bodies by the sound they make" as we can hear in a watch

the beating of the Balance, and the running of the Wheels, and the striking of the Hammers, and the grating of the Teeth, and Multitudes of other Noises; who knows, I say, but that it may be possible to discover the Motions of the Internal Parts of Bodies, whether Animal, Vegetable, or Mineral, by the sound they make, that one may discover the Works perform'd in the several Offices and Shops of a Man's Body, and thereby discover what Instrument or Engine is out of order.[3]

Hooke further imagined "that in Plants and Vegetables one might discover by the Noise the Pumps for raising the Juice, the Valves for stopping it, and the rushing of it out of one Passage into another, and the like." He tried to keep himself from going further because "methinks I can hardly forbear to blush, when I consider how the most part of Men will look upon this," but he could not stop.

Despite the derision of "the Generality of Men," Hooke thought that believing such "foolish and phantastick" things possible "may perhaps be an occasion of taking notice of such things as another would pass by without regard as useless."[4] Hooke noted that "I have been able to hear very plainly the beating of a Man's Heart, and 'tis common to hear the Motion of Wind to and fro in the Guts, and other small Vessels, the stopping of the Lungs is easily discover'd by the Wheesing, the Stopping of the Head, by the humming and whistling noises, the sliping to and fro of the Joynts in many cases, by crackling, and the like." Because he can hear such motions that are "quite invisible as to the Eye," Hooke imagines "an Artificial Timpanum" that could be made "more nice and powerful to sensate and distinguish them," which he proposes "as Opportunity is offer'd to make Tryal of, which if successful and useful, I shall not conceal."[5] Here Hooke seemed poised to invent the stethoscope a century and a half before Laennec. Though Hooke had the means and skill, there is no record of any such "Tryal," one suspects that he was simply distracted by other projects or duties, of which he had so many. Had he succeeded, the course of modern medicine might have been greatly advanced.

Having made such powerful use of the microscope, Hooke quite naturally realized the power that such an "Artificial Timpanum" might offer as a microscope for sound. As Hillel Schwartz put it, "Having explored worlds that became more populated under lenses microscopic and telescopic, Hooke was prepared for a world filled with noises at every scale, noises that people were already, unawares, listening through, noises that would eventually deserve to be listened *to.*"[6] Though Hooke never brought forward such a device, fifteen years later he was still thinking about the possibilities. In his *Lectures of Light* delivered in 1682, Hooke considered that the minute vibrations of small bodies might elude the eye: "But then where the Eye is unable to assist us any further in distinguishing the swiftness of Vibrations, there the Ear comes in with its assistance, and carries us much further."[7] Noting that "the smaller the Bell, the sharper and more shrill its Sound," Hooke conceives "that there may be yet beyond the reach of our Ears infinite shriller and shriller Notes, which may be distinguished by Ears or Organs of Hearing adapted by their lesser Bulks and finer Parts, to distinguish those quicker Vibrations," namely, the ears of "lesser Creatures that we discover." Though "possibly they cannot hear those Sounds which we hear," those creatures "may have as great variety in the differences of Sounds wholly imperceptible to us, as we have within the reach of our Ears."

Hooke's extrapolation of "quicker Vibrations" than the sounds we hear waited centuries before decisive evidence emerged. The first step came about a century later in the work of Lazzaro Spallanzani, most often remembered for his incisive experiments refuting the notion of spontaneous generation by showing that small, living organisms could not arise from inorganic matter. He also was the first to demonstrate that digestion involved the action of gastric acids, supplanting the ancient concept of digestive "concoction," understood as a kind of internal cooking. Spallanzani considered himself not a "scientist" (a word coined decades later) but a natural philosopher, trained by his renowned aunt Laura Bassi, the first woman to receive a university degree in physics and a noted exponent of Newtonian science.[8] Thus, Spallanzani's work should be understood not only within older traditions of natural history—descended from Aristotelian observational and classificatory practice—but also within the larger stream of experimental investigation that reached from Newtonian physics to influence the formation of biology as we now know it. In that sense, Spallanzani was arguably one of the first modern biologists, whose investigations had immense importance for the later work of Louis Pasteur.

Spallanzani's third remarkable discovery came near the end of his life, as described in his "Letters on a Suspected New Sense in Bats" ("Lettere sopra il sospetto d'un nuovo senso nei pipistrelli," 1793).[9] His letters presented themselves within the context of physics, as he praised a text titled *Elements of Physics* and took that opportunity to recount a "very curious discovery": a bat that had been blinded was "just as clever and expert in his movements in the air as a bat possessing its eyes. . . . In contrast, I have found that other animals, such as birds, quadrupeds, amphibians, fish, and insects behave in their respective movements after having been blinded as if really blind."[10] Throughout these and the following experiments by him and others, it is disturbing to contemplate the sufferings of these bats subject to all kinds of intensely invasive experiments, beginning with their blinding. Suspecting that some other sense took the place of sight, Spallanzani conducted many experiments with negative results, tentatively concluding that "in the absence of sight there is substituted some new organ or sense which we do not have and of which, consequently, we can never have any idea."[11]

Spallanzani systematically tested each of the remaining senses, especially touch, which seemed to him the most obvious possible replacement for the bats' lost sight. He had the bats fly in a large, vaulted room that made a right angle midway down its length, which the bats navigated perfectly while flying several feet distant from the walls. He also filled the room with fine silk threads hung with weights vertically from the ceiling, spaced far enough apart that the blind bats could pass between them and against which they brushed no more frequently than did sighted bats. In contrast, covering blind bats with coats of varnish to deaden their sense of touch did not seem to affect the accuracy of their flight once they got used to the difficulty of flying in that condition.[12] Stuffing the bat's nostrils to impede the sense of smell did not seem to affect its flight, though blocking the nostrils caused the bats great harm: they could no longer breathe, which for them requires both mouth and nostrils. Indeed, tying their mouths shut made them collapse; Spallanzani noticed "a slight but very audible whistle" as the animal struggled to breathe.[13]

Spallanzani initially interpreted his experiments as indicating that hearing was not the compensatory sense he sought. He blocked the blinded bat's ear canals with bird lime, which did not seem to impede its navigation; one of his correspondents used sealing wax with similar results. His failure to find a single sense capable of compensating for sight drew Spallanzani to assert that finding that power would require "long and continuing researches with the greatest reflection and most careful consideration," especially if indeed this were "a new sense, a sense we do not have."[14] The next step came when his results were considered by members of the natural history society in Geneva, whom he asked to check his findings. Among them, Louis Jurine, "a surgeon and a great entomologist, ornithologist, and botanist," studied the anatomy of the bat and found "the organ of hearing very great in proportion to that of other animals and [having] a considerable nervous apparatus assigned to that part."[15] He tried to put a small hood on the bat or stop its ears with cotton, which the animal quickly tore off. After Jurine poured "a mastic of turpentine and wax into its ears," the bat reacted with "a great deal of impatience, and flew afterwards very imperfectly." A long-eared bat whose ears were bound up "flew very badly"; another whose ears had been filled with "liquid pomatum" was "much affected by this operation; but when the substance was removed, it took flight." It seems that Jurine was able to block hearing much more effectively than Spallanzani was; Jurine concluded that "the eyes of the bat are not indispensably necessary to it for finding its way" so that "the organ of hearing appears to supply that of sight in the discovery of bodies, and to furnish these animals with different sensations to direct their flight and enable them to avoid those obstacles which may present themselves."[16]

In 1794, after learning about Jurine's experiments, Spallanzani repeated and confirmed them, then reversed his initial opinion to conclude that "the ear of the bat serves more efficiently for seeing, or at least for measuring distances, than do its eyes."[17] He could not, however, identify the source or nature of the sounds involved, speculating that perhaps they were the reflections of the sound of the bat's wings or body, though he did not find this convincing. He did notice that covering the bat's mouth impeded its flight, though when its mouth was uncovered, he heard no sound other than the "slight whistle" noted above, which did not seem significant to him except as an indication of the animal's distress.

At this point, further advances were obstructed by the celebrated anatomist Georges Cuvier, who in 1795 rejected Spallanzani's sixth-sense hypothesis in favor of a novel form of touch sensitivity in the bat's skin that he thought might arise in response to air movements. Cuvier also rejected the experiments indicating hearing, arguing that the operations involved "were extremely cruel and have done much more than simply to lessen their [the bats'] power of hearing," as if that were more cruel and invasive than blinding them.[18] Though he performed no experiments to demonstrate his hypothesis, Cuvier's fame and commanding reputation effectively sidelined the hearing hypothesis for over a century. To be sure, it was difficult for natural historians to imagine sounds lying outside human hearing. As George Montagu sarcastically put it in 1809, to believe "that the ears of bats are more essential to their discovering objects than their eyes, requires more faith and less philosophic reasoning, than can be expected of the zootomical philosopher, by whom it might be asked, Since bats see with their ears, do they hear with their eyes?"[19] Though in 1906 Arthur Whitaker continued to prefer the "now generally accepted" touch theory, he noted the sensitivity of bats to very high pitches on the piano and concluded that "a bat's sense of hearing is adapted to sounds of a much higher pitch than our own and than that of most animals."[20] In 1908, though he noted the importance of the bat's hearing, Walter Hahn concluded that "the presence of a sixth sense, that of direction, will explain all of the facts."[21]

The sinking of the *Titanic* in 1912 stimulated efforts to find ways for ships to locate and avoid obstacles, for which Hiram Maxim turned to the "blind" navigation of bats. The inventor of the Maxim gun, one of the first automated weapons, had apparently begun thinking about the "sixth sense" of animals long before the maritime disaster. In his published account, Maxim took as his starting point Spallanzani's work on the "sixth sense" of bats. Though acknowledging Cuvier's defense of touch, Maxim went back to Spallanzani's idea that bats use for navigation the vibrations from wing motion, which Maxim interpreted as causing sounds *below* the threshold of human hearing (infrasound, as it would now be called). The bat's wings beat about twelve to fourteen times per second, which, "of course, produces an extremely low note that does not appeal to our ears, but it travels after the manner of sound—or light, for that matter—strikes all the surrounding objects, becomes modified by their character and size and is reflected back." Thus, "the bat obtains its knowledge of surrounding objects by sending out certain atmospheric vibrations and receiving back, in a fraction of a second later, the reflected and modified vibrations."[22]

Though Maxim did not use the modern term for such sound vibrations below the threshold of human hearing (which he considered to be about 16 Hertz), he essentially proposed to use infrasound to locate seagoing vessels. He envisaged powerful sirens of about "two or three hundred horsepower" sending out vibrational waves "that have as amplitude and energy at least three hundred thousand times as great as those sent out by the bat" yet inaudible to humans and hence "cannot be considered as sound." Their reflections would be detected by "an apparatus which might be considered an artificial ear [figure 13.1], . . . always able to vibrate freely in response to the waves of the echo, and its vibrations are made to open and close certain electrical circuits which ring a series of bells of various sizes," depending on the dimensions of the distant object. "This apparatus gives an audible notice if anything is ahead of the ship."[23] In addition, the artificial ear might also "produce a diagram of the disturbances in the air," generating a "wavy line" in a paper recorder that might be used to produce more detailed estimates of the object's distance and speed. Thus, a ship might be turned into a kind of gigantic artificial bat capable of detecting other ships or icebergs through the thickest fog.

As evidence, Maxim referred to "astonishing results" found by the well-known physicist John Tyndall (whom he called "the poet of science") confirming that "the reflection of sound from a solid body like a ship is very great," as well as from "acoustic clouds" observed in clear weather, though never in fog.[24] Though Maxim's particular method using infrasound did not prove practicable, it was a significant example for other kinds of artificial echolocation that led to sonar. In 1912, the physicist Lewis Fry Richardson also submitted patents for airborne and underwater echolocation systems but proposed using high-pitched sounds around 100,000 Hertz underwater; his work went no further because he could not find commercial backing, and his pacifism deterred him from turning to the War Ministry for support.[25] In this context, it is worth noting that in 1875, Wilhelm Preyer claimed to have made tiny tuning forks that produced audible frequencies of 40,960 Hertz, though by 1897, Carl Stumpf and others disproved their audibility. That same year, though, Rudolph Koenig made an (inaudible) tuning fork producing 90,000 Hertz, which he verified by visual tests.[26]

As Richardson's work indicated, the search for effective detection had moved to using high-frequency sound rather than low, for physical reasons. As he noted (referring to Lord Rayleigh's *Theory of Sound*), for high-frequency sounds, "diffraction effects are reduced and it is possible to concentrate the sound into a beam which diverges but slightly and to receive the echo from a fairly definite direction, so that the apparatus is in a crude way the acoustical analogue of the search light and telescope."[27] Richardson did not have any detailed proposal for generating or detecting such sounds, but in 1913, Reginald Fessenden patented an oscillator that was basically an underwater loudspeaker that could also function as a microphone. Using it, he could detect icebergs about two miles distant. The British Navy installed "Fessendens" on its submarines during World War I (figure 13.2).[28] The device's usefulness was limited by its rather low frequency (at first 540, later 1,000 Hertz, both quite audible pitches in the octave above A440), which caused the sound beam to be rather large and limited in resolution. In addition, any neighboring ships could easily detect when the device was in use simply through listening by a hydrophone (underwater microphone).

Addressing these problems required much shorter wavelengths that were ultrasonic and hence not directly audible by human ears, meaning frequencies above 20,000 Hertz.

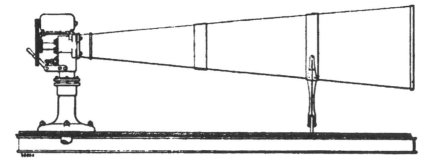

Fig. 2.—Side Elevation of the Siren or Vibrator.

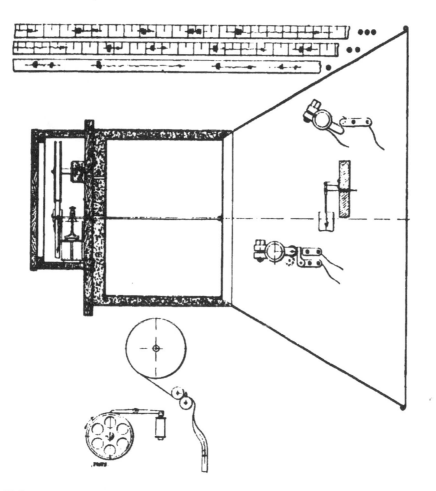

Figure 13.1
Hiram Maxim's "siren or vibrator" generating sounds whose frequency lies below the threshold of human hearing and his "artificial ear," a "receiver for recording the vibrations" reflected back from distant objects at sea. From his article, "The Sixth Sense of the Bat" (1912).

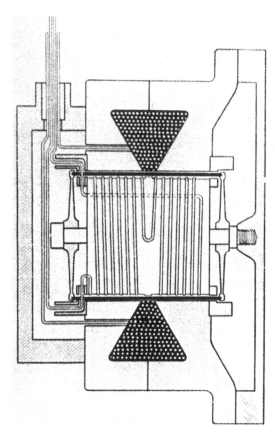 

Figure 13.2
Reginald Fessenden's sonic oscillator (1913).

Generation of such high frequencies became practically feasible after the discovery of piezoelectricity by Pierre and Jacques Curie in 1880, who noted that certain kinds of crystal (such as quartz and Rochelle salt) were capable of producing electricity under pressure. Conversely, application of electricity to such crystals could then induce deformations and even vibrations of the crystal.[29] This became the first practical way to generate ultrasonic vibrations artificially, which Paul Langevin used in his 1917 "asdic" (later called "sonar") device, capable of ranging distant objects up to two kilometers by using frequencies around 40,000 Hertz.[30] Langevin made his quartz crystals resonate at 150,000 Hertz, which he tested on fish in a laboratory tank, with startling results: "Fish placed in the beam in the neighborhood of the source were killed immediately, and certain observers experienced a painful sensation on plunging the hand into this region."[31] We will return to the medical consequences of such disruptive powers.

After World War I, these discoveries about artificial echolocation informed reconsid-eration of the bats, which had been the impetus for Maxim's ideas. In 1920, Hamilton Hartridge (a physician who had served as an experimental officer at a British airship sta-tion) instanced "experiments on 'sound ranging' apparatus during the war" as part of his argument that "bats during flight emit a short wave-length [ultrasonic] note and that this sound is reflected from objects in the vicinity. The reflected sound gives the bat informa-tion concerning its surroundings."[32]

Experimental verification of this theory required use of the very technologies that had enabled ultrasonic sonar as a tool of naval warfare and also radar. Though operat-ing through radio (electromagnetic), rather than sound, waves, the first discoveries that led to radar were essentially sonic observations. In 1922, A. H. Taylor and L. C. Young were studying high-frequency radio communication and happened to be working along-side the Anacostia River in Washington, DC. They set up a transmitter on one side of the river, a receiver on the other, and generated a steady sound tone.[33] They were puzzled that this tone would sometimes unexpectedly swell in volume from its initial normal level to nearly double in volume, then die away to almost nothing. After a few moments, this pro-cess would reverse, swelling from nothing to double volume, then back to normal. They noticed that this sequence coincided with the passage of a steamer along the river between their transmitter and receiver, inferring that the increase in volume corresponded to the constructive interference of the radio waves reflecting back from the bow and stern of the passing ship. Taylor and Young worked for the US Navy and proposed that radio might help detect the passage of ships in and out of harbors.

Then in 1930, Young and L. A. Hyland were working with short-wave direction finding, again using a steady tone now transmitted over several miles. The receiving antenna had a very narrow blind spot in one direction: when the antenna was aimed directly toward the transmitter, the tone could scarcely be heard and helped fix that direction. As they worked, the tone would sometimes mysteriously get loud and fluctuate violently, then return to steadiness. They realized that this "dancing" tone occurred when an airplane was flying overhead and proposed that radio could be used to detect passing aircraft. In subsequent years, these discoveries underwent intensive development.[34]

The techniques and instruments that emerged were crucially important for ultrasound as well as for radar, leading back to the study of bat navigation and Hartridge's ultra-sound hypothesis. In 1938, Donald Griffin, an undergraduate biology student who was fascinated by bats, approached George Washington Pierce, a physics professor at Harvard much involved in wartime research on radio and sonar. He was responsible for the "Pierce circuit" widely used to stabilize a radio frequency oscillator using a piezoelectric crystal. Using techniques we will shortly consider, he had developed almost the only apparatus that could detect and generate ultrasounds in the range 20,000 to 100,000 Hertz.

Already in 1936, Pierce had begun using his ultrasonic detector to study the sounds of insects, so that though Griffin approached him "with some trepidation," Pierce was in fact eager to work with bats to test Hartridge's hypothesis.[35] When Griffin first brought a cage full of bats to Pierce's lab "and held the cage in front of the parabolic horn we were surprised and delighted to hear a medley of raucous noises from the loudspeaker." Because even ordinary sounds have overtones above the range of human hearing, some care was needed to check that the bats really "were talking in ultrasonics."[36] They turned off the loudspeaker and relied instead on a paper chart recorder, which allowed them to confirm that the sounds were correlated with the bats' behavior. Because a thin layer of almost any solid strongly absorbs ultrasound, interposing such a layer of material between the bats and the microphone extinguished the signal, thereby confirming their discovery. They were surprised that the maximum response of the detector was around 50,000 Hertz, more than double the highest frequency available to human hearing.

The work of Griffin and Pierce had some unexpected twists that depended on the particular properties of the sounds they were using. When they released the bats to fly around the room, the sounds seemed to stop except when the animals flew toward the microphone. Could it be that the bats only occasionally emitted ultrasounds? This setback led them at first to doubt their findings, so that their first publication only claimed that the bats' ultrasounds were a kind of call rather than their principal means of navigation. After Robert Galambos joined their collaboration, they realized that this discrepancy could be explained by the highly directional sensitivity of the microphone they were using, which was greatest along the axis of its parabolic horn. Indeed, the high directionality of short-wavelength sounds probably underlay the evolutionary selection of their use by the bats no less than for human sonar. Once the microphone was pointed at the bats, it always detected the ultrasounds (figure 13.3). From this, Griffin drew the lesson that "excessive caution can sometimes lead one as far astray as rash enthusiasm."[37]

Let us pause to consider the underlying sonic principles that underlay Pierce's detector, so essential for these discoveries. He used piezoelectric crystals to generate and detect ultrasounds, as the Curies' work had first suggested. Yet there remained the problem of how to render ultrasounds audible to human ears. This was also a problem for radio, whose waves (though electromagnetic rather than acoustic in nature) in general have very high frequencies. For instance, in 1900 Fessenden achieved the first radio transmission of his voice using a frequency that may have been as high as 5 million Hertz.[38] If a voice signal is imposed on such a high-frequency wave, how could it be presented to a human ear limited to frequencies less than 20,000 Hertz?

To deal with the general problem of modulating the radio signal and rendering it readily intelligible, Fessenden developed a technique he named *heterodyning* that ultimately rests on the phenomenon of "Tartini tones," understood as the difference tones produced by two relatively close frequencies." In 1714, the violinist, composer, and theorist Giuseppe Tartini observed that if "a violinist plays simultaneously, with strong and sustained bowing, the following intervals [shown as whole notes in figure 13.4], perfectly in

Figure 13.3
Robert Galambos with a bat and an ultrasonic detector (about 1960).

Figure 13.4
Giuseppe Tartini's examples of "third sound" (*terzo suono*), showing the two notes to be bowed simultaneously as whole notes, producing a "third sound" shown as a quarter note below them (♪ sound example 13.1). From his *Treatise on Music* (*Trattato di musica*, 1754).

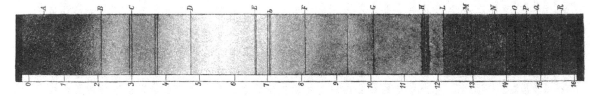

Figure 13.5
Hermann von Helmholtz's photograph of the solar spectrum, from his *Treatise on Physiological Optics* (1867), including the dark lines seen in that spectrum. The range from *F* to *G* shows blue and indigo, *H* to *L* violet, *L* to *R* ultraviolet. Helmholtz noted that "the intervals are arranged on the principle of the musical scale, because this seemed to be the best method for physiological reasons."

tune, a third sound [*terzo suono*], entirely distinguishable, will be heard," indicated by the filled-in notes.

These came to be called *Tartini tones* or combination tones.[39] Because they reflect the precise tuning of the two bowed notes, they have long been used to refine the intonation of string players.[40] Around 1800, Thomas Young argued that these were *beat tones*, physically produced at the difference of the two directly sounded frequencies (therefore called *difference tones*), a view that Hermann von Helmholtz confirmed later that century. More recent research indicates that there is no actual physical vibration of some body at this difference frequency, so that it is really a *phantom tone*, a kind of acoustic illusion.[41] Other such "virtual" sounds had long been known; for instance, the apparent pitch heard in many large bells—often its apparently "loudest" note—is not actually one of the bell's observable modes of vibration but is entirely constructed by the hearer's mind.

By heterodyning, Fessenden meant the use of such a tone intentionally produced as the difference between a signal (say, a voice) and a nearby "carrier" wave frequency that is internally generated by the electronics. The resultant small difference between these two very high-frequency waves is devised to give a wave whose much lower frequency range can drive a loudspeaker producing sounds audible to the human ear. As Fessenden put it in his 1901 patent application, "My invention consists in the production of electrical 'beats' analogous to sound-beats and their utilization in receiving-conductors tuned to the sending-conductors for wireless selective signaling."[42] Early accounts of this new principle also emphasized its sonic grounding, for instance, by comparing it to the beats heard from two organ pipes of nearby frequencies.[43] In this case, the beat theory was gradually replaced by an understanding of the Tartini tones as "phantom tones," in which the brain effectively supplies a missing pitch that is never physically sounded. In contrast, the heterodyne-generated difference tone actually drives the loudspeaker of the radio or ultrasound detector, thus transforming what had been an acoustic illusion into sonic reality.

With all this in mind, we return to Pierce's detector, which used Fessenden's hetrodyne principle to render the ultrasonic cries of insects or bats audible as difference tones within the frequency range of human hearing. In so doing, this detector brought together all the diverse strands of the story that began with bat echolocation. The very detectors that eventually "heard" the bats' elusive sounds had grown from devices designed to detect ships and submarines—using the ultrasonics the bats had first demonstrated. This curious imbroglio of biology, music (Tartini tones), physics (via Young and Helmholtz's beat theory), electronics, and military technology grew from—and circled around—the quest for a new sense that turned out to be a new kind of sound and hearing.

Such was the early history of ultrasound, strikingly different from that of ultraviolet light, discovered in 1801 by Johann Wilhelm Ritter in chemical reactions he observed as caused by "chemical rays" beyond the violet part of the spectrum. He had searched for such reactions past the violet by explicitly seeking a symmetrical complement to William Herschel's discovery (1800) of "heat rays" below the red part of the spectrum (infrared) that also caused chemical reactions. In both cases, observable chemical reactions were the proxies interpreted to have been caused by new, invisible forms of light. The paradox of invisible light was somewhat mitigated by the visible reactions that it generated; chemical reactions provided the needed intermediaries that made visible what otherwise would have remained invisible. Indeed, decades later, Hermann von Helmholtz noted that if the more luminous part of the spectrum is blocked off, "ultraviolet rays are visible without difficulty. . . . At low intensity, their color is indigo-blue, and with higher intensity bluish gray" (figure 13.5).[44] Evidently Helmholtz's eye did not register ultraviolet frequencies with a separate "color" other than shades of blue. Today we dare not check these observations because of the damage ultraviolet light can do to the unprotected eye, which Helmholtz did not realize. Safety standards now prohibit any such observations.

The path to understanding ultrasound was far less direct. The human ear has no access to the bat's sounds beyond the highest audible range in which we can still hear its squeaking. The bat's ultrasounds lie at far higher frequencies, requiring—and calling forth—a whole new technology of electronics, not having recourse to known chemical reactions. In "What Is It Like to Be a Bat?" (1974), the philosopher Thomas Nagel argued that bats were "a fundamentally *alien* form of life" whose "sonar, though clearly a form of perception, is not similar in its operation to any sense that we possess, and there is no reason to suppose that it is subjectively like anything we can experience or imagine."[45] We will return to this provocative assertion in the context of listening to neurons. For the moment, we may wonder whether the experiments of Griffin, Pierce, and Galambos might not give us access to some aspect of the subjective experience of listening to ultrasounds.

Having begun with the mysterious senses of bats, it seems appropriate that the study of ultrasound circled back to illuminate the bats themselves and then went on to enable new modes of diagnosis that "heard" the interior of the body in ways that went far beyond percussion and auscultation. The study of bats flying in the dark transformed the practice of war, whose instruments in turn made audible the bats' unhearable sounds and then opened powerful new ways to diagnose human illness.

14

Ultrasounding Bodies

The paradoxical discovery of unhearable sounds that started with animal navigation led to important human applications. Using ultrasound to locate ships and submarines drew on several sets of techniques that had to be reshaped when put to medical use. Ultrasounding the body required sustained reconsideration of its inner acoustics, especially how the multitude of echoes could be intelligibly heard or registered, then connected to what had been a predominantly visual understanding of anatomy and pathology. In this process, not only the equipment but the needed skills had to be invented in order to connect medicine, engineering, and physics. The story of how this happened involved many combinations of individuals whose varying experience and expertise gradually converged as they synthesized war-surplus parts and homemade audio equipment into sensitive diagnostic tools.

The *Titanic* and the Great War stimulated enormous activity to develop ultrasound into a practical technology. If, as Friedrich Kittler opined, "war spawns most technological inventions," medicine then tries to repurpose those tools.[1] At first, though, ultrasound seemed suspect. In the 1930s, ultrasonic heating of tissue (alarmingly demonstrated by Langevin's experiments) became so widely used that it smacked of being a cure-all.[2] That somewhat unsavory reputation may then have inhibited further medical developments for a time, at least those involving the interventional possibilities of ultrasonic heating. When attention turned to using ultrasound as a new diagnostic method, early efforts tried to locate brain tumors, which are hard to detect using X-rays because highly absorbent bones in the skull tend to shield the brain and absorb the rays. Also, the difference in X-ray absorption between normal and pathological tissue was generally too small to register a clear distinction between them. Therefore, it was speculated that ultrasound might be able to surmount these problems. In the initial attempts, the "pulse-echo" method used by bats and imitated by sonar came into conflict with the dominant example provided by X-ray imaging, in which the rays transmitted through the body darken a photographic plate.

Among the early pioneers in diagnostic ultrasound of the brain, Karl Theodore Dussik rejected the pulse-echo method because it "would us give complicated curves, which would be inconvenient for clinical use." Instead, he chose to use transmission because "I thought that visualization of differences of attenuation could bring us real pictures, in some sense similar to the way in which we can read an X-ray picture."[3] Likewise, André Denier initially opted for transmission because the problem of separating the echo from the direct signal seemed "like listening for a cricket in a bombardment."[4] In the late 1940s, Dussik claimed to have made "hyperphonograms" that imaged the cerebral ventricles, even (on one occasion) showing a tumor, though in 1953 those images were shown to be artifacts produced by the varying density of the bones even of an empty skull.[5]

Their decisions reflected contemporary industrial applications of ultrasound to detect flaws in metal machinery using the transmission method: cracks would reveal themselves as sonic shadows on the detector, including fine cracks not detectable by X-rays. Airplane wings and ships' hulls needed even higher standards because the finest hairline cracks eluded detection by the ultrasound transmission method. Accordingly, in 1942 efforts led by Donald Sproule at the British firm Henry Hughes Ltd. turned to using the pulse-echo method, originally suggested in 1928 but not feasible until the wartime development of radar provided oscillators capable of generating high-frequency, short-duration pulses of ultrasound.[6] Here again the repeated alternation between military and civilian uses of ultrasound enabled advances that otherwise would probably have come much later. Biological and nautical parallels underlied Sproule's description of applying "echo-sounder principles to sounding in a solid sea."[7] Sproule's invention could image the sea floor as a black outline and could detect schools of fish; in 1935, it was used to locate the wreck of the *Lusitania*.[8]

In 1943, Sproule and his colleagues demonstrated a pulse-echo flaw detector operating at 2.5 million Hertz, using one probe to emit the ultrasound while another receiving probe was moved across the test object. The results appeared as spikes on an oscilloscope, the height of each spike showing the amplitude of the reflected wave, its horizontal position along the screen showing how deep the flaw was inside the object. This mode of display became known as *A-scan* or *A-mode*. In 1942, the American researcher Floyd Firestone independently received a secret patent for his "Supersonic Reflectoscope," generally similar to the Sproule design but needing only a single probe to emit and receive the ultrasounds. Firestone used to point the probe of his Reflectoscope on his leg and say, "See, you might be able to find things in the body," though he and his associates did not follow this up.[9]

After World War II, this work was declassified; for about $100, electronics hobbyists could get a war-surplus pulse generator and probe their own bodies, encouraged by a June 1948 article in *Audio Engineering*, "Applications of Ultrasonics to Biology," by S. Young White, a well-known inventor and engineer.[10] Because pulse generators were so cheap and readily available, White thought that "it is hardly worthwhile to build one." He advised readers to try these generators on their leg, "and you will have a nice tool for work on the body," capable of detecting bullets, "an ideal type of load." Only three years after the war, he seems to have expected that some of his readers would have had bullets still left in their body. The article showed how to listen for the sounds of the sphincter muscle and pointed out that at ultrasonic frequencies, some of the sounds near the vocal chords are "very amusing" and "could be used as sound effects on radio." Taking a lesson from the bats, White speculated that the blind could use ultrasonic devices attached to their cane for echolocation. The same issue of *Audio Engineering* also contained articles about record wear, loudspeaker placement, high-fidelity amplifiers, and reviews of classical music recordings, showing the close interplay of all these domains of audio engineering. White invited professionals and hobbyists to apply their audio expertise to ultrasound in order "to solve some general problems and give medicine some of the tools it needs."[11]

By the early 1950s, several research groups were already engaged in that endeavor. In Denver, the radiologist Douglas Howry used Air Force surplus amplifiers, a hobby-kit oscilloscope, and the power source from his record player to produce A-scan images from within the body.[12] His diverse tool kit showed that even serious researchers used a bricolage of audio components, hobby electronics, and war-surplus ultrasonic pulse generators. Howry was dissatisfied with the complex patterns he saw on the oscilloscope and wanted instead to produce images like those he knew from *Grey's Anatomy* or at least like "a better X-ray." To achieve a two-dimensional image, Howry collaborated with engineers whose experience in radar included the technique of plan position indicator (PPI), which used a rotating radar beam that showed the beam's echoes as bright dots corresponding to their points of origin in a two-dimensional plane.[13] Such images were familiar from World War II radar screens, whose sweeping beam showed their targets as glowing dots. In the medical context, rather than having a rotating signal from a stationary source, it was more convenient for the ultrasonic transducer itself to move around the scanned area, which then (using a technique similar to PPI) could give a two-dimensional cross-sectional image later known as *B-scan* (or *B-mode*). Howry's work, however, required that the body part under examination had to be immersed in heavily salted hot water (sometimes for as long as forty-five minutes) in order to effect sufficient transduction of ultrasounds into the body, akin to sonar examining a ship. Though he could produce recognizable images of simple inanimate test objects or foreign bodies in tissue, as well as notable images of the cross-section of the neck, his insistence on refining the resultant ultrasonic image toward what he knew from X-rays kept him from applying ultrasound successfully to hospital diagnostic work.

In contrast to Howry's frustration with A-scan images, John J. Wild thought they were "a bloody miracle" and was more comfortable with exploiting the specific possibilities of ultrasound rather than requiring that they conform to expectations formed from X-ray imagery.[14] Wild began by using an ultrasound radar trainer, developed at the MIT Radiation Laboratory so that pilots could practice radar skills on the ground by using an ultrasonic probe "flying" over a model landscape, here again cross-fertilizing radar and ultrasound. With considerable assistance from a skilled engineer, Wild replaced the model with layers of dog intestine sections in water and noticed different A-scan echoes when the sections were arranged in varying numbers of layers. Wild felt "the challenge to get something out of them" by deciphering the A-scan signals. He and the engineer John Reid received a National Institutes of Health (NIH) grant that enabled them to go beyond what they could accomplish with the radar trainer, though they were still obliged to build their equipment out of "primarily war-surplus parts from the basement."[15] Following advice from one of Reid's teachers, they used a moving ultrasound probe to generate what they called "two-dimensional echographs," essentially B-scans. By 1953, they were able to use these to diagnose cancerous lesions in breast tissue, sometimes lesions so small that they could not be detected by palpation but were later confirmed by biopsy (figure 14.1). Wild claimed a success rate of 90 percent in detecting such cancers. Despite these successes, though, Wild's results did not meet with widespread acceptance. No one else was working at 15 million Hertz, so his results were not confirmed. The NIH was critical of his direct use of these techniques on human subjects, bypassing what they considered necessary trials on animals. Finally, he was embroiled in a legal battle for the control of his lab equipment that resulted in his equipment being seized and his laboratory closed. As Reid later noted,

Imaging in medicine was done with X-rays and people did not know what to do with us. . . . We had difficulties getting papers accepted for publication because the only available reviewers were apt to be radiologists who knew nothing about what we were trying to do, or physicists who knew too much about low frequency sound waves. . . . [We] found ourselves treated as misguided or demented investigators at most of the scientific meetings.[16]

The development and wide application of fetal ultrasound finally brought these new techniques to wide acceptance, especially through the efforts of the British physician Ian Donald. His mother had serious ambitions to be a concert pianist, and Donald himself studied music as an undergraduate. Like many other scientists, he experienced severe illness in childhood, losing both parents and a sibling to diphtheria; rheumatic fever damaged his own heart in ways that gave additional impetus to his research, as we will see.[17] As a medical student in London during the 1930s, he was attracted to obstetrics, a field then considered "the Cinderella of the clinical sciences," treated with disdain and even contempt by many elite physicians as a "messy" specialty, tainted by their distaste for "feminine complaints." Separate departments of obstetrics were lacking in several of the major London hospitals; general practitioners could graduate without ever having delivered a baby, hence were far less skilled than midwives. In consequence, as infant mortality in the 1930s reached abysmal levels, worse than fifty years before, reformist voices called for renewed attention to obstetrics.

Having a mechanical turn of mind, Donald was involved in the development of several mechanical devices, including some to alleviate breathing problems of newborn infants. His sister was the assistant to Robert Watson-Watt, a pioneer of radar, and during World War II, Donald became closely involved with echolocation through radar and sonar. As a military physician, he was tasked with showing that the high-frequency oscillations used in radar did not (as Nazi propaganda had it) cause erectile dysfunction; he flew several

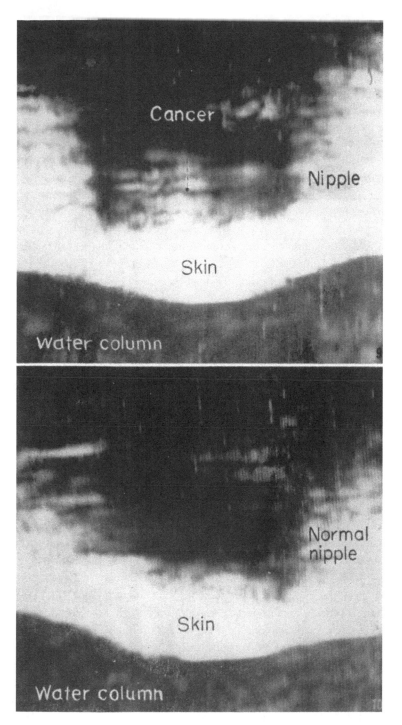

Figure 14.1
John Wild and John Reid's 1953 ultrasonic image of a breast malignancy using B-scan.

missions to demonstrate personally the safety of the radar onboard. Though he did not have to do so, Donald served as an extra radar operator and spent time familiarizing himself with the new developments of electronics and sonar.

As a physician at St. Thomas's Hospital in London, he had been trained by patrician clinicians who emphasized a clinical holism they called "neo-Hippocratism." They tended to prefer the highly trained judgment of generalist diagnosticians to overreliance on laboratory tests; some of them scorned the electrocardiograph, for instance, preferring an educated ear attuned to the most subtle stethoscopic sounds.[18] Though trained in this sensitive milieu, Donald had a penchant for engineering and wanted to bring to clinical medicine the technology he learned in the Royal Air Force.[19] After the war, he met with John Wild and had a long conversation about ultrasonic imaging. Because the ultrasound he himself had been using used was at much higher frequencies, Wild suggested that the lower frequencies used by ultrasonic industrial flaw detectors might be suitable for uterine investigations.

Just after this, Donald moved to Glasgow, where he had accepted the Regius Professorship of Midwifery (obstetrics), a major step in his career. At that point, his work on breathing apparatuses for newborns had reached an impasse that led him to seek new directions in research. By fortunate coincidence, the British firm of Henry Hughes Ltd. (for whom Sproule had worked) had relocated to Glasgow as Kelvin Hughes, so Donald now had direct access to the industrial ultrasound equipment that Wild had suggested. In 1955, Donald visited the Babcock & Wilcox factory, where he could experience their ultrasonic flaw detector, used in work on boilers for nuclear power stations. When he arrived, Donald noticed that the "ultrasound technicians were in the habit of testing their equipment by bouncing the beam off the bone in their thumbs."[20] Thus, he saw the flaw detector applied to biological materials, however crudely, by technicians who evidently treated ultrasound as wholly benign, posing no health hazard. On a subsequent visit, he brought an assortment of biological samples from the operating room, including "recently excised fibroids, small, large and calcified, and a huge ovarian cyst"; Babcock & Wilcox provided "an enormous lump of steak by way of control material."[21] Donald applied the ultrasonic probe to his samples, struggling without the help of a technician to make acoustic contact with the probe; the factory artist made sketches, no camera being available. The results "were beyond my wildest dreams and clearly showed the difference between a fibroid and an ovarian cyst. . . . I could see boundless possibilities in the years ahead."[22]

Despite Donald's optimism, he did not grasp the magnitude of the difficulties that remained before diagnostic ultrasound was accepted. Malcolm Nicolson and John E. E. Fleming's 1996 reenactment of his experiment showed that with other settings of the sensitivity of the detector and other samples, the results could have been far less convincing: they concluded that "Donald did not understand what he was seeing. . . . [He] was evidently still a naive user of ultrasound," fortunate to be a gynecologist examining abdominal cysts and fibroids, not brain tumors.[23] Still, "he had stumbled upon application of ultrasound for which the material characteristics of the target object happened to be very favorable."[24] This sustained him through the discouragements of the subsequent phase in his research, in which so many difficulties emerged that the local medical establishment started calling him "Mad Donald" for his stubbornness and odd stratagems (such as using water-filled condoms to make better acoustic contact between probe and tissue).[25]

In this struggle, he received crucial help from his fellow clinician John MacVicar and especially from Thomas Brown, a talented young engineer at Kelvin & Hughes. Brown realized that they were working with out-of-date equipment and applied his own experience working with both radar and sonar. By this point, Donald was experiencing increasing cardiac deterioration from his childhood rheumatic fever and needed major cardiac surgery, still risky at that time; his sister died during a similar procedure in 1957. His mortal hurry spurred on Brown and MacVicar.[26]

They worked to resolve difficult cases in which clinical and ultrasonic diagnoses were at odds; over the course of hundreds of cases, they made many mistakes in order to gain the needed experience. In the process, they "earned great mockery from our colleagues who naturally suggested that digital palpation was surely good enough."[27] In Donald's recollection, there finally came "a great stroke of luck" when he was asked by his colleagues to offer a demonstration. As subject, a sixty-four-year-old woman was chosen; she had a grossly distended abdomen, seemingly ascites—the same abnormal buildup of fluid in the abdomen that was one of the first objects of percussion even before Auenbrugger, as noted in chapter 9. Edward Wright, Regius Professor of the Practice of Medicine, had diagnosed gastric cancer. The paitent was vomiting and losing weight rapidly. Because her condition was considered inoperable, she was receiving nursing care and essentially waiting to die. At the bedside, Donald's own physical examination (presumably including percussion) supported the diagnosis of ascites. They then applied the ultrasonic flaw detector to her abdomen, with his medical colleagues gathered around:

I had explained that I expected to find bowel echoes floating up towards the center [as would be expected for ascites from gastric cancer]. You can imagine my dismay when on applying the probe to the most protuberant part of the patient's abdominal wall there were no bowel echoes to be seen in this area but simply a large clear space presumably of fluid with a very strong echo at great depth, almost off the screen. Just at this moment Dr. MacVicar came looking for me, poked his head over the screen and commented, "Seems a large cyst." I apologized, of course, to my medical colleagues and modestly pointed out that the machine might not be working very well. I also felt like apologizing for Dr. MacVicar. However, he tenaciously stuck to the point of view that if it wasn't a cyst it ought to be.[28]

During the ensuing discussion, Wright was modest enough to admit that he wasn't certain of his diagnosis; no malignant cells had been found in the abdominal fluid, and the radiological evidence was equivocal. Donald continued: "On the strength of the new provisional diagnosis of ovarian cyst we arranged to take the patient into my own department for laparotomy. On removing a truly massive mucinous cyst of ovary her vomiting cleared up and the radiological appearances in her stomach were declared to be an artefact. This discovery undoubtedly saved this woman's life." This story stands in his recollection as a kind of turning point in which ultrasound overturned a well-considered diagnosis and saved a life by offering a new kind of evidence. Mere percussion of the abdomen—which confirmed the presence of abnormal quantities of abdominal fluid—gave way to a sonic finding that went beyond the abilities of the human ear, going from sounding bodies to ultrasounding them.

This demonstration moved Donald's colleagues to invite him to consult on a difficult cardiac diagnosis about a palpable mass in a patient's heart that surgeons feared was a myxoma (a benign tumor of the heart's fibrous tissue), though other physicians thought it was a clot. The ultrasonic probe determined that the mass was indeed a clot, as was confirmed during a subsequent surgery, which unfortunately the patient did not survive. As Nicolson and Fleming observed, Donald's ultrasonic techniques could succeed in both realms of gynecological and cardiac medicine because "the heart has some of the characteristics of a cystic structure, consisting as it does of fluid-filled cavities containing solid structures."[29] This was only the beginning of the systematic application of ultrasound to the heart; though Donald did not publish this case, that same year (1957) a German group published a similar case. Though its later development goes beyond the scope of this book, echocardiography became "the technique of choice in diagnosing structural abnormalities of the heart."[30]

As Donald prepared to use his probe to investigate the wombs of pregnant women, he felt the need to defend the safety of ultrasonic procedures in that most sensitive and important arena. Well aware of Langevin's findings, he noted in his group's first published paper (1958) that "there can be no doubt that intense ultrasonic energy can produce damage." Nevertheless, the existing literature confirmed that ultrasound could only cause structural damage were it applied at energy levels many times greater than those Donald intended to use for obstetrics. Still, the delicacy of the fetus called for even higher levels of safety than those expected for older patients. Donald arranged for experiments that exposed kittens to one hour of ultrasound, far longer exposure than any used clinically. Comparison with unexposed control animals showed that "exposure of the kittens to more than thirty times the dose of ultrasound necessary in its diagnostic use produced no detectable neuropathological change." For Donald, this demonstrated "beyond all shadow of doubt that susceptible tissues would not suffer even from these dosages," whose very small energies he calculated.[31] In light of that, Donald argued that being exposed to diagnostic ultrasound was "no more dangerous than listening to Beethoven's Fifth Symphony," perhaps the first comparison of that celebrated work to the effects of ultrasound.[32]

The images in this first paper contrasted "A-scope" (A-scan) images with a new kind of "B-scope" (B-scan) images that had been developed through Brown's efforts. Though clear and informative, the A-scan images required considerable interpretation to go from the visible trace to the underlying physiology (figure 14.2). For Brown, "it seemed plain,

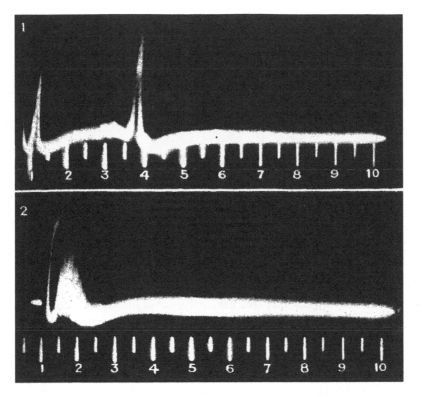

Figure 14.2
A-scan images. (a) "Acute retention of urine, showing bladder walls separated by gap representing urine." (b) The same bladder "after having been emptied by catheter, showing bladder walls no longer separated by gap." From Ian Donald, J. MacVicar, and T. G. Brown, "Investigation of Abdominal Masses by Pulsed Ultrasound" (1958).

as the nose on one's face, that what was needed . . . was some kind of system which would display echoes in the position from which they were coming."[33] He had some trouble persuading Donald, whose experience of radar had been limited to locating a target in the air or on the ocean's surface. To make ultrasound more diagnostically effective, they needed to address acoustic issues that were not so important in the much less reflective setting of aircraft or surface ships. Donald was not familiar with the more comparable problem of locating submarines underwater using sonar, compared to which ultrasonic probing of the abdomen or womb evoked more confusing echoes.

As had been the case for percussion and mediate auscultation, this new way of "hearing" the interior of the body required a whole new set of auditory techniques. But where users of stethoscopes had to develop their recognition of unfamiliar and subtly different sounds, ultrasound required correlating the acoustic with the visual and spatial. Ultrasonic listening to the body needed to be translated into *seeing* in order to map echoes to

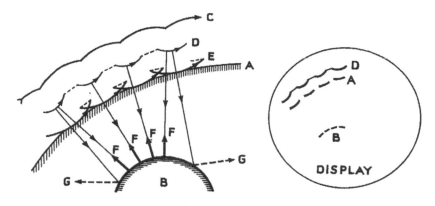

Figure 14.3
"Diagram of method combining B-scope and P. P. I. presentation: *A*, patient's skin; *B*, reflecting mass; *C, D, E*, paths traced by probe spindle, transducers, and probe face respectively; *F*, paths of echoes returning to receiving transducer (where reflecting surface is at right-angles to incident ultrasonic beam); *G*, paths of useless reflections (where beam strikes surface obliquely)." From Donald, MacVicar, and Brown (1958).

anatomy. The sonic turn that prepared the emergence of ultrasound was thus reconciled with the human (and clinical) need to see, not just hear. Subsequently, the analog displays of early ultrasound results gave way to digital versions; much later we will return to the significance of this further turn from analog to digital that essentially reduced visual as well as auditory displays to binary (rather than continuous analog) formats.

Where Wild had held his probe in a constant direction, Brown synthesized the B-scan with the plan position indicator (PPI) method used in radar in order to produce a more useful ultrasound image. Rather than having the probe mimic the scanning made by a fixed but rotating radar antenna, Brown moved the probe over the skin surface, thereby "'collecting' echoes on the one picture from as many angles as possible, registering simultaneously not only the echoes and their strength but also the position of the probe and the angle of the incident beam," as their paper explained (figure 14.3). Using this movable scanning that Brown called "compound B-scope," they could maximize the chances of the ultrasound hitting the inner surfaces at optimal angles (near ninety degrees) that would create a strong echo. The trace of a "brightness modulated display" could form a representation of those inner surfaces if the oscilloscope screen had a sufficiently long persistence time (during which the trace remains visible), as it had when used in PPI radar.[34]

The resultant procedure remained fundamental for subsequent ultrasound imaging, familiar to anyone who has ever witnessed a fetal ultrasound: the technician moves the probe over the surface of the mother's belly, using a special jelly to facilitate the transduction of ultrasound into and out of the body. At the beginning, though, Brown just used a generous application of olive oil and assembled his equipment by bricolage, still relying on war-surplus electronics and "chains, sprockets, and so forth, from a Meccano set," rigged up on a movable hospital bed-table borrowed from the infirmary (figure 14.4).[35]

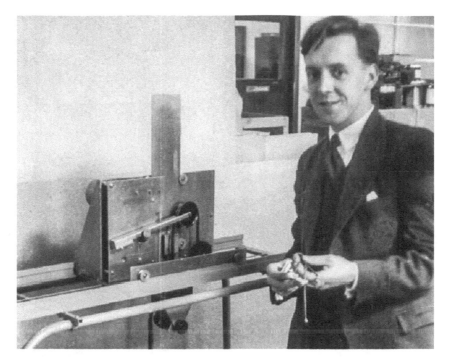

Figure 14.4
Thomas Brown with his "bed-table scanner" (1956).

Though generally hard to grasp visually as still images (as in figure 14.1), some of Wild's oscilloscope images had shown much more detail when filmed, which recorded the results of moving the probe in real time. Brown had not seen these films before developing his new B-scan pictures, in which the moving probe (along with long-persistence screens) was able to create an even stronger record, as in the images shown in their 1958 paper (figure 14.5).[36] In these images, clinicians could much more clearly recognize anatomical features and pathology. The image of a cyst now showed a dark space in its interior (figure 14.5a) compared to the more schematic indication of a mere gap between echo spikes in the A-scan image (figure 14.2). Most striking, they could image the fetus only fourteen weeks after gestation (figure 14.5b), allowing a reliable diagnosis of pregnancy far earlier than ever before.

Still, "the challenge posed by the interpretation of the early two-dimensional images can hardly be overstated" not only because of their fuzziness but even more because they represented thin cross-sectional slices of the body that had never been represented or studied in detail.[37] To explain these to his medical colleagues, Donald compared these cross-sections to slices of bread (the body represented by the whole loaf), but (unlike bread) ultrasound "slices" were not only transverse but could be longitudinal or oblique.

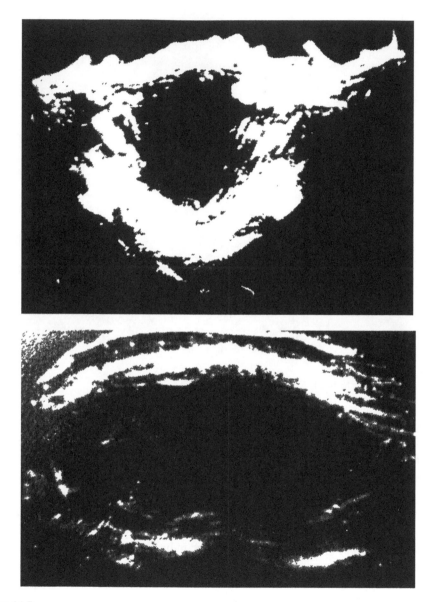

Figure 14.5
Ultrasonic images of (a) a "unilocular ovarian cyst of moderate size" (b) and a "uterus at 14 weeks' gestation, showing echoes from foetus towards left half of uterus. Provisional clinical diagnosis had been that of fibroid." From Donald, MacVicar, and Brown (1958).

There were no anatomical atlases for such slices, and the first people who tried to learn Donald's techniques struggled to recognize the most basic features, such as the parts of the fetus imaged in figure 14.5b. As one of them recalled, "We didn't know at that time what we were looking at! . . . 'Oh look,' he said, 'There's a round thing there,' he says, 'I wonder if that could be a head?' And that more or less illustrates what we were trying to do at the time."[38]

Admittedly, this was spoken by a physicist, one of the first to train under Donald, rather than a medical professional. Yet even physicians had problems, especially radiologists used to reading X-ray transmission images produced by a technician who had the patient hold a static pose. In contrast, "to an obstetrician, ultrasonic diagnosis was, unlike other forms of imaging, essentially interactive: an understanding of the contents of the abdomen emerged in the course of the scan and was the direct product of physical as well as visual inputs."[39] This interactive quality was enhanced by the development of real-time scanners in the 1970s, pioneered in the United States and then adopted also in Glasgow. "Real-time scanners proved far easier to learn on and to employ clinically than the static B-scanners" because the simultaneous motion of probe and image "increased the visual and proprioceptive feedback for the operator during the scanning process. Conceptualization of the abdominal contents was thereby facilitated. The need to wait for a photographic image to develop was removed."[40] In this process, the introduction of Doppler techniques allowed the ultrasound also to display the exact direction and rate of motion of blood flow.[41]

These interactive, real-time processes were the key to making ultrasound intelligible and even compelling to the general public as well as practitioners. The moving probe did not just educe a static image but produced something like a movie unfolding in real time. Consequently, one could then "see" the baby in the womb, moving and alive, so that "parents on seeing their fetus moving on the screen were informed and delighted and indeed the ultrasound session became a family event," as Stuart Campbell (one of Donald's protégés) put it.[42] By Donald's retirement in 1976, diagnostic ultrasound was firmly established as an essential component of modern clinical routine not only in obstetrics but many other fields of medicine, notably echocardiography.[43] Even the destructive powers of ultrasound, carefully controlled, have important applications, such as dissolving cataracts (phacoemulsification) or kidney stones (lithotripsy).[44]

As its techniques matured, ultrasounding bodies became more like musical performances, interactive and responsive to sonic cues rendered visible rather than contemplation of static images. Through the special properties of its high frequencies, ultrasound afforded medicine an enormous extension of the diagnostic powers first manifested in percussion and auscultation by visualizing in four dimensions (time as well as three-dimensional space) the dynamic process of listening to the body.

15

Tuning the Nerves

The final chapters of this book concern ways in which sound became a tool for observing the electrical activity of the nervous system. In that project, the details of the observing apparatus prove to be important because they reveal the increasing use of sonic devices in connection with electrical instruments that measured the details of nervous action, first described by Luigi Galvani. Conversely, as Emil du Bois-Reymond vividly showed, contracting a muscle leads to an observable electric current. To explore nerve action, Hermann von Helmholtz developed new tools for studying the nerves by amalgamating tuning forks with electromagnetic devices, which he also used to synthesize vowel sounds and listen to the sounds made by muscles. His student Julius Bernstein improved this equipment to "tune" the firing of nerves even more closely and thereby illuminated the underlying chemical processes. Throughout, vibrating bodies like springs and tuning forks were essential tools for controlling the stimulation of the nerves and timing their response. Careful hearing could register the response of muscles, followed by the further listening techniques explored in the next chapters.

We begin by setting eighteenth-century views about "nerves" and "nervousness" within the larger context of vibrating fibers that gradually replaced ancient ideas of the body in terms of humors. Chapter 7 considered fibers in the bodily fabric first as purely material constituents, then as functional elements capable of dynamic activity and vibration, which seemed particularly relevant to understanding the higher functions of living beings. Newton even speculated that nerves transmitted vibrations using electrical forces.[1] Among bodily fibers, the nerves took on a particularly important role. As Penelope Gouk put it, "During the eighteenth century learned medical explanations for health and disease gradually moved away from the balance of humours, and increasingly focused on the state of the nerves."[2] This new emphasis took different forms, some of them noting a connection between music and the nerves. For instance, the Newtonian David Hartley's *Observations on Man* (1749) linked the laws of bodily action to the mind through the musical vibrations of a fine medullary substance stored in the brain.[3] The physician Richard Brocklesby's *Reflections on Antient and Modern Musick, with the Application to the Cure of Diseases* (1749) promoted the use of music to treat mania and related nervous disorders, which were notably resistant to other therapies. David Campbell's *Inaugural Dissertation on the Effect of Music on Soothing or Banishing Pain* (*Disquisitio inauguralis de musicis effectu in doloribus leniendis aut fugendis*, 1777) held that pain "appears to consist in an over-violent shock to the nervous fibers out of their natural state, due to any cause."[4] Campbell treated his own case of inflammatory fever by having a fiddler play to him for twelve hours over the course of three days. In these ways, long-standing applications of music to regulate emotions found new justification in terms of the sympathetic vibration of the nerves. The whole discourse of "nerves" also affected the self-understanding of sufferers, not merely their medical handlers, ranging from statements by the "mad" King George III (1788)—"I am *nervous*, he cried, I am not *ill*, but I am *nervous*"—to T. S. Eliot's "The Waste Land" (1920): "My nerves are bad to-night. Yes, bad."[5]

Music consoled and soothed much as it had since antiquity, though now directed to alleviate "nervousness." Because the use of music to quiet the nerves has been well studied by Gouk and others, we will not treat it further here. Instead, we consider a newer (and much more physical) "tuning" of the nerves that emerged in response to emerging physiological discoveries. In 1791, Luigi Galvani observed that electricity applied to a frog's leg would make it twitch (figure 15.1). From this began a whole series of developments that connected these physiological findings with contemporary developments in electricity and magnetism. We will trace these developments in order to understand the context in which sound became an important tool for investigating the electrical activity of nerves.

Early on, Galvani became embroiled in a controversy with Alessandro Volta (the discoverer of the "voltaic pile" or battery) about whether dissimilar metals or materials (such as those used in the action of batteries) were necessary to induce the muscle to twitch. Clarifying these issues required measuring instruments capable of registering minute electric currents. In fact, Galvani's frog was effectively the first such device, though capable only of a brusque, impulsive response, not a quantitative or detailed measurement.[6]

Going beyond this "frog galvanoscope" depended on the discovery by Hans Christian Ørsted that an electric current generates magnetic effects. As I have discussed elsewhere, Ørsted's discovery emerged from his search for overarching connections among music, electricity, magnetism, and light, synthetic ideas he shared with another radical pioneer, Johann Wilhelm Ritter.[7] Both of them were deeply interested in Galvani's work; Ørsted thought that the action of positive and negative electricity respectively explained "the striking form of vegetation" and "the internal form of the plant."[8] Their search for the unity of Nature implied a newly electric concept of life emerging from deep connections among electricity, magnetism, and sound. All this preceded and prepared Ørsted's 1820 discovery that a magnetic compass near a current-carrying wire would be deflected in proportion to the current, Thus, Ørsted's achievement represented a profound reaction to Galvani's work, part of a daring attempt to extend the relation between biology and electricity through its magnetic effects, which in turn would reveal the electrical operations of physiology.

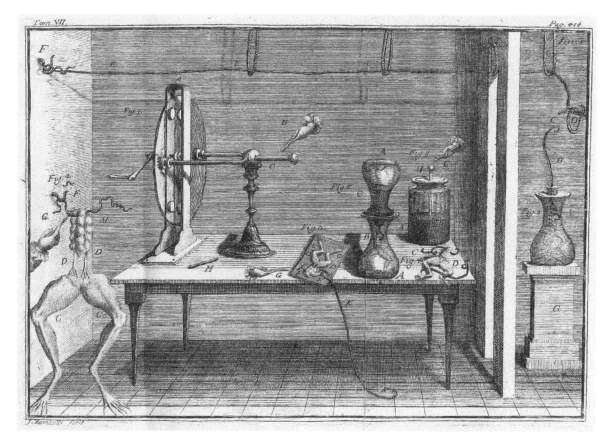

Figure 15.1
Luigi Galvani's engraving of his experiment, from *On the Effect of Electicity on Muscular Motion* (*De viribus electricitatis in motu musculari*, 1791). A hand-cranked static electricity generator stands on the left of the table; on the far right is a Leiden jar that could store the electricity.

Ørsted did not develop his discovery into a practical measuring instrument, probably because he was more interested in what he considered its philosophical implications. In the hands of others, it became the basic principle of all galvanometers, whose name memorialized the connection of electricity, magnetism, and Galvani's work. Each successive development sought to increase the accuracy with which the needle registered the electric current. For instance, in 1825 Leopoldo Nobili presented his "astatic galvanometer" (figure 15.2), which used two paired magnets (their poles oppositely oriented) in order to cancel out the magnetic field of the Earth. With this instrument Nobili was able to observe the tiny currents from the cut surface of a frog muscle, yet such was the influence of Volta that Nobili ascribed his currents to thermal effects rather than animal electricity.

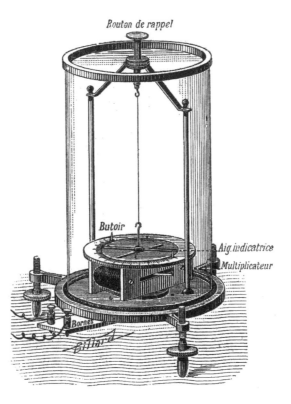

Bouton de rappel

Butoir

Aig. indicatrice

Multiplicateur

Bornes

Figure 15.2
Leopoldo Nobili's galvanometer, its indicator needle visible on the top plate (1825).

Carlo Matteucci in Pisa further improved Nobili's galvanometer by winding 3,000 turns of wire to augment the magnetic effect of the current by that factor. He also devised other ingenious experiments; in 1842, for example, he placed the cut end of one frog's sciatic nerve on the intact thigh muscle of another, then observed that the first frog's nerve made the second frog's muscle twitch. Not relying on any metals, this clearly demonstrated an electrical discharge or "action current" (as it came to be known) associated with muscle contraction. He also showed that this "proper current" (later called "demarcation current") could be found in warm-blooded animals like pigeons and rabbits. Yet his galvanometer revealed an apparent contradiction: the ensuing muscle contraction momentarily seemed to cancel out the very demarcation current that had initiated it.

These perplexing experiments attracted the attention of Johannes Müller, who is often regarded as the founder of modern physiology. His *Elements of Physiology* (*Handbuch der Physiologie des Menschen*, 1833–1840) brought comparative anatomy, chemistry, microscopy, and physical science into a new synthesis. Müller's most memorable contribution to neurophysiology was his "law of specific energies," namely, that each particular nerve of sense has special powers or qualities which the exciting causes merely render manifest. Sensation, therefore, consists in the communication to the sensorium, not the quality or state of the external body, but of the condition of the nerves themselves, excited by the external cause—We do not feel the knife which gives us pain, but the painful state of our nerves produced by it. The probably mechanical oscillation of light is itself not luminous; even if it could itself act on the sensorium, it would be perceived merely as an oscillation; it is only by affecting the optic nerve that it gives rise to the sensation of light. Sound has no existence but in the excitement of a quality of the auditory nerve; the nerve of touch perceives the vibration of the apparently sonorous body as a sensation of tremor. We communicate, therefore, with the external world merely by virtue of the states which external influences excite in our nerves.[9]

This summation established the various specific kinds of nerves as the link between mind and world, not through philosophical speculation but as a physiological foundation. Though Charles Bell had already argued in 1833 that "each filament or track of nervous matter has its peculiar endowment, independently of the others which are bound up with it," Müller installed this as a fundamental principle of neurology.[10] The quotation above exemplifies his equal treatment of the various senses, among which sound now had an equal place next to vision, which (from Aristotle on) had tended to be regarded as the principal sense.[11] Then too, until this time, otology (the anatomy and physiology of the ear) had been "the runt of the anatomic litter," as Jonathan Sterne put it, reflecting the lower status of hearing relative to sight and also the notable difficulties of dissecting the ear.[12]

In his eulogy of Müller, his student Emil du Bois-Reymond noted that despite the "hopelessly dark nature" of the anatomy and physiology of hearing, Müller's musical ear "naturally" destined him to penetrate its secrets.[13] He was not merely an ordinary music lover, as was relatively common among his educated contemporaries; du Bois-Reymond thought his "self-absorbed listening into his sense organs, their doubling, as it were," had led to his nervous breakdown in 1827.[14] After years of debilitating self-experimentation, Müller's "rich imagination" and peculiar hypochondria had collapsed, yet as a result left him with heightened abilities to address the darkest recesses of sense-experience, especially of hearing.[15] Müller went on to give a detailed anatomic description of what he called the "auditory apparatus," including extensive comparisons between that apparatus in different species, in order to develop what Sterne called "an entirely functional and mechanical theory of hearing, one that separated it from the other senses and defined it as a complex mechanism."[16]

As part of his argument, Müller presented evidence of the interrelation between electricity and hearing, going back to the foundational experiments: "Volta states that, when the poles of a battery of forty pairs of plates were applied to his ears, he felt a shock in his head, and a few moments afterwards perceived a hissing and pulsatory sound like that of a viscid substance boiling, which continued as long as the circle [that is, the galvanic circuit] was closed."[17] Müller also noted that Ritter too heard the G above middle C when he attached his head to a strong battery, reminding us that both of them shared alarming practices of self-experimentation that tested on their own bodies the connections between electricity and physiology. That kind of dangerous experiment led to Müller's observation that electrical stimulation of the auditory nerve resulted in the perception of sound no less than when that nerve was stimulated by an external vibration.

With this in mind, Müller assigned du Bois-Reymond to resolve Matteucci's paradoxical findings about muscle action seeming to cancel its own current. In 1843, du Bois-Reymond found that the nervous impulse was an oscillatory current marked by what he called "negative variation" (*negative Schwankung*), a characteristic swing to negatively charged current after an initial positive current, hence explaining Matteucci's observation.[18] In order to observe this in detail, du Bois-Reymond devised a "multiplier," a highly improved version of Nobili's galvanometer using over 24,000 coils of wire to achieve the needed sensitivity (figure 15.3). This required careful attention to weed out extraneous effects (such as contact problems and skin conduction) that would interfere with measuring the nerve impulse precisely.

At length, in 1847 du Bois-Reymond was able to perfect his most famous experiment. Having attached the leads of the galvanometer to platinum plates resting in "conducting vessels" filled with saline solution, he then immersed his fingers in the electrolyte and waited for the galvanometer needle to come to rest. Suddenly he tensed one of his arms and the needle jumped. As Galvani had shown that electricity could make a frog jump, du Bois-Reymond's simple demonstration showed conversely that a twitching muscle gives rise to an electrical impulse, a "voluntary tetanic current," using that term to refer to the strong, almost convulsive character of the muscle contraction. He went on to perform this publicly to astonished observers in Berlin and London, though he faced considerable skepticism in Paris.[19] His published version shows an idealized muscular youth producing the tetanic current (his right hand immersed in the solution) while gazing at the needle registering the effect of his own impulse (figure 15.4). Representing the way his volition moves the galvanometer, arrows on his naked torso illustrate the direction of the invisible current he is producing. The drapery over his loins decorously veils yet also thereby hints at the universality of electricity in every aspect of his activities, even those not directly connected to the galvanometer.

The very nature of du Bois-Reymond's demonstration, however, indicated the limitations of what he could observe. His galvanometer, though sensitive, could only register the tetanic current as a fleeting impulse, not present any details of its course beyond its negative variation or make any permanent record. His fellow student Hermann von Helmholtz found ways to make such a record.[20] Leading up to this, in 1850, Helmholtz determined the actual speed of this impulse (about 30.8 meters per second), a feat his teacher Müller considered impossible.[21] Helmholtz went on to measure the speed of nerve impulses in the human forearm, which was slightly faster than in the frog.

To do so, Helmholtz also used a galvanometer, but it is important to differentiate the different modalities of observation he used. The galvanometer itself operates purely through electric and magnetic effects: the changing electric current causes a magnetic effect that

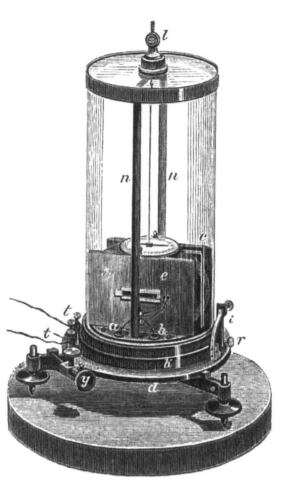

Figure 15.3
Emil du Bois-Reymond's multiplier, with 24,000 coils of wire wound around the plates *e*, while the indicator needle hangs from *l*.

then moves a magnetized pointer (essentially a compass needle that detects the magnetic field). As du Bois-Reymond experienced, all that an observer sees would be a very swiftly transient deflection of the needle, too fast for accurate recording by a stopwatch, especially given the vagaries of human sensory response controlling the clock. After all, any human observer's nerves would themselves exhibit some kind of time delay—their "personal equation"—that would have to be taken into account.[22] Helmholtz's initial 1850 experiments had coped with this by having the nerve impulse automatically turn on and off a chronometer, using contemporary ballistic methods that could (for instance) measure the speeds of bullets as they pass through artillery barrels.[23]

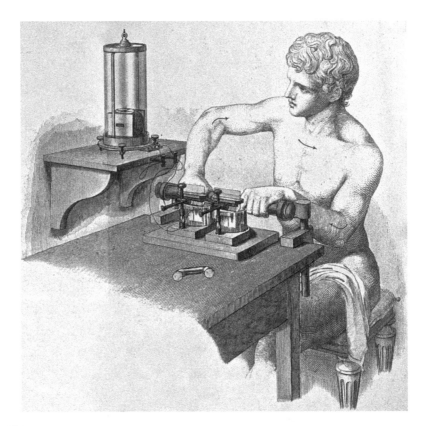

Figure 15.4
Du Bois-Reymond's demonstration of a voluntary tetanic current, whose direction through the body is indicated by arrows, from his *Investigations on Animal Electricity* (*Untersuchungen über thierische Electricität*, volume 2, 1848). The galvanometer rests on a separate shelf in order to avoid disturbance.

In order to measure these short time intervals, Helmholtz needed to control very precisely the electrical stimulation of the nerve. A simple spark (such as Galvani had used) was too uncontrollable to be suitable. For his purposes, Helmholtz adapted Christian Ernst Neeff's "interrupter" (1847) to be an electromagnetic tuning fork (figure 15.5) that combined acoustic, electric, and magnetic elements. In this hybrid device, an electromagnet makes a tuning fork vibrate continuously by magnetizing the metal of the fork in pulses. Because one end of the fork is attached to a wire that dips down into a cup filled with mercury, as the fork vibrates, the contact with the mercury (a good conductor of electricity) is designed to be interrupted when the wire rises above the surface of the mercury. The current resumes when the wire reenters the surface. This creates an intermittent electric current at the fork's frequency of vibration. This current is then fed back into the electromagnet, causing it to keep the fork vibrating without any further

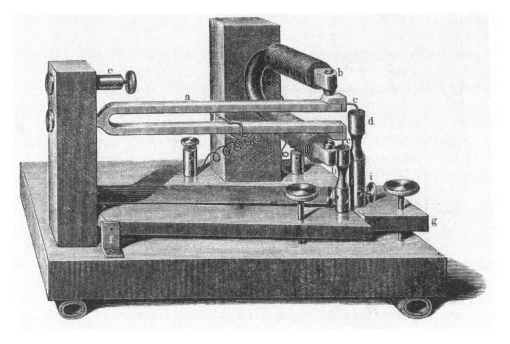

Figure 15.5
Helmholtz's electromagnetic tuning fork *a*, which is fixed horizontally between the poles *b b* of an electromagnet. The wire *c* dips into the mercury cup *d*. The vibrations of the fork then cause the wire periodically to
emerge from and then reimmerse in the mercury cup, thus creating an intermittent current that feeds back into
the electromagnet, as illustrated in his *Tonempfindungen* (1863).

external mechanical excitation. In designing his apparatus, Helmholtz noted that Neeff's interrupter used a spring whose vibrations unfortunately "communicate themselves to all adjacent bodies and are for our purposes both too audible and too irregular. Hence the necessity of substituting a tuning-fork for the spring."[24] Even more important, the fork is designed to produce a more precisely calibrated frequency than the spring because the fork was made to tune pitches with high accuracy, which then could tune the nerves with that same accuracy.

I pause to note that during that same period, tuning forks also became essential parts of physics experiments, such as Albert Michelson's 1880 measurement of the speed of light, which used an "electric tuning fork" to regulate the rotation of the mirror at the heart of the apparatus and thereby achieve an accuracy ten times greater than previous measurements.[25] As Daniel Kleppner noted, Michelson achieved this accuracy by using his "good ear" to calibrate his tuning fork by counting beats to within 0.01 to 0.02 beats per second.[26] Michelson also found an "acoustical criterion" to regulate the turbine that rotated the mirror; as he noted, "when the adjustment is perfect, the apparatus revolves without giving any sound."[27]

By 1851, Helmholtz incorporated his electromagnetic tuning fork into what he called a myograph (from the Greek for "muscle writing") in order to provide a larger and more detailed picture of the nerve impulse.[28] He still used his electromagnetic tuning fork to stimulate the nerve. To record its response, the needle of the myograph's galvanometer moved against a rotating drum with smoked paper (a "kymograph"), thereby recording its changing position on the paper, leaving a permanent image for later study (figure 15.6).[29] In this way, the frog's nerve inscribed its own characteristic firing pattern. In order to provide an accurate time standard against which the frog's impulse could be compared, another tuning fork inscribed its vibrations on the drum. The two forks book-end the process, one tuning the nerve's stimulus through its audible vibration, the other calibrating its response via a silent trace on the rotating drum.

This was only the beginning of Helmholtz's involvement with tuning forks. He went on to an extensive series of research into the nature of sound and hearing that he summarized in *On the Sensations of Tone as a Basis for the Theory of Music* (*Die Lehre von den Tonempfindungen als physiologische Grundlage für die Theorie der Musik*, 1863). In this seminal work, Helmholtz often used tuning forks to test various aspects of hearing as well as to give a time standard, using the same method of inscribing its variations on a rotating drum (figure 15.7), for "this wavy line once drawn, remains as a permanent image of the kind of motion performed by the fork during its musical vibrations."[30] The tuning fork thus was a bridge between music and time, otology and physics. These were crucial connections: sound can register temporal phenomena (especially very brief ones) with a clarity and immediacy difficult, if not impossible, for instruments reliant on purely visual display.

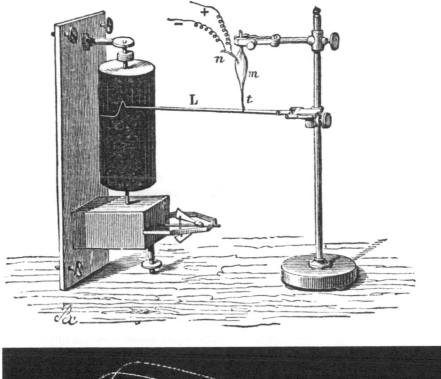

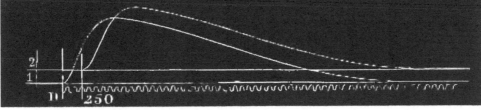

Figure 15.6
A schematic depiction of a myograph such as Helmholtz used in his later work measuring nerve and muscle impulses over time. (a) A galvanometer records the changing current from a frog's muscle as a tracing on a rotating drum (kymograph), compared to the vibrations of a tuning fork (not shown). (b) The drum's tracing of a human muscle response shows the changing current (vertical) against time (horizontal), traced by a tuning fork at 250 Hertz. Curve 1 shows the response when the nerve was stimulated very near the muscle, curve 2 when stimulated 30 centimeters farther away. From E. J. Marey, *La machine animale* (1873).

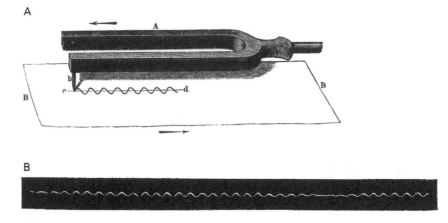

A

B

Figure 15.7
Helmholtz's illustrations of the vibrations of a tuning fork (above) and their trace on a rotating drum (below), from his *Tonempfindungen* (1863).

In the course of his work on physiological acoustics, Helmholtz became particularly interested in the problem of understanding how various vowel sounds differ, even if sung on the same fundamental pitch.[31] Initially, he sang into a piano, the instrument he had played since youth, noting that the different vowels would excite different strengths of overtones among the strings. He noted that this experiment "succeeded with my wife better than with myself." If you try it (with the sustaining pedal held down), you will experience for yourself the eerie sense that, even after you stop singing, the piano keeps resounding your specific vowel.[32] It is beguiling to imagine the scene, he and the talented Olga comparing vowels through the piano's sympathetic strings. Indeed, the piano became for him the model for the operation of the inner ear itself, whose hair fibers he analogized to sympathetically resonating strings.[33]

In order to understand vowel sounds fully, Helmholtz judged that he would need to produce them artificially, for which he again used the electromagnetic tuning fork. Instead of using it to stimulate a frog's nerve, he adapted it to sustain and magnify its vibrations through a nearby hollow cylindrical body, its size designed to give the greatest resonance to the fork's pitch (figure 15.8). In order to set this whole apparatus into vibration, Helmholtz used one of his original electromagnetic tuning forks (figure 15.7) at the same frequency, which then would generate an intermittent current that would be fed into the electromagnet of the resonator. The resonant cavity was stationed as close as possible to the fork; its opening could be more or less exposed through the action of a movable cover, which allowed him to regulate the volume of the resonance.

This resonator would produce one single pitch, but Helmholtz needed a broader range of overtones in order to try to produce vowel sounds, for he had already demonstrated that timbre in general (and vowel quality in particular) was determined by the particular

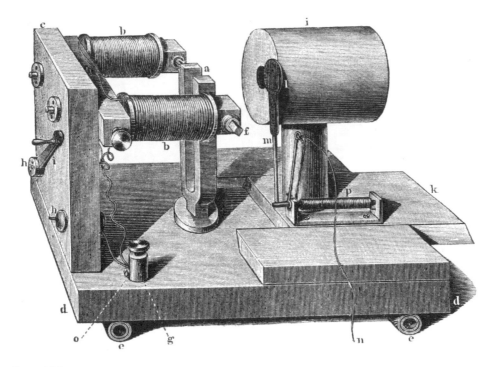

Figure 15.8
Helmholtz's resonator. The fork *a* is kept vibrating by the electromagnet *b b*, which itself is driven by an intermittent current generated by another electromagnetic tuning fork (at the same frequency), as shown in figure 15.7. Here, the fork's vibrations excite the resonance of chamber *i*, whose movable cover *m* allows its volume to be regulated.

Figure 15.9
Helmholtz's large apparatus for synthesizing the timbres of ten harmonics, as built by Rudolf Koenig in Paris (about 1865; ♪ sound example 15.1).

mixture of the various overtones and the fundamental pitch. In essence, he constructed the first sound synthesizer by assembling several resonators so he could produce and regulate ten overtones of a fundamental pitch (figure 15.9).[34] This required virtuoso craftsmanship, as beautiful to look at as it was beguiling to hear, thanks to the workshop of Rudolf Koenig, a violin maker who went on to produce many such intricate acoustic instruments for Helmholtz and others.[35] Helmholtz used a subvention from the king of Bavaria to pay for his expensive apparatus, whose fundamental tone was a B-flat "in the deepest octave of a bass voice" and included its next seven overtones up to the B-flat three octaves higher, in "the highest octave of a soprano."[36] In Koenig's version, a keyboard with eight keys allowed the different overtones to be included or not, their relative volumes also adjustable through the movable cover of each resonator. By altering this relative mixture of overtones, Helmholtz claimed that he had reproduced the basic vowel sounds; for instance, producing the vowel *U* needed a strong first octave overtone (B-flat), while the vowel *E* needed a strong F a fifth above that. To be sure, these were the vowels of a singing rather than a speaking voice, striving for the purity of timbre achieved by a trained singer. The demonstrations in ♪ sound example 15.1 give a sense of this instrument's sound, its mechanically pure vibrations giving an eerie intimation of a sung vowel. Though synthesized sound now seems commonplace, one can imagine how striking and surprising Helmholtz's synthesizer must have seemed to his contemporaries.

The same year his *Tonempfindungen* first appeared, Helmholtz also published a new (and far lesser-known) series of experiments on the sounds of muscles. First described in print by Francesco Maria Grimaldi (1665), these sounds came to wider attention in the writings of William Hyde Wollaston (1810) and Paul Erman (1812). These observations came even before Laennec's stethoscope and involved the fringes of ordinary experience, such as the sound heard when you press the end of your finger into your ear. "A sound is then perceived which resembles most nearly that of carriages at a great distance passing rapidly over a pavement," as Wollaston put it.[37] He added that this resemblance arises "not so much from the quality of the sound as from an agreement in frequency with an average of the tremors usually produced by the number of stones in the regular pavement of London, passed over by carriages moving quickly." He then calculated that frequency as roughly twenty to thirty vibrations per second. If the carriage were going about eight miles per hour (twelve feet per second) over paving stones about six inches long; it would vibrate about twenty-four times in every second (because it bumps twice over the edge of the stones twelve times per second). Wollaston also confirmed this estimate by rubbing a pencil for a given time (five or ten seconds) along a board in which regular notches had been cut, in order to try to replicate the frequency of the sound he heard in his ear, which was almost infrasonic, just at the lower limit at which the ear can perceive vibrations as sounds (about 32 Hertz).

If you try to hear these sounds for yourself, I found helpful the description given in 1812 by Paul Erman (who does not seem to have read Wollaston's account): in the absence of other sounds (ideally at night), place your head against a hard cushion and then strongly and sustainedly contract your maxillary muscles by tensing your jaw. As Erman observed, "one has the subjective sensation of a series of separate sounds (for one finds no characteristic note or pitch). These follow each other in equal rhythm as long as the contraction persists. If the will to contract becomes stronger, the series of pulsations takes on an accelerating rhythm" and contrariwise for a weakening contraction.[38] Erman's description of these as rhythms underlies the felt sensation of regular pulsation, for he is describing a sound that is felt as vibration, not a pitch in the normal sense. He seems to imply that muscular action must be rhythmic, not static or constant, but does not make this explicit.

Wollaston was careful to note that this sound came not from the tympanum (eardrum) itself but from the muscles of the head, an inference he tested by moving the thumb in his ear so that it did not rest on the head at all, only the ear. He wondered whether he might be hearing an octave, a 2:1 ratio of tremors that "coincided only at alternate beats," which he tested by changing the spacing of the notches on his board, noting that the ensuing rumbling sound still matched the muscle sound. He also listened to the muscles in his foot by resting his notched stick on it and then putting his ear on a cushion at the other end of the stick.

Wollaston drew deep implications from this curious phenomenon. Whereas Grimaldi thought this sound came from the "agitation of [animal] spirits," Wollaston attributed it to the physical operation of the muscles themselves. In general, he deduced that "each [muscular] effort, apparently single, consists in reality of a great number of contractions repeated at extremely short intervals," those contractions then resulting in the rumbling sound he perceived.[39] Though he began as a physician, Wollaston became a famous chemist, the discoverer of the elements palladium and rhodium, who made pioneering observations about atomic theories of matter. In this less well-known physiological work, he suggested an atomic theory of muscular effort, breaking seemingly continuous actions into discrete contractions made perceptible as low-frequency vibrations.

When Helmholtz turned to this phenomenon, he had tools to determine the frequencies of these vibrations more precisely.[40] By then, the stethoscope had become well established in other realms of physical diagnosis, especially the examination of the thorax, lungs, and heart, as we have seen. Helmholtz's 1864 "Experiments on Muscle Noise" ("Versuche über das Muskelgeräusch," translated in the appendix to this book) began by referring to the "well-known and often questioned muscle noise" as observed by stethoscope. His word *Geräusch* spans an interesting and complex terrain of meanings from rustling to roaring and thundering; in this context, it corresponds to the modern medical term *bruit* denoting a variety of whispery sounds heard through a stethoscope, thereby setting them aside from more ordinary noises.[41] Helmholtz's allusion to the controversy about muscle sound indicates the difficulty of characterizing these near-infrasonic vibrations. He surely knew the famous Schubert songs that evoked *Rauschen* (the sound of rushing or rustling) in brooks and in the wind (figure 15.10a,b; ♪ sound examples 15.2 and 15.3), whose accompaniments variously depict the lulling or ominous vibrations of water or wind.

Helmholtz began by comparing Wollaston's estimate of the pitch of muscles to more recent determinations by the Irish physician Samuel Haughton (1863), who found with "great surprise and pleasure" that the pitch of "muscular susurrus" happened to be several octaves above the tinnitus he experienced after a fever. When Haughton independently surveyed his "musical friends" about the pitch of the susurrus, they divided into two groups, one choosing C, the other D, two octaves below the bass staff (♪ sound example 15.4), pitches so low that "they are found on the new pianos only."[42] This would place the susurrus between 32 and 36 Hertz. Taking Wollaston's calculations a bit further, Haughton measured the "intervals of the Guernsey granite pavement" and calculated that a carriage moving at 8 miles per hour should make 35.2 vibrations per second, thereby confirming his result.

Helmholtz repeated these observations himself and generally agreed with their results but wanted "not to produce the contractions with my own will," which raised the difficulty of the same subject both producing and observing his own contractions. Then too, Helmholtz sought a higher standard of accuracy than the estimates presented by Wollaston and Haughton or even his own self-observation. Instead, he used a form of his induction apparatus with an adjustable spring (rather than a single tuning fork) to stimulate his muscles at a known (but variable) frequency, being careful to put the apparatus in another room behind two sets of closed doors "so nothing of its pitch could be heard directly. As soon as I put the electrodes on my masseter and brought it into strong contraction, I heard the pitch of the spring of the induction apparatus. When I changed the adjustment of its [the spring's] screw, then I heard the change [in the muscle]."[43]

Figure 15.10

Two Schubert songs about *Rauschen*: (a) "Wohin?" from *Die schöne Müllerin* (1824, D975); Text: "I hear a little brooklet rushing from its source in the rocky spring" (♪ sound example 15.2); (b) "Herbst" (1828, D945); Text: "The winds are rustling, so autumnal and cold" (♪ sound example 15.3). Note the difference between the piano accompaniments, (a) a lulling circular figure depicting the brook's gentle motion and (b) the more insistently repetitive rhythm of a cold wind.

By placing the source of the excitation in an objective source outside himself, Helmholtz could show that the muscle sound he heard was in fact a direct response to its electrical stimulation, synchronized to its frequency. This amplified Wollaston's hypothesis that muscular action was composed of discrete contractions by showing that their frequency was given by that of the current exciting them. Surprisingly, the effective frequencies went up to 130 Hertz, quite a bit higher than those Wollaston, Haughton, or he himself had observed earlier. To check this, Helmholtz noted that the muscle sound "only became audible when the current strength was increased sufficiently to give a contraction of the muscle. Likewise, the sound was also heard, though less strongly, through a stethoscope on a young man's arm muscles, which were brought into contraction by induction currents flowing through them. In this case, the ear and auditory nerve of the observer were not affected by electrical currents."[44] When he reduced the electric current sufficiently (or moved it slightly away from the forearm muscles), the muscle sound ceased, no longer vibrating in response to the driving frequency. Thus, the muscle sound was directly linked to its contraction, eliminating all doubt about the sound's existence.

At this point, Helmholtz switched from using an adjustable spring to a tuning fork of 120 Hertz, about the frequency he had noticed earlier (sounding about D below middle C). He then heard "relatively strongly" the muscle sounding the pitch an octave higher (240 Hertz). By that era, the conventions of scientific papers (and perhaps also his own reticence) would not allow any expressions of personal feeling, but one can sense Helmholtz's surprise at having heard in a muscle the primal musical interval of an octave: these fibers really were vibrating like strings. He also tried using even higher-frequency tuning forks (around 600 Hertz, an octave higher) to induce tetanic contractions in frog legs, though he could not hear their sound. Evidently he was trying to see how many overtones he could hear from muscles, as if his investigations of vowel timbres had informed his hearing of muscle tone, no longer a mere figure of speech (*tonus* as the general condition of muscle "tone") but its audible sound.

The title of his subsequent paper, "On Muscle Tone" ("Ueber den Muskelton," 1866, translated in the appendix to this book), shifts from describing muscle sound as "noise" (*Geräusch*) to pitch (*Ton*), the term for a musical note.[45] To do so, Helmholtz returned to using a variable spring in order to explore very low frequencies that would be difficult or impossible to achieve with tuning forks. Such low frequencies required huge tuning forks, but even Koenig's mammoth fork could reach only 32 Hertz (figure 15.11). Helmholtz wished to go lower still because he suspected that the muscle frequencies heard around 36 to 40 Hertz by Wollaston, Haughton, and himself were actually octaves of a still lower fundamental frequency. By tuning the variable spring carefully, he was able to determine that in fact, the muscles were vibrating in the range 18 to 20 Hertz, well below the range of human hearing. He and the others had in fact been hearing the pitch an octave higher, which they mistook as being the fundamental pitch. This indicated, however, that the

Figure 15.11
A large tuning fork built by Rudoph Koenig (1890), 1.35 meters long, with an adjustable slider that allows it to give pitches from 32 to 50 Hertz.

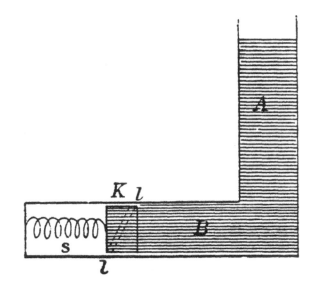

Figure 15.12
Eduard Pflüger's mechanical model (1859) of nerve and muscle mechanics. The spring *s* and piston *K* represent
the molecular forces opposing the potential energy stored in the nerve, which he represented by the hydrostatic
pressure in the tube *AB*. The vibrations of the muscle would correspond to those of column of fluid when dis-
placed from equilibrium.

nervous system operates at infrasonic frequencies, which would have to be considered in
interpreting the muscles' audible overtones.

Helmholtz himself did not carry these investigations further, facing limitations on the
frequencies he had available to simulate the muscles. The work later went forward under
his assistant Julius Bernstein, who had studied medicine with du Bois-Reymond and
hence was a second-generation descendant of Müller's scientific lineage. Both Helmholtz
and du Bois-Reymond had suggested that nerves and muscles operated through biochemi-
cal processes, which became the focus of increasing attention. The vibrational character
of nerve action also suggested mechanical models, some of which represented nerves
and muscles in terms of actual vibrating springs (figure 15.12).[46] Testing the applicability
of this and other models called for significant improvements in measuring all aspects of
nerve transmission.

With that in mind, Bernstein set about improving the accuracy of Helmholtz's results,
beginning with the determination of the speed of transmission of nerve impulses for which
he devised new apparatus. Bernstein also took up the measurement of muscle sounds and
was able to carry it to frequencies as high as 1,380 Hertz by using an improved induction apparatus or "inductorium" (figure 15.13). This new device used a spring of variable
length whose frequency of oscillation was determined by comparing it with the pitch of
a tuning fork. With it, he confirmed Helmholtz's fundamental findings and argued that

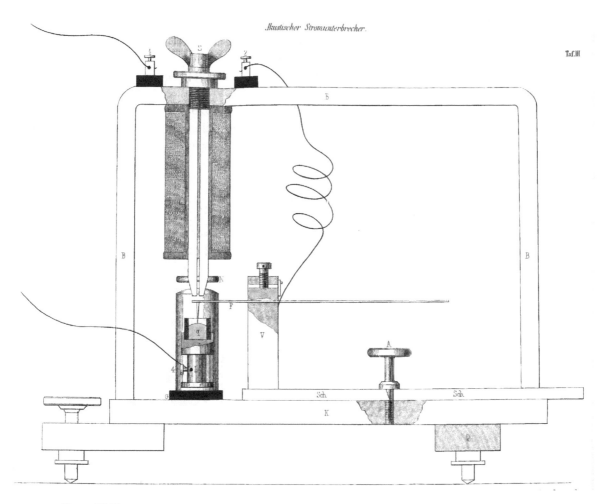

Akustischer Stromunterbrecher.

Taf. III

Figure 15.13
Julius Bernstein's acoustic interrupter (1871). Current from a battery enters at terminal 1, then passes through the coils of the electromagnet to terminal 2, continuing to spring F, then to mercury bath q, returning to the battery from terminal 4. The spring's house can slide along the track Sch so that the length of the spring can be varied. When the circuit is closed, the electromagnet breaks the contact between F and q.

that muscle action during contraction is discontinuous, as Wollaston and others had surmised: separate muscle fibers contract either completely or not at all, and never gradually. Further, muscular action seemed to have an atomic or molecular basis. Indeed, Bernstein interpreted the equality of the pitch of the muscle sound to that of the interrupter's electrical frequency as indicating a one-to-one correspondence between the stimulus pulses and individual "capacitors" (on the molecular level) inside the muscles.[47]

Ultimately, from this and many other experiments Bernstein developed a membrane theory of bioelectric potentials that explained nerve and muscle action in terms of the activity of semipermeable membranes. Nerves and muscles operate through the gradients of concentration of ions across such membranes, which then generate bioelectric potentials, a basic understanding that remains in place today. Bernstein initially identified potassium and sodium ions as responsible, though he opted for potassium alone in his final version.[48] His arguments involved many aspects of contemporary chemistry and physics, including the amounts of heat generated by these molecular actions of the electrochemical ions, alongside the evidence from the speeds and frequencies of nerve and muscle action. Bernstein's work offered the first quantitative theory of nerve action that revealed its atomic basis, crucial to the twentieth-century developments that went even further into these microscopic details and led to fundamental changes in medicine as well as in physiology. Though only one strand in this complex tapestry, sonic investigations proved an essential element to understanding its unfolding design.

16

Telephonic Connections

As much as the work of Helmholtz, du Bois-Reymond, and Bernstein revealed about the operation of nerves and muscles, carrying that project further required understanding how an individual nerve worked, which was greatly aided by rendering its electrical activity audible. The first efforts in this direction used the technology of the recently invented telephone to listen to nerves and muscles, even activating them by voice telephonically. Soon after Alexander Graham Bell made public his telephone in 1876, physiologists began experimenting with it. As Axel Volmar has shown, "The potential of the telephone for acoustic observation of electrical muscle and nerve currents became known very early and was used productively."[1] We will situate Volmar's findings in the larger story of neural signals.

In November 1877, only a few days after he had received a Bell instrument, du Bois-Reymond performed an experiment with his usual flair for the dramatic. Having attached the receiving end of the telephone electrode to the nerve of a frog's thigh muscle, when he "said, sang, whistled" *zucke!* ("twitch!") into the microphone, the muscle would twitch.[2] This startling scene did not involve communication of the word's meaning, however, for "even setting down a funnel on the table somewhat strongly" would suffice if the microphone picked up that sound. He noted that the frog's muscle remained quiet in response to *liege still!* ("lie still!"), not through understanding his meaning but because "sounds with deeper characteristic overtones are more effective than those with higher," such as the *u* in *zucke* compared to the *i* in *liege*. Here du Bois-Reymond drew on Helmholtz's work with tuning forks imitating human vowels in order to understand the electric signals that connect vowel and frog. Du Bois-Reymond connected this with other experiments that used a violin bow to set a magnet into vibration, thereby causing an electric current that made a frog's muscle contract in "acoustic tetanus."[3] For him, the telephone was one of many sonic interventions in nervous action, sharing with Helmholtz's induction apparatus the use of electric currents, whether generated by tuning fork or by voice.

Shortly after, in 1878 Bernstein and Ludimar Hermann (another of du Bois-Reymond's students) used the telephone to receive, not just send, neural signals. After another researcher questioned the accuracy of his acoustic interrupter (see figure 15.13), Bernstein used the telephone to demonstrate that his device indeed produced currents whose frequencies were given by its vibrating spring. Having attached a telephone to receive the signal from the spring set to an arbitrary frequency (in this case, about 180 to 190 Hertz), "immediately one would hear the same pitch through the whole room, though mixed with a rattling noise often heard when the telephone plate [in the receiver], driven by the strong currents necessary for the interrupter, hit the iron core."[4] Despite the rattling, the pitch of the vibrating spring driving the nerve came through loud and clear.

In 1878, Hermann tried to use a telephone to detect the very small currents generated by nerves. Like Bernstein, he too observed that the telephone could pick up the vibrations of the inducing apparatus so audibly that he heard the vibration frequency "many feet away from the receiver," which itself could be a meter away from the inducer.[5] He therefore hoped that hearing the exact pitches of muscle sounds through the telephone would clarify their physiology. Yet he could not hear the induced nerve currents through the telephone beyond a "rattling noise" because the Bell instrument was not sufficiently sensitive.

That same year, interest in using the telephone to investigate the nerves spread rapidly across Europe and beyond, as Volmar points out.[6] In Paris, Jacques-Arsène d'Arsonval (who had been the assistant of the famous physiologist Claude Bernard) used the more sensitive telephones made by Siemans and Halske to argue that "the worst constructed telephone is at least a hundred times more sensitive than the nerve in detecting weak electric variations."[7] The telephone was also far superior to du Bois-Reymond's galvanometer, which (despite its sensitive coils) "lacked instantaneity and whose needle, because of its inertia, could not show rapid electric variations such as take place (for example) in a tetanized muscle." To test the presence of a weak current from a muscle, d'Arsonval proposed "to send the current through the telephone and, in order to obtain variations, I interrupt the current mechanically with a tuning fork. *If no current is flowing through the telephone, it will stay silent, but if on the contrary there is the weakest current, the telephone will vibrate in unison with the tuning fork*."[8] Yet again the tuning fork emerged as a valuable instrument, capable of modulating the delicate nervous currents. D'Arsonval had no doubt that the use of the telephone "will furnish interesting results in the study of animal electricity that I will study in this new way."

Later that year in St. Petersburg, the Georgian physiologist-physician Ivan Tarchanow (Tarkhnishvili) confirmed d'Arsonval's assertions about the sensitivity of the telephone. Tarchanow used a 100 Hertz tuning fork (about G-flat two octaves below middle C) to create the electrical induction stimulating a muscle. Then, using "two simple Bell telephones, one only has to put telephones on both ears to perceive a clear tone that repeats the pitch of the tuning fork."[9] He performed this experiment on the muscles of both a frog thigh and a human arm: "As long as the hands remain in complete rest, there is not the least trace of tone in the telephone. As soon as the muscles in one hand were set into contraction, at once a clear tone comes out that dies away when the muscles begin to relax."[10] As Volmar notes, Tarchanow's choice of words showed that he thought hearing no less than vision capable of producing a "display [*Anzeiger*]" of nerve currents.[11] To do so, Tarchanow emphasized the greater clarity achieved by using both ears, probably influenced by the long-standing use of the binaural stethoscope (dating at least from the 1850s) as well as Alexander Graham Bell's experiments on "stereoscopic" telephony.[12]

When Bernstein returned to this field in 1881, he too heard "a distinct rattle" when stimulating frog muscles with his acoustic interrupter using currents whose frequencies went as high as 700 Hertz.[13] In describing these frequencies, he used musical terminology to give the pitches of the tuning forks he employed to produce these currents, showing how much he viewed these experiments in terms of the underlying musical apparatus. This sensitivity to musical sound also marked his description of experiments in which the muscle was dosed with strychnine to induce a different kind of tetanus, which produced "in the telephone a very clearly audible deep singing tone."[14]

During this period, the telephone fascinated the music-loving physiologists in Berlin. In June 1881, Franz von Mendelssohn (a cousin of the composer) telephoned the Helmholtzes and played one of Joseph Joachim's compositions for them over the phone. Anna, Hermann's second wife, reported that "the sounds could be heard clearly, right up to the finest nuances of the violin's sound."[15] Joachim, who was also listening, was able to judge that he was hearing "the Amati, not the Stradivari," owned by Mendelssohn. Anna thought that "this omnipresence of the distant is uncanny and ghostly."

Only a few years later, such uncanny moments had become routine. As Volmar notes, by 1888 du Bois-Reymond's "telephone experiment" had become so familiar that William Stirling included it as a laboratory exercise in his introductory *Outlines of Practical Physiology*.[16] In the subsequent decade, researchers sought ways to make permanent records of the data obtained through the telephone, which Hermann called "phonophotographic studies" and Bernstein "phototelephonic."[17] These hybrid devices involved an optical apparatus that transduced the telephone signal into a moving light beam, which then could be projected onto a rotating drum covered with photographic paper. Though other investigators still used galvanometers to measure the currents, telephones and tuning forks often proved helpful parts of these devices. For instance, in 1895 J. Burdon Sanderson noted that "when a muscle is stimulated by leading alternating currents of great frequency through its nerve the effect is the same as if the excitation is continuous. The simplest method of testing this experimentally is to employ telephone currents. When for instance the note c ([notated by the] space c in the treble [clef, i.e., an octave above middle C]) is loudly sung to a telephone, of which the terminals are connected with the exciting electrodes, the nerve receives over a thousand excitations per second by currents in alternately opposite directions." The resultant image (figure 16.1), taken with a moving photographic plate, shows the initial "negative variation" along with the nerve's subsequent response to the sung c.[18] In this experiment, by fortunate accident the telephone was registered by the galvanometer even before the nerve received the telephone current, so that the photograph shows both the form of the stimulating current as well as that of the muscle's vibration in response.

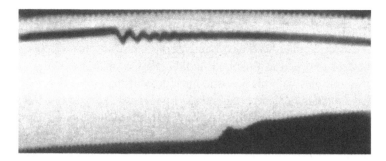

Figure 16.1
J. B. Sanderson's 1895 photograph of an experiment in which the note c was sung through a telephone to a frog's nerve. The upper line shows the baseline response of the galvanometer, which received the telephonic current even before the nerve (middle line). The nerve's response shows the initial "negative variation" as well as its continuing vibration at the frequency of c, gradually fading out over time.

In 1900, the Russian physiologist Nicolai Wedensky described "the telephonic method" as "virtually irreplaceable" because "we can trace *any arbitrary point* in the nerve *without interrupting its connection with the muscle*."[19] Taking a larger view of the ensemble of instruments, he concluded that "in the study of every complicated process in the nerve fiber it is necessary to use the muscle, the telephone, and the galvanometer. Every one of these devices is speaking in its own language and appears to be a good witness under certain conditions and a weak one under others."[20] His complete apparatus (figure 16.2) included all these multiple modalities to give fuller insight into nerve and muscle action, for which he found the timbral possibilities of the telephone particularly revealing: "In some studies it was easy to distinguish one physiological tone from another through their particular timbres. When I stimulate [the muscle] with currents that already have begun to be heard in the telephone, first I hear the muscle tone and then finally, after muscle fatigue from the stimulation has set in, the quiet tone of current cycles with a different timbre, compared to what until now had for a long time been perceived without any modification."[21] He even could hear through the telephone various phases of the action of poison on the nerves, which evoked an initial tone characteristic of "very *strong* as well as very *moderate* stimulating currents" changing to "a *weak, muffled tone complicated by nerve noises*" after the poison had begun to act.[22] Like Laennec and Bernstein, Wedensky drew on his sonic experience to describe what he heard, such as the deep noise "like a distant waterfall" during a voluntary contraction.

Alongside these developments involving sonic modes of observation, the galvanometer itself underwent much development as an instrument for physiology. The Dutch physicist Willem Einthoven devised a particularly powerful version about 1901 in order to measure the very small electric currents of the heart.[23] The Einthoven galvanometer (figure 16.3a) was devised to perform the earliest electrocardiograms, leading the cardiac current through an extremely fine filament of gold- or silver-covered quartz thread passing

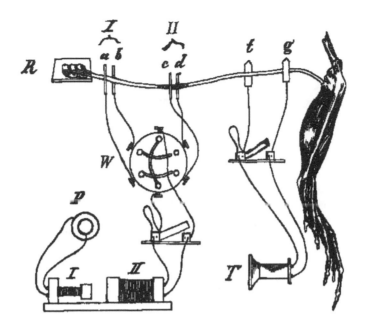

Figure 16.2
Nicolai Wedensky's diagram of his experimental setup (1900). An induction apparatus (*P*, with coils *I, II*) generates currents that are regulated in frequency by the rotating wheel *W* and then flow through the wire at the top into a frog muscle; telephone *T* can be switched into the circuit to hear the incoming current as well as the muscular response. In other experiments, Wedensky used an electromagnetic tuning fork to regulate the frequency.

between strong electromagnets, called a "string" galvanometer because of that conducting thread. The current passing through it would then cause it to move in the magnetic field and cast a shadow on a moving roll of photographic paper, whose rate of motion registered the time coordinate of the image, thus replacing Helmholtz's tuning fork. The resultant electrocardiogram looks very much like those taken today (figure 16.3b). But the replacement of a sonic observation—the vibration of the string—by its visual trace along the paper tended to obscure and even hide the sonic turn that underlay the whole apparatus.

For this discovery, Einthoven received the 1924 Nobel Prize in Physiology or Medicine, but his instrument was a behemoth weighing 270 kilograms (because of the water cooling for the electromagnets) that required five people to operate it. Though much more sensitive than earlier galvanometers, its quartz thread was easily broken, so that the instrument was delicate and hard to use. Many measurements of physiological currents particularly challenged instruments like the galvanometer, which had been developed for much larger currents. Though Wedensky, among others, made some experiments involving amplification of the tiny nerve currents, substantial advances came only after the development of multistage amplifiers needed to produce usable radio transmissions.

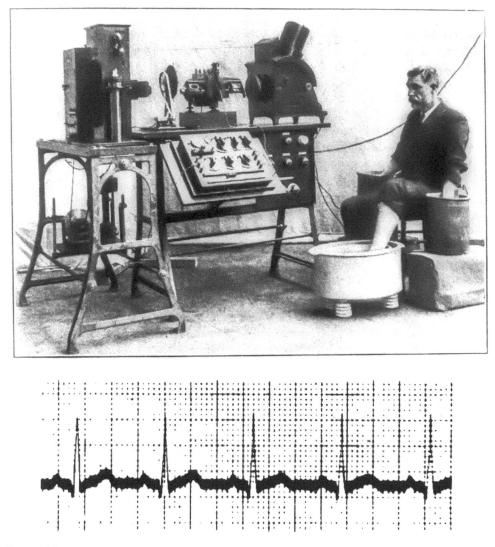

Figure 16.3
(a) An Einthoven string galvanometer manufactured in 1911; the patient sits with an arm and a leg in strong salt (conductive) solutions. (b) One of the first electrocardiograms made on the string galvanometer (1902).

Here again military necessities provided the stimulus. As Rudolf Höber noted in 1919, the Great War stimulated the development of technology that could "amplify weak currents to an almost arbitrary degree."[24] One of the new three-stage vacuum tube amplifiers allowed him to redo the observations of Bernstein and Wedensky by amplifying the action currents more than a millionfold. Höber noted that nerve and muscle sounds "so greatly amplified could easily be demonstrated in favorable circumstances, heard through a telephone with attentive observation in a quiet workroom."[25] Because of this greater volume and clarity, he could distinguish between different frequencies of stimulation (still provided by an electronic tuning fork) more readily than Wedensky could. Through his telephone, Höber heard a vivid spectrum of pitches and timbres from a human forearm muscle successively stimulated at "50 Hertz, loud humming tone; 100, tone like a cello; 160, higher, louder tone, unison with the stimulus frequency; 256, tone a bit lower than the stimulus frequency mixed with noise; 512, loud noise; 850, loud scraping noise."[26] Höber evidently considered such timbral descriptions a significant part of his experimental record and used them in clinical settings to assess the sounds he heard from the muscles of patients suffering from spastic contractions after apoplexy or encephalitis, as well as the very different muscular sounds of catatonic patients.[27]

In an afterword, Höber applied these techniques to make the sounds of the human heart audible even in a large auditorium: "If one passes through the amplifier the current a microphone picks up from the heart sounds and puts a sound horn [*Schalltrichter*] on the telephone, from far away one hears clearly the rhythm, as well as any possible arrhythmias that might be present and the accentuation of the first or second [heart] sounds."[28] By putting a horn or megaphone over the telephone, Höber evidently created a simple loudspeaker of the sort that was already beginning to be used for sound reproduction from radios or phonographs. In this way, sounds that previously could only be heard by paying close attention in a quiet room now could fill an entire auditorium. Höber regretted that the loud background noise "as yet still disturbs finer observation" of more subtle sonic details and planned "as soon as possible to seek methods to overcome" these problems.

Responding to Höber's call, by the mid-1920s the Austrian electrophysiologist Ferdinand Scheminzky had developed improved techniques incorporated in what he called an electrostethoscope, the next iteration in the sonic technology that had begun with Auenbrugger and Laennec. To be sure, Scheminzky's work was only one of many attempts to apply the new electronic amplifiers to the long-standing problems of listening to heart sounds.[29] We single it out because of the way Scheminzky synthesized several aspects of Höber's work, including the sonic amplification of heart sounds as well as muscle sounds, along with measurements of the amplified currents using the string galvanometer.[30] Scheminzky's 1927 paper showed his compound apparatus, a state-of-the-art instrumentation for the audible investigation of cardiac status in real time (figure 16.4). This photograph's odd resemblance to scenes of Dr. Frankenstein's laboratory in old movies should not distract from closer reading of this image. Against a wall filled with diagrams

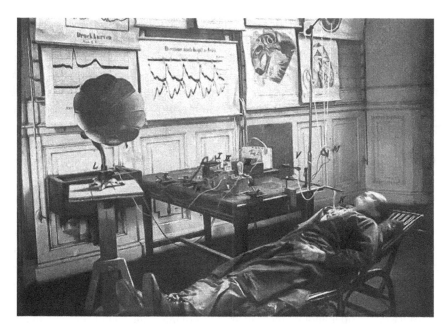

Figure 16.4
Ferdinand Scheminzky's 1927 picture of his apparatus to amplify heart sounds. The microphone *M* suspended via the recording capsule *K* over the patient's chest produces a current passing through a three-stage tube amplifier (*Tk, I, II, III*) on the table next to him that then drives the loudspeaker *L*.

and graphs of the heart, the loudspeaker projects the patient's amplified sounds, giving mediate auscultation immediate, public availability. He took special care to use the best available carbon microphones to achieve "an appreciably greater fidelity of reproduction, especially at low currents," thus applying contemporary concerns with sound quality in order to elucidate what he called the "tone character" of heart sounds, especially the fine nuances of cardiac pathology.[31] These microphones gave much better results than those in normal telephones and also enabled enhanced audition and discussion of heart sounds even in a large auditorium, useful for medical education and clinical consultation.

Scheminzky also used his setup to make permanent visual records, as well as to hear those sounds in the moment. He used an Edelmann string galvanometer to produce a graphic display of the microphone current next to a time marker from a clock. To monitor the sound driving these images, he used a telephone, further integrating sonic and visual modes of observation. His graphs include sound records of the tricuspid valve of the heart and of the facial muscles (figure 16.5). The electromagnetic tuning fork remained his source to stimulate the masseter muscles, using currents generally at low frequencies (20 and 50 Hertz) falling near or below the threshold of human hearing. Scheminzky particularly noted the ability of the carbon microphone to register these sounds even when human ears could not, thus indicating the usefulness of these sonic methods in overcoming the limitations of unassisted hearing.[32]

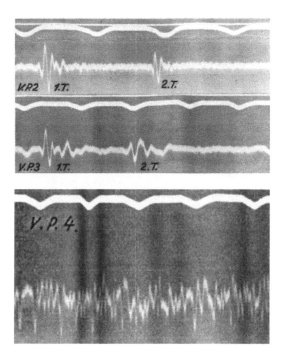

Figure 16.5
Ferdinand Scheminzky's 1927 images of microphone signals from (a) unamplified heart sounds from the tricuspid valve of two healthy patients (V.P.2, V.P.3), showing first and second heart sounds (1.T, 2.T) and (b)
amplified muscle sounds from the masseter muscle in contraction stimulated by an electromagnetic tuning
fork. At the top of each image, a time signal (from a Jacquet clock) shows intervals of ⅓ second.

Bernstein and those who preceded him had found suggestive evidence of the underlying electrical and chemical mechanisms of nerve action through inferences from phenomena such as muscle sounds. Yet confirming, clarifying, and deepening these explanations
required investigating the "atom" of nerve action, the individual neuron and its exact
behavior. Because the electrical currents involved were so small, measuring instruments
were challenged to respond adequately to observing an individual neuron at work, even
with amplification. More fundamentally, the kind of listening to nerves and muscles we
have been considering in the past two chapters (and indeed since Laennec) had been
essentially *analog*, in the sense of hearing the interior soundscape of the body made audible by devices like stethoscopes or telephones that amplify small but continuous sounds.[33]
As we will see in the next chapter, amplifying the signals of single neurons gave access to
a new realm of *binary* sounds, not continuous—either on or off, nothing in between. The
sonic turn thereby led to a new realm of hearing the frequencies of such discrete binary
pulses, which in general are perceived as rhythms.

17

Listening to Neurons

When Edgar Adrian in 1928 began using a loudspeaker to play the amplified output of single neurons, he provided far easier and more intuitive access to their activity than purely visual inspection of galvanometer or oscilloscope traces. This technique became an important part of his experimental repertoire and played a significant role in several of his discoveries. After reviewing the history of attempts to measure single nerves, we will trace Adrian's investigations and the role sonic methods played in them. Amplifying nerve signals through telephones or loudspeakers not only gave immediate access to the overall character of those signals but proved to be a valuable tool in fixing on one neuron and locking on to its individual response. As such, it became a significant technical innovation that contributed to seminal discoveries about the all-or-nothing discreteness of nerve impulses throughout the nervous system, disclosing information embedded in the impulses' varying frequencies, often so low that they were perceived as rhythms rather than pitches.

Though the previous two chapters have taken place mostly in Europe (particularly Germany), the developments we consider now mostly happened in Britain and the United States.[1] This shift began to occur even before the Great War wreaked havoc on German institutions. The first measurements of the action potential of a single nerve were made in 1888 at Oxford by two brothers-in-law, Francis Gotch and Victor Horsely. Their interest in "the nature of the excitatory processes of the epileptic convulsion in the spinal cord" led them to investigate how a nerve would respond to a single electrical impulse, as opposed to sustained stimulation (for instance, using an electromagnetic tuning fork).[2] They generated an impulse by breaking the current passing from a single Calland cell battery through an induction apparatus such as du Bois-Reymond used. As a result, "the break shock produced in the secondary coil by this means was so feeble as to be barely perceptible on the tip of the tongue."[3] Though weak, this brief but controlled electric shock sufficed to activate the sciatic nerve of a toad; they went on to stimulate this nerve in anesthetized cats and monkeys in order to assess the generality of their findings across different species.

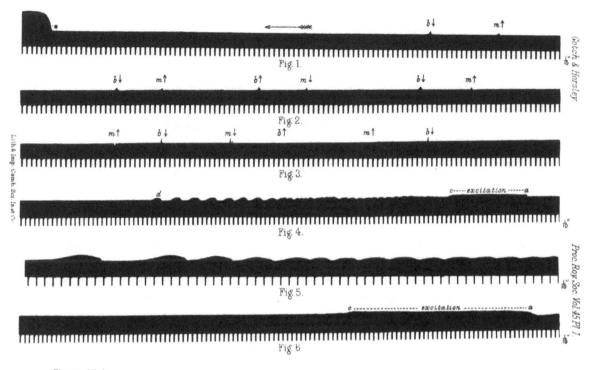

Figure 17.1
Gotch and Horsely's 1888 measurements of the electrical response of the sciatic nerve to electric impulses, with facsimile drawings of the original photographic plates, to be read right to left. The regular lines under each figure show the time in tenths of second as given by a vibrating shutter. The dark area shows the meniscus (interface) of the mercury; *m* denotes making a current, *b* breaking it; figures 4 and 6 show the nerve's response to sustained excitation (as indicated). Figures 4 and 5 show rhythmical response that "corresponds to the clonic stage of the epileptic convulsion."

To observe the response of the nerve, Gotch and Horsely used a recently devised (1873) form of galvanometer, Lippmann's capillary electrometer (illustrated below in figure 17.5). This device could register brief and weak currents, which would cause a column of mercury to rise inside a thin glass tube, though so slightly that a microscope was required to observe the change of mercury level. Because of the thinness of this vessel, even a small change of current would quickly cause a shift in the level of the mercury, on top of which rode a layer of sulfuric acid; the interaction of the mercury and the acid created a special effect of capillary action that gave this electrometer its extreme sensitivity.[4] As they made their measurements of a nerve's reaction to stimulus, Gotch and Horsely watched through the microscope for displacements of the capillary surface on the scale of 1/400 of a millimeter of actual motion. They were able to record these displacements using a rapidly moving photographic plate, against the time standard given by a vibrating shutter (figure 17.1). Their results show "a well-marked difference or demarcation current," which registered sharp responses to making or breaking the stimulus.

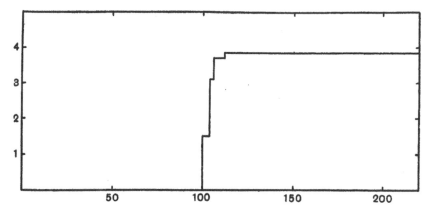

Figure 17.2
Keith Lucas's 1909 diagram of the response of a frog muscle to increasing stimuli, showing the "steps" characteristic of an all-or-none response. The graph plots the height of contraction against strength of stimulus (where 100 arbitrarily designates a current just barely able to excite the motor nerve).

Gotch and Horsely did not go further with these observations, becoming more interested in the physiology of epilepsy and the development of the new discipline of neurosurgery.[5] Sixteen years passed before Keith Lucas at Cambridge began his investigations into individual nerve impulses. His engineering background helped him find new ways to approach the problem of individual nerve impulses, in particular how varying stimuli could cause a gradation of response in many-fibered muscles. As he put it in 1905, "the contraction in the whole muscle might conceivably be graded either by gradation of the contraction in each several fibre, or by variation of the number of fibres called into play, or by a combination of both these methods."[6] He began with frog muscles comprising 150 to 200 fibers, which he gradually peeled away, one by one. At each stage, he tested the remaining fibers with a brief direct stimulus to find out how the response depended on the number of fibers, as measured by a mirror attached to the muscle. He concluded that

the contraction of a frog's striated muscle excited directly does not increase continuously from zero to its maximum with increase of the exciting current. The contraction increases in abrupt steps to which further increase of the exciting current will add little or nothing, until a new step is reached. Since the number of steps was never found to exceed the number of fibres composing the muscle, I drew the inference that each step represented the addition of a fresh fibre or group of fibres to the number previously excited.[7]

In 1909, he extended these experiments by stimulating a muscle through its motor nerves rather than by direct stimulus to the muscle fibers. He found a certain cutaneous frog leg muscle that had about eight or nine nerve fibers, which he could gradually tease apart. He noted that "the contraction does rise in steps when increasing stimuli are applied to the nerve, just as it does when the stimuli are applied directly to the muscle."[8] These steps are shown in figure 17.2. Lucas cautiously inferred that his data supported what he called "the property of 'all or none' contraction": each muscle fiber either is or is

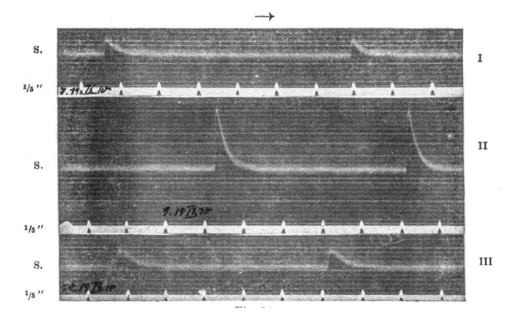

Figure 17.3
Siegfried Garten's 1910 images of the response of a frog sciatic nerve, using a string galvanometer. The time signal at the bottom of each set of images shows ⅕ of a second.

not maximally contracted, nothing in between, so that "the 'submaximal' contraction of a skeletal muscle is the maximal contraction of less than all its fibres."[9] If so, muscle—and presumably nerve—action is "quantized," discrete like the quantized energy Max Planck reluctantly introduced into physics in 1900, though Lucas did not mention this curious parallel.[10]

Lucas drew back from an outright assertion of the truth of the all-or-none principle, judging that "the question can hardly be settled absolutely until we are able to experiment with a single cell," meaning a single nerve fiber.[11] Here the limitations of the available instruments became increasingly important in registering such weak currents that lasted only about 0.1 of a millisecond (0.0001 second). To measure such short times would require an instrument having a natural frequency of at least 50,000 vibrations per second, far higher than the natural frequency of string or capillary galvanometers.[12] For example, Siegfried Garten's 1910 recordings of a frog sciatic nerve (made with a string galvanometer) are difficult to interpret in detail (figure 17.3) because they push so far past the capacity of the instrument he used.[13]

Lucas's student Edgar Adrian eventually performed the decisive experiments. Though they had begun collaborating in 1911, their work was interrupted when Adrian went to medical school in 1914; Lucas volunteered as an airman during World War I and died in a training accident in 1916 at the age of thirty-seven. Lucas's work was a triumph of

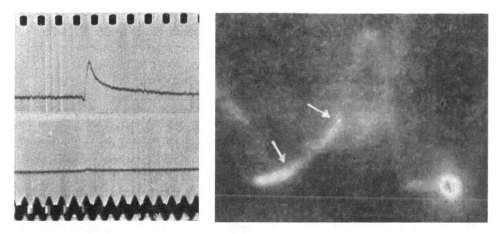

Figure 17.4
(a) Alexander Forbes and Catherine Thatcher's 1920 recording of a frog sciatic nerve using a one-stage amplifier and a string galvanometer, showing an extraneous fine vibration at 330 Hertz. (b) Joseph Erlanger and
Herbert Gasser's 1922 image of a bullfrog sciatic nerve response as shown on an oscilloscope. They note that
because of "diffuse fluorescence and the dispersion of the light," this photo "gives a poor idea of the actual
clearness" of their images.

ingenuity, patience, and immense skill; his cellar laboratory "filled with water after rain so
that people had to walk about on duckboards in wet weather."[14] He made no particular use
of sonic techniques beyond the occasional use of an electromagnetic tuning fork as a source
of continuous stimulation for nerves or muscles. In contrast, Adrian's work introduced sonic
techniques that remain in use even today and that helped him locate and measure single
neurons reliably. Looking back in 1932 over his work, he observed that "the history of electrophysiology has been decided by the history of electrical recording instruments."[15] As we
will see, that history included the ways those instruments could be rendered audible.

When Adrian resumed his work on nerve action in 1918, he proceeded in light of new
initiatives elsewhere. In 1920, Alexander Forbes (Ralph Waldo Emerson's grandson) and
Catherine Thatcher at Harvard were the first to use a one-stage amplifier to study the
action potentials of a frog sciatic nerve. They passed the amplified potentials through a
string galvanometer whose shadow was recorded on moving film. Ironically, their work
was plagued by excess vibrations that their tuning fork or even the sound of their voices
evoked in the physical structures inside the vacuum tubes; these extraneous vibrations
caused an extra jittering visible in some of their images (figure 17.4a), eventually removed
by isolating the amplifier in a soundproof box.[16] In 1922 at St. Louis, Joseph Erlanger and
Herbert Gasser measured the response of a bullfrog nerve using an oscilloscope, which
had been invented in 1890. Because the oscilloscope's natural vibrational frequency was
about 200 million vibrations per second, it could give far more precise results than the
string galvanometer, though on those older instruments, the resultant images did not persist long enough on the screen to be clearly photographed (figure 17.4b).[17]

Though Adrian was working with a capillary electrometer, his detailed description of the action potential was submitted slightly before Erlanger and Gasser's oscilloscope images.[18] His apparatus, like so many of the others we have been considering, also included a tuning fork (at 200 Hertz) to calibrate the velocity of the nerve impulses measured.[19] In 1912, he and Lucas had begun to measure the "refractory period" of a nerve, the time during which it does not respond to further stimulation, and its subsequent "supernormal period" of heightened response.[20] Now, in 1921, he offered further details and related these periods to Bernstein's membrane theory of nerve action in terms of the ionic potentials.[21]

Shortly after, Adrian met Gasser, who generously gave him the circuit design for a three-stage amplifier; just before his death, Lucas had discussed with Adrian the possibilities of amplification. After the war, Adrian built a one-stage device. As he wrote to Forbes in August 1920, "I have got the capillary electrometer going again during the last two months & it seems to me that an amplified electrometer would be about the last word in recording nerve impulses."[22] Forbes sent Adrian a gift of vacuum tubes and came to Cambridge to collaborate and set up a string galvanometer there. In this joint work, Adrian continued to extend the range of the "all-or-nothing" principle (as he now called it) to include all kinds of sensory nerve fibers in both frogs and mammals.[23] The cathode ray tube of the oscilloscope continued to be too faint to be photographically recorded.[24]

By about 1924, Adrian was unhappy about the progress of his research.[25] Many years later, he wrote that he had begun working with the string galvanometer

in the hope of being able to find out exactly what was coming down the nerve fibers when the muscle contracted. We knew then that nerves sent down nerve impulses as signals, but we didn't know anything about the way in which the impulses would follow one another. We didn't know whether they came at a high frequency, or at a steady frequency. We didn't know whether the frequency varied or not. In fact, we didn't know at all how the nervous signals were controlled. Alexander Forbes had been working with me in Cambridge and I had learned a great deal from him about string galvanometers and about mammalian preparations, but the experiments I had started became more and more unprofitable. You know the sort of thing that happens—they become more and more complicated and the evidence more indirect, and after a time it was quite clear that I was getting nowhere at all.[26]

In 1925, he finally had a working three-stage "valve amplifier," using the contemporary term for what later were called vacuum tubes. Adrian noted that though the string galvanometer was indeed highly sensitive to small currents, it had a limitation "which no amount of amplification can overcome, and that is the limitation imposed by the inertia of the moving system. Owing to the mass of the string, the record of its movement does not give a true picture of the changes of electromotive force applied to it and the distortion, though of little account in the record of a muscle action current, is quite enough to obscure the true form of the much briefer response of a nerve fiber."[27] After all, the heart of the string galvanometer is a vibrating string a few centimeters long, with all its attendant mechanical limitations yet tasked to record the frequencies of nerves firing on the

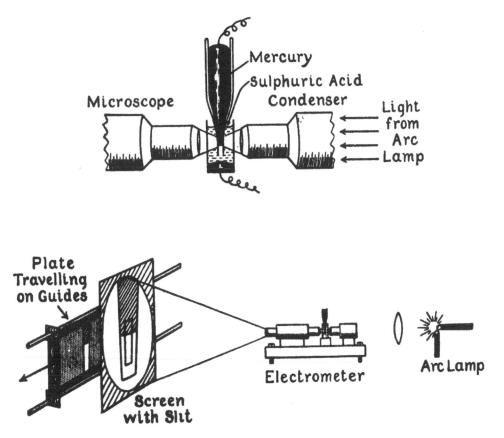

Figure 17.5
Edgar Adrian's capillary galvanometer, from *The Basis of Sensation* (1928).

timescale of milliseconds, enormously shorter than the string's natural frequency. Then too, fluctuations in the amplified current all too often blew out the gold-covered quartz string, causing trouble and expense. Adrian chose to keep working with the capillary galvanometer (figure 17.5), whose "inertia factor is extremely small," applying corrections to its readings by an ingenious method Lucas had worked out.[28]

Using his new three-stage amplifier with this galvanometer in 1924 and 1925, Adrian conducted a routine calibration of his equipment, set up to check the baseline reading from a frog's leg muscle.[29] He later recalled that "the amplifier had to be treated with great respect, as in those days the valves [tubes] were terribly microphonic," meaning vulnerable to unwanted vibration.[30] He was "distressed, but not very greatly surprised, to find that the base line wasn't a bit steady. It was oscillating rapidly all the time, . . . a fine, rapid affair" he ascribed to "picking up an artifact from somewhere," which he would need to eliminate. When he tried readjusting the apparatus, "this little oscillation was only there when the muscle was hanging down quite freely, from the knee joint of the frog's nerve-muscle preparation. If the muscle was supported on a glass plate there was no oscillation at all and the base line was quite steady." Suddenly it dawned on him that "a muscle hanging under its own weight, ought, if you come to think of it, to be sending sensory impulses up the nerves coming from the muscle spindles. . . . When you relax the stretched muscle, when you support it, those impulses ought to cease." Within a week, he was nearly certain "that many of them came from single nerve fibres and that by some extension of the technique it ought to be possible to find out exactly what was happening in single nerve fibres when the sense organs attached to them were stimulated."[31] If so, Adrian had reached the long-sought goal Lucas had envisioned.

Nevertheless, Adrian realized he needed to clarify his methods and findings so that they could be reliably repeated and their import assessed. In early 1926, he published images of his "leg-pull" experiment (figure 17.6a) showing the dependence of the frequency output of the nerve on the weight the muscle was bearing.[32] This was particularly important because (in contrast to the many earlier experiments using artificial stimulation) here the impulses were being generated by the nerve itself rather than through external electrical stimulation. Noting that the potential changes he measured were a thousand times smaller than from the whole sciatic nerve (comprising 3,000 to 4,000 individual fibers), he deduced "that only a very few fibres are concerned in producing the isolated responses—almost certainly less than ten and probably only one."[33] He sealed this argument in a subsequent 1926 paper with Yngve Zotterman, a visiting Swedish researcher. To do so, they used the sterno-cutaneous muscle of a frog, which Lucas had noted had only twelve to twenty-five fibers and hence could more easily be pared down to a single nerve fiber. Using this cunningly chosen specimen, they argued that "these responses are the product of a single end-organ," as shown in figure 17.6b.[34]

The significance of these findings (along with others completed that year) was summarized by Alan Hodgkin, a student of Adrian who later became a Nobel laureate himself: "In a single nerve fibre the electrical impulse is invariant and does not change with the nature or strength of the stimulus; the intensity of a sensation is controlled by the frequency of the discharge and the number of fibres in action, whereas the quality depends on the central connections of the type of nerve fibre that is being stimulated."[35] Adrian himself represented these ideas in a diagram (figure 17.7) he made for lectures he gave in 1928 as *The Basis of Sensation*:

The stimulus is represented as appearing suddenly and remaining at a constant value. The excitatory process in the receptor declines gradually, and as it declines the intervals between the impulses in the sensory fibre become longer and longer. The impulses are integrated by some central process, and the rise and decline of the sensation is a fairly close copy of the rise and decline of the excitatory process in the receptor. The quality of the sensation seems to depend on the path which the impulses must travel, for apart from this there is little to distinguish the message from different receptors.[36]

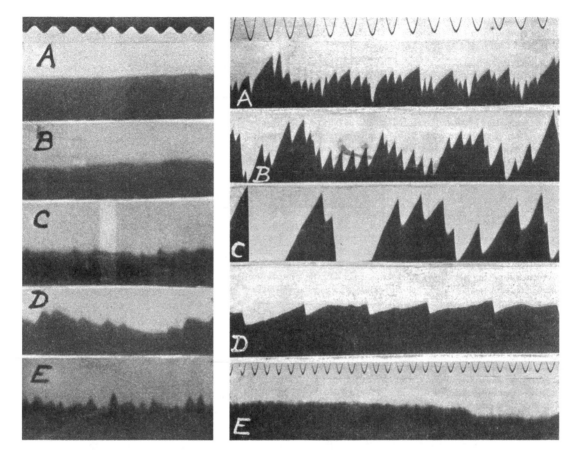

Figure 17.6
(a) Adrian's 1926 images of the response of the sensory nerve coming from a frog's leg muscle bearing various weights ranging from 10 g (C) to 100 g (E). The tuning fork signal at the top shows 200 Hertz. The greater the weight the muscle bears, the higher the frequency at which its nerve fires. (b) Adrian and Zotterman's 1926 images of the sterno-cutaneous muscle of a frog bearing 2 g or 1 g weight, with an increasing number of strips removed (B–E). The firing pattern becomes most regular in D. In E, the last remaining fiber was evidently destroyed because the pattern disappeared.

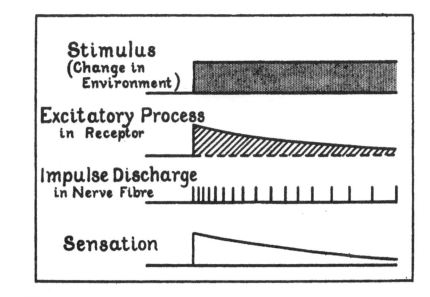

Figure 17.7
Adrian's depiction of the relation between stimulus and sensation. From *The Basis of Sensation* (1928).

In essence, a nerve fiber transduces a stimulus into pulses whose frequency gradually decreases as the sensation fades from the immediacy with which it had begun.

Because those pulses are always equal in amplitude and duration, nerve signals encode their response via their varying frequency in time, their *rhythm*. That musical term took an increasingly large place in the papers Adrian wrote with Rachel (née Eckhard) Matthews between 1926 and 1928, the first investigation ever of the impulses in the optic nerve of vertebrates. This required substantial changes in technique because recording optic nerve impulses was more difficult than recording impulses from peripheral sense organs: the optic nerve is short (about 15 mm) and delicate yet comprises many fibers. As they noted, "The greater the number of fibres in a nerve and the smaller the distance between the leading off electrodes, the more difficult it will be to detect the electric response of a single fibre, yet if a record of sensory action currents is to give much information the activity must be confined to a few nerve fibres to allow the individual impulses to be distinguished."[37] To overcome this, they built a four-stage amplifier as well as new optical equipment for stimulating the eye, recording the capillary galvanometer readings on 4,000 feet of photographic film (rather than plates).[38]

Though they did not comment on it in their papers, Hodgkin noted that they "listened for impulses with telephone earphones," at the suggestion of Bryan Matthews, a research student of Adrian who became Rachel's husband.[39] They were doubtless aware of the use of telephones by the German physiologists, as discussed earlier. Yet whereas the Germans used the telephone simply to hear the raw signals of nerve impulses, Adrian and Rachel

Matthews also listened through the earphones to help them find the signal of a single neuron in order to avoid the trial and error of trying to cut through a thin and delicate bundle of fibers. This required a new and different kind of listening we shall gradually explore.

Working with eels, Adrian and Matthews established that the optic nerve, like other sensory nerves, works on the all-or-nothing principle; the optic impulses did not differ in time relations or groupings from other sensory nerves. As with those other nerves, the size of the action current did not depend on the strength of the stimulus, but the frequency of impulses did: when the eye is illuminated, the discharge of impulses rises to a maximum frequency, then declines rapidly to zero, the phenomenon of adaptation also characteristic of other sensory nerves (see figure 17.7). Here the change from darkness to light seemed to be the decisive factor; likewise (though more surprising), when the light was switched off, there was a renewed outburst of impulses, the "off discharge." These were fundamental findings that eventually led to what Hodgkin considered "almost an industry" of research that was already producing a thousand papers annually on this subject in 1979.[40]

After another 1927 paper on flashes of light and the reaction time of the optic nerve, in 1928 Adrian and Rachel Matthews turned to the interaction between the neurons in the retina itself and those in the optic nerve.[41] In this paper, the term *rhythm* and its derivatives appeared 128 times to describe the types and changes in frequency of nerve impulses. It is not clear at what point in their work they began listening to the nerve impulses through the telephone earpiece, whether they had done so since the beginning (late 1926) or had only just begun using this technique in 1928, as evidence we will consider below seems to confirm.[42] Yet this notable emphasis on rhythm suggests that they were actually hearing those rhythms through their telephone. To be sure, a paper by Sybil Cooper and Adrian in 1924 used *rhythm* nine times to describe patterns they presumably were not hearing but seeing in electrometer recordings.[43] Still, Adrian and Matthews's much more extensive usage of *rhythm* is consistent with what they would have heard: the varying frequencies of the impulses they were observing were not heard as varying *pitches* but as varying *rhythms* just because the ear perceives the low frequencies involved (5–15 Hertz, for an eel eye) in that way.[44]

Whatever may have been the exact chronology of their use of telephone earpieces, Adrian and Matthews used rhythm to describe many aspects of the functioning of the eye because "most preparations of the eel's eye will sooner or later give the rhythmic type of discharge."[45] First, "the waves in the optic nerve discharge of the eel are caused by a rhythmical waxing and waning in the number of impulses in the nerve fibres. It follows that the ganglion cells of the retina must all be working in unison with alternating periods of rest and activity."[46] Thus, the rhythms of impulses manifest rhythmical coordination of retinal cells. Such coordination could be caused by flickering lights (which present a rhythmical stimulus) or by nerve connections between retinal ganglion cells. Because strychnine promotes synchronous discharge of neurons in the spinal chord, they hypothesized that it should also favor rhythmic response in the retina, a prediction they confirmed. Therefore, they concluded that "the rhythmic discharges from the retina are due to nervous interconnection between the [retinal] ganglion cells and not to intermittent stimulation of the rods and cones."[47] Thus, their observations about rhythm gave crucial insight into the functioning of the eye.

This connection may seem less surprising when one realizes that Adrian (educated at an elite school) knew ancient Greek so well that "when someone at Trinity High Table quoted a line from a Greek play . . . Adrian was able to continue for a dozen lines or more."[48] Thus, he probably knew that in Greek, *rhythmos* meant not only musical rhythm but also patterns in dance, sculpture, visual art, and architecture, for which "*rhythmos* gave rational order to motion," as mentioned in chapter 1.[49] Still, one can feel the wonder and surprise in the way he and Rachel Matthews captioned the visible records of the rhythms that they probably first experienced by hearing, especially that turning a light *off* (A in figure 17.8) causes a rhythmic discharge no less than turning it on. Their assemblage and presentation of these images highlight the striking patterns of rhythmic regularities, as well as the way the rhythms slow down as the eye adapts to the stimulus. We have here the interesting situation of researchers using a new technique (at least for them) silently, as it were, not yet commenting on or even mentioning it. Yet even so, their descriptions and probably their actual procedures seem to have changed significantly, compared to before they started using the telephone to listen to the optic nerve.

This "silent" phase found its voice in Adrian's next papers. Though (in Hodgkin's opinion) Adrian "could have worked on the retina for the rest of his life," he said that the eye was so "beastly complex" that he wanted a change.[50] After this long series of papers on sensory (afferent) neurons, which carry impulses inward to the brain, with the visiting American researcher Detlev Bronk he now turned to motor (efferent) neurons, along which the brain sends impulses outward to muscles and organs. Their conclusions affirmed the all-or-nothing doctrine for these neurons as well; indeed, they found that "the discharge of a motor nerve cell can scarcely be distinguished from the discharge of a sense organ. In both, the frequency of the discharge controls the intensity of the effect that the message produces and is itself controlled by the intensity of excitation. The sense organ may give a series of impulses with a nearer approach to a completely regular rhythm, but in both, the frequency varies over much the same range under average conditions of excitation."[51]

Near the beginning of their first paper (1928), after describing the particular problems of trying to record from a single motor neuron, Adrian and Bronk made explicit their new sonic technique:

The use of the telephone for investigating nerve or muscle action currents is well known, for the telephone has much the same electrical sensitivity as the string galvanometer. We have been accustomed for some time to use a telephone connection in the output of the amplifier system before making photographic records. In the present work, however, in addition to the telephone we have found it a great convenience to have a continual sound record of the nerve discharge made by a loud speaker.[52]

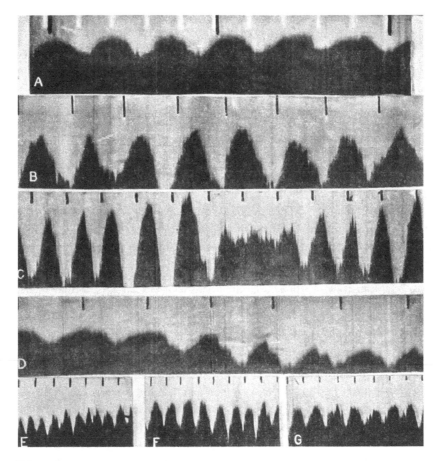

Figure 17.8
Edgar Adrian and Rachel Matthews's 1928 figure illustrating "rhythmic action current discharge in eel's optic nerve": (A) "off discharge" when the eye was in darkness for ½ sec after long exposure; (B) "on discharge" 45 sec after beginning of exposure; (C) 40 sec after beginning of exposure: "regular beat in the rhythm"; (D) "increase of frequency when light is increased. . . . half-way through the record the light is moved nearer the screen and the frequency rises"; (E–G) "decline in frequency as exposure continues."

To be sure, the use of a telephone may earlier have seemed so "well known" to Adrian that he never mentioned it in his descriptions of methods in earlier papers.

Still, the addition of a loudspeaker seemed significant enough here that they highlighted it as a new part of their method and connected it to the particular problems they faced in isolating a single neuron, for which they had proposed a new technique, splitting the connective sheath in order to expose a small stalk of intact fibers in the middle (figure 17.9a): "The object of the dissection is to leave a few fibres in the middle of the nerve intact but separated as far as possible from one another. If this is accomplished, the remaining fibres are cautiously divided until only the smallest strand is left joining the two ends of the nerve."[53] At that point, though, they did not usually try to carry the dissection to leave only a single undivided fiber because it could well be afferent (sensory) rather than motor. Then too, "The electrical records usually provide sufficient evidence of the number of motor fibres in action, and the actual number of undivided fibres can be checked by microscopical examination after the experiment is over." They connected this new technique of dissection to their new sonic technique (figure 17.9b):

When only a few fibres are in action the electrometer excursions may be too small to detect on a screen, but they produce a series of faint clicks in the loud speaker, and it is thus possible to control the dissection, to expose a plate at the moment when the discharge is at its height, etc., without the inconvenience of wearing telephones.[54]

Thus, Adrian and Bronk could "control" their dissection using either telephone earpieces or the even more convenient loudspeaker. Their language supports our earlier inference that Adrian had in fact already been controlling earlier dissections (such as in his work on the optic nerve) by "wearing telephones," inconvenient as they were. The difference may not only have been shedding cumbersome headgear but the more enveloping and intense aural experience of hearing the neurons throughout the room, helping facilitate the arduous process of isolating a single neuron and also allowing easier collaboration between the researchers running the various pieces of equipment. Essentially they were making a kind of movie while at the same time doing microsurgery; hearing the "soundtrack" was evidently helpful to coordinate these activities.

By the time of Adrian and Bronk's 1928 work, loudspeakers had become so familiar—and even fashionable—that it was natural to think of using them in this new context.[55] The earliest type of loudspeaker had appeared in 1921; by 1927, the Orthophonic Victrola advertised its new sound "as radically different as the modern motor-car in comparison to the horseless carriage," its records having "a *character of tone* that is pleasing beyond description."[56] In 1928, Leon Theremin and his students performed before 12,000 people using "massive" loudspeakers; as Emily Thompson notes, stadium and movie audiences "were accustomed to hearing 'radio sound' emitted from loudspeakers."[57] The BBC had been founded in 1922, and radio news had increasingly become the norm, along with radio music and other programs. In fact, during the newspaper strike of 1926, the radio was the only source of news in the United Kingdom.[58] Thus, Adrian and Bronk's adoption

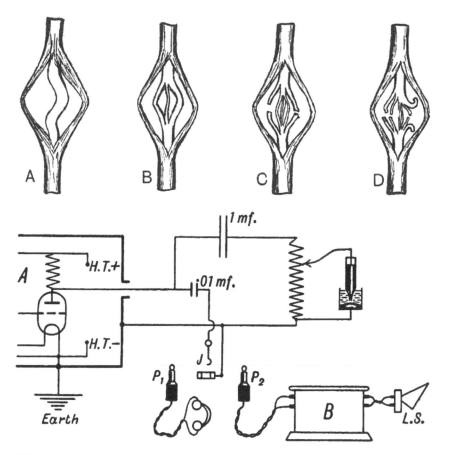

Figure 17.9
(a) Adrian and Bronk's 1928 "method of dissecting nerve so as to divide all but two or three fibres." (b) "Connections for telephone or loud speaker used simultaneously with capillary electrometer." The amplifier *A* leads to a jack *J* whose plugs P_1 and P_2 can connect to telephone earpieces or a loudspeaker *L.S.*

of loudspeakers took place during the flowering of radio, so that their amplified experiments were a kind of live radio program accompanying and guiding their dissections and filming.

As Sterne has argued, "Mediate auscultation was the first site where modern techniques of listening were developed and used," leading the way for the development of other listening practices used in telegraphy and various emergent media.[59] To these we can now also add the new listening practices that "unit recording" evoked, to use the usual term now for electrophysiological work monitoring single neurons. Interestingly, there was originally no name for the use of loudspeakers or headsets in this process, though Hodgkin noted in 1979 that they "have been widely used ever since," along with the special technique of dissecting nerves just discussed.[60] We return in the next chapter to the particular listening processes involved in unit recording.

Later in 1928, Adrian's student Bryan Matthews took up the loudspeaker as an observational instrument and carried it further still, as seems appropriate for the person who suggested that Adrian use the telephone in the first place. Because of its innate lack of inertia and quick response to rapidly varying currents, Matthews believed that a loudspeaker was capable of replacing the capillary galvanometer, which was sorely taxed to respond to rapidly changing impulses whose onset and decay may overlap confusingly. To that end, he designed a special loudspeaker whose frequency response was as flat as possible (responding equally to all frequencies), then attached a mirror to the central vibrating iron structure that drives the paper cone and projects the sound outward (figure 17.10a and 17.10b). The mirror could then reflect a light beam that would track the movements of the loudspeaker; that beam could be recorded on film to produce a permanent record. As he noted, "The system is in fact the same as that used for moving the diaphragm of a telephone" except now the "voice" could be seen as well as heard.[61] This Matthews oscillograph became the instrument of choice for physiological measurements of nerve impulse currents, producing beautifully clear images (such as figure 17.10c) yet never breaking down. It was only superseded by improved versions of the cathode-ray oscilloscope after World War II.[62]

Adrian's work with Bronk continued in a second paper (1929) that began with studies of motor neurons involved in the reflex movements of a cat's hind limbs.[63] They noted that "by listening to the nerve discharge throughout the experiment we have been able to save a great deal of time and photographic material and to detect various points which would almost certainly have been missed had we relied entirely on photographic records."[64] They also used the loudspeaker in a way that is strikingly similar to the Matthews oscillograph: to the iron core of a loudspeaker, they attached a pointer that could be seen in the microscope that was also observing the capillary galvanometer reading the values of the nerve currents. This new system "was adequate to signal the beginning and end of the discharge in the entire nerve. The loud speaker movement forms a convenient signal for electrical stimulation."[65] Thus, they adapted Matthews's innovation as a kind of adjunct or viewfinder to help in their measurements.

Adrian and Bronk also studied the motor nerves activating the human triceps muscle. For this, the subject was Adrian himself, disclosed only by his initials attached to the relevant figures. Instead of the complex and difficult dissection of the nerve to get down to a few fibers, they used a thin needle (figure 17.11b), a technique that remains standard to this day.[66] Still, it is daunting to contemplate the combination of physiological skill and physical endurance involved in sticking a needle into your own triceps, then seeking its fine electrical response as you contract it. This scene takes its place in a long series of electrical self-experimentation that reaches back at least as far as Ritter and Humboldt applying electricity to their own bodies. But those early investigators did not achieve

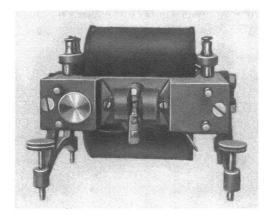

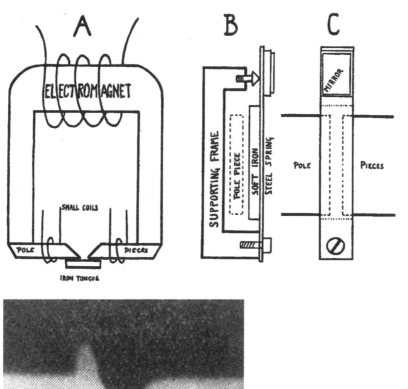

Figure 17.10
The Matthews oscillograph (1928). (a) General view. (b) Detail of the iron core and mirror. (c) Image it produced of the action currents from a frog sciatic nerve (time signal at bottom shows 200 Hertz).

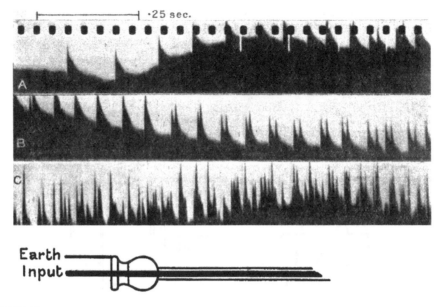

Figure 17.11
Figures from Adrian and Bronk (1929). (a) "Action currents in human triceps (E. D. A.) recorded with concentric needle electrodes during gradually increasing voluntary contraction. *A*, beginning of contraction; *B*, follows on *A*; *C*, powerful contraction." (b) Concentric needle electrode.

anything like the exquisite accuracy of Adrian's experiment or the insights it disclosed. In his data, one can see the nerve beginning from rest and firing ever more rapidly as he willed his muscle into increasing contraction (figure 17.11a). As Adrian pointed out, this exhibits voluntary action, hence the electrical impulses are the visible—and audible—traces of the human will itself.

As Hodgkin noted, this became one of Adrian's favorite demonstrations for his students, teaching them both physiology and sangfroid.[67] When he and Bronk described it in their 1929 paper, they made clear that the demonstration centered around amplifying motor impulses through the loudspeaker, noting that this "method was demonstrated to the Physiological Society on October 13th, 1928."[68] To be sure, at the time there was no way to project the visual display of the galvanometer so that it would be visible to an audience. Yet the loudspeakers were not just a compensation for the lack of a proper visual display, a mere dramatization of the effect after the fact, as if the nerve impulses had been found purely visually. On the contrary, *it was through sound that he found the electrical signal*, as he and Bronk noted: "The individual rhythms can often be followed much more easily in the sound records made by leading the amplified action currents into the loud speaker, for the ear can pick out each new series by slight differences in intensity and quality which are hard to detect in the complex electrometer record."[69] We will later consider what this special listening technique entailed as it passed into common use.

Almost immediately, this series of discoveries by Adrian became central parts of neurophysiology, recognized in the citation for his 1932 Nobel Prize in Physiology or Medicine: "Edgar Adrian developed methods for measuring electrical signals in the nervous system, and in 1928 he found that these always have a certain size. More intensive stimuli do not result in stronger signals, but rather signals that are sent more often and through more nerve fibers." His Nobel lecture, "The Activity of the Nerve Fibres," did not include his favorite demonstration of the electrical signals in his triceps, which probably would have been hard to bring off on such an occasion, even for Adrian. Yet he evidently felt the need of a sonic demonstration, which he provided using another popular sonic technology of the time, not mentioned so far: "Since rapid potential changes can be made audible as sound waves, a gramophone record will illustrate this, and you will be able to hear the two kinds of gradation, the changes in frequency in each unit [single nerve fiber] and in the number of units in action."[70] This was the first occasion on which gramophone records were played to illustrate a Nobel lecture, along with "lantern slides." Though the slides were probably versions of figures in his papers, one would be more curious to hear the records; their disappearance from the Nobel archives shows how poorly preserved sonic artifacts tend to be compared to texts or images.[71] Their use on this festive occasion, as well as during his 1928 presentation to the Physiological Society, indicate the significance of such sound examples for his discoveries, their justification, and their public demonstration.

18

Sonic and Rhythmic Knowledge

Listening to single neurons enabled a profoundly new kind of physiological understanding, whose character we now consider in detail. Adrian and his collaborators showed that each neuron conveyed its signal purely through the frequency of its firing. But "frequency" is simply a general term for rhythm.[1] The all-or-nothing principle showed that neural rhythm is the universal language of every nervous system, its binary "machine code" (to borrow a term from computer science). This neural code underlies our thoughts, feelings, perceptions. Among our senses, those that are fundamentally rhythmic—haptic (tactile) awareness and hearing—perceive the neural code as rhythm—its native idiom, as it were. All the other senses represent those binary impulses in other guises, such as visual or olfactory perceptions.

As noted earlier, humans generally perceive vibrational frequencies less than about 20 Hertz as rhythms rather than distinctly audible pitches. Changes in an experimental animal's neural rhythms, when amplified and made audible, can directly affect the experimenters' own neural rhythms and thereby their mental state. These changes then manifest themselves in the experimenters' emerging "awareness" that something happened in the experimental animal. As we will see, both experimenter and subject are joined together by these shared neural rhythms, whose changes mark every aspect of their interaction.

This *rhythmic knowledge* gives access to the phenomena of life *as they unfold in time* more directly than by visual representations. Though the historical path to rhythmic knowledge was opened by *sonic knowledge*, these two forms of knowing should be distinguished as binary is from analog, discrete from continuous. Thus, the sonic turn we have been following since the eighteenth century led to the significantly different rhythmic turn of the twentieth. Subsequently these binary rhythms became digitized data manipulated by computers, themselves designed to mimic the neural networks of animals. This digitization reflected the inherently binary character of neural responses.

My choice of terminology shows the difficulty of finding words to contrast auditory and rhythmic experience with visual or purely numerical thinking. The anthropologist and ethnomusicologist Steven Feld coined *acoustemology* to "suggest a union of acoustics and epistomology, and to investigate the primacy of sound as a modality of knowing and being in the world."[2] Similarly, the term *auditory epistemology* has gained favor with some scholars.[3] Though these terms are helpful and apt, I feel the need to discriminate between sonic and rhythmic modes of knowledge because these words underscore the historical shift between analog and binary phenomena. This has a special importance because the discovery of the all-or-nothing code was not just one "medium" among others but proved to be the common foundation on which all nervous systems operate. Rhythmic knowledge reflected—and revealed—the fundamental rhythms embedded in all neural activity on Earth.

To assess these modes of knowledge, we will consider two episodes from Adrian's later work: his sonification of brain waves and a curious experience he had when experimenting with the retina of a toad. Each turned on neural rhythm, especially its changes at critical moments. Adrian's technique of audio monitoring became common in neurophysiology labs throughout the world. In particular, it played a significant role in the Nobel-prize-winning work of David Hubel and Torsten Wiesel on the visual system. Through these examples, we can compare the fundamental character of rhythmic knowledge to the kinds of objectivity (and subjectivity) Lorraine Daston and Peter Galison described in their analysis of the ways in which vision can mediate knowledge.

The preceding three chapters presented the unfolding evidence that nerves communicate via all-or-nothing impulses encoded purely in the frequency of their firing. Trying to understand this fundamental physiological fact, Helmholtz and du Bois-Reymond had already begun comparing the nerves to a telegraphic system.[4] Though at times he too used this telegraphic metaphor, in 1930 Adrian contrasted neural signals with ordinary codes and telephonic communications:

A nerve fibre cannot conduct all manner of changes like a telephone wire; it cannot even conduct the dots and dashes of a Morse code. As a rule, with a steady stimulus, the impulses in a sensory discharge are evenly spaced and the only variable is their frequency which gives a measure of the intensity of excitation in the sense organ. This is all the information which a single fibre can convey, but as there are a million or more sensory fibres entering the spinal cord in man, we need not be surprised at the range of our sensations.[5]

Adrian considered frequency to be the true means by which neurons transmit what he called "information," a term he used in a new sense, compared to earlier discourse about the nerves conveying "impulses," "reactions," or "activity." As Justin Garson pointed out, "By using the language of information, nerve physiologists like Adrian posed questions that could not be asked previously, namely, questions about the abstract relationship between impulse and stimulus."[6] Fundamentally, this new approach turned on the concept of frequency, which describes both sound and electric signals, inviting comparisons between them.

To illustrate the sonic investigation of these frequencies, consider Adrian's pioneering work on listening to brain waves. This research he and Bryan Matthews conducted shortly after Hans Berger published his discovery (1929–1933) of the basic brain rhythms:

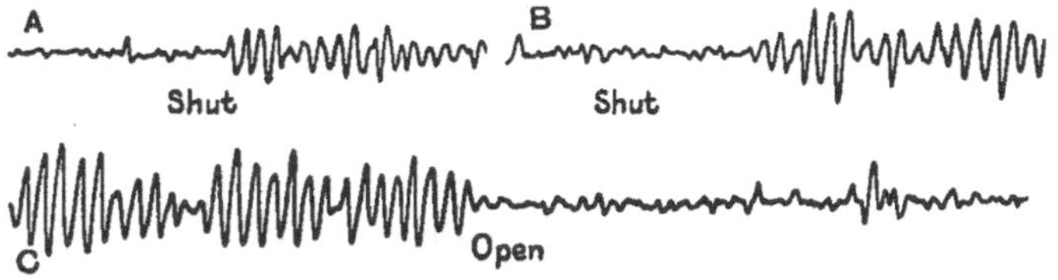

Figure 18.1
"The development of the [alpha] rhythm in the absence of visual activity. (a) E. D. A. The rhythm appears when the eyes are closed; (b) B. H. C. M. Ditto; (c) E. D. A. The rhythm disappears when the eyes are opened." From Adrian and Matthews, "The Berger Rhythm: Potential Changes from the Occipital Lobes in Man" (1934).

alpha (7–14 Hertz) and beta (15–30 Hertz).[7] Adrian and Matthews confirmed Berger's work and brought it to wide attention; though they differed with his interpretation on some points, Adrian nominated Berger for a Nobel prize.[8] In their initial 1934 work on animals, they used no fewer than three Matthews oscillographs to record brain waves, noting that "a loudspeaker was not often used, as the cortical potentials often rise and subside too slowly to give clearly audible sounds."[9] Evidently these very low frequencies were not rendered usefully by ordinary loudspeakers. In their subsequent work on the human brain, they again used the Matthews oscillograph but added that "it is sometimes an advantage to be able to hear as well as see the rhythm; although its frequency is only 10 a second it can be made audible by using a horn loud speaker with the diaphragm set very close to the pole pieces and a condenser in parallel to cut out high frequencies."[10] In this way, they rigged up a sort of subwoofer and filter that could do better with the low frequencies they were hearing from their own brains (figure 18.1). They also noted the differences between them: "One of us (E. D. A.) gives the [alpha] rhythm as soon as the eyes are closed, and maintains it with rare and brief intermissions as long as they remain closed. The other (B. H. C. M.) is better in the role of observer than of subject, for in him the rhythm may not appear at all at the beginning of an examination, and seldom persists for long without intermission."[11] Such striking rhythmic changes could readily be heard even when the experimenter's eyes were closed.

In 1941, Adrian described further refinements of his listening to the electrical impulses coming to the cerebral cortex from peripheral sense organs:

An essential feature of the method has been the use of a loud-speaker and amplifier system giving a faithful reproduction over a wide range of frequencies. The electrical changes in the cortex include both the very brief axon potentials due to impulses in nerve fibres and the much slower waves which are the characteristic product of the cortex. An optical system making a photographic record can be adapted to show one or the other but can rarely do justice to both simultaneously, whereas with a good loud-speaker it is easy to detect both impulses and waves.[12]

In this paper, Adrian used the term *noise* both to describe what he heard from the loudspeaker as well as to characterize the neural impulses themselves, one of the earliest uses of "noise" in this context.[13] He also characterized the impulses as "signals," well before others used that term in this context.[14] Thus, his listening informed his categorization of the nerve signals in relation to noise. To do so, he attended to "faithfulness" or "fidelity," which Sterne has called "sound's own 'dismal science'—it was ultimately about deciding the values of competing and contending sounds."[15] By devising more "faithful" loudspeakers, Adrian could directly hear the interplay of global brain waves and local impulses in real time, as he or Matthews produced them. This important discrimination depended on hearing's ability to counterpoint the local and global without blurring them together.

Adrian continued to use loudspeakers as an essential tool of his work, registering the signal and helping him find a single neuron. For instance, in his very next paper, after introducing this technique (1929), Adrian mentioned using "the additional amplifier and loud speaker arrangement" along with a Matthews oscillograph, an instrument he apparently viewed "with some considerable awe."[16] In a 1938 paper, he noted that "a loudspeaker was particularly valuable and discharges could often be detected by ear although the oscillograph record was too confused to show them."[17] By 1942, Adrian's paper on the olfactory response of hedgehogs described sonic monitoring "in the usual way," as if the technique had become so routine as to be unremarkable.[18]

Still, in 1947 he recounted an incident that happened "not long ago" that indicated how listening to neurons calls for further thought. He began by noting that though neural response is "best studied" through the photographic record, "the timing of the impulses can be made more intelligible by reproducing them as a series of sounds," for each impulse is "brief enough to be turned into a sharp click in a telephone or loudspeaker or to be recorded permanently on a gramophone disk." Through hearing the impulses, "the changes in frequency are reproduced directly and there is no need to infer them from a still picture."[19]

Then Adrian contemplated the curious "vicissitudes through which the sensory message must pass when we listen to such a recording," beginning with the experimental animal's muscle twitching, whose neural impulses he then recorded on a gramophone disk:

When a copy of the disk is played, the sounds fall on the ear-drum of the listener and set up a succession of impulses in the auditory nerve corresponding to those set up originally by the stretched muscle-spindle. Thus the message ultimately reaches a central nervous system, though not the central nervous system which was its original destination, and it has been curiously changed *en route*.[20]

Through hearing, the animal's nervous system had been "curiously" connected to Adrian's with a kind of directness not possible for a visual reading of the galvanometer. This directness meant that Adrian was hearing—and thus feeling—what the animal saw:

I had arranged electrodes on the optic nerve of a toad in connection with some experiments on the retina. The room was nearly dark and I was puzzled to hear repeated noises in the loudspeaker attached to the amplifier, noises indicating that a great deal of impulse activity was going on. It was not until I compared the noises with my own movements around the room that I realized I was in the field of vision of the toad's eye and that it was signaling what I was doing, for the slight shadow movements which I was making on the toad's retina were (and ought to have been) quite enough to send messages to the brain, if it had been there to receive them.[21]

Having previously removed the toad's brain, Adrian realized that in effect, he had put his own brain in its place. In that dark room, Adrian heard the toad seeing *him*. The hail of nerve impulses in the toad's optic nerve responded to Adrian's movement, thereby evoking an exactly similar hail of impulses in his own brain, registered as his sudden realization. This moment of rhythmic crisis opened a passage from one brain to the other. Compared to vision, hearing generally connects more primally to the brain; as Nietzsche put it, the ear is "the organ of fear," always open to alert an animal to imminent danger.[22] For scientists, Daston and Galison assert that "all epistemology begins in fear"—the fear of succumbing to delusions, bias, or the deceptive allure of pet theories.[23] Because of its exquisite sensitivity, hearing can offer critical help at the moment of discovery.

In particular, perception of rhythmic changes marked the work of David Hubel and Torsten Wiesel on how the brain processes visual perception. During the late 1950s, they investigated how visual signals are encoded by the visual system.[24] They projected various shapes onto a screen that would be viewed by an animal and monitored the response of a single neuron in the animal's brain. As Bevil Conway, one of their collaborators, explains, "One listens to the neurons with limited filtering of the signal, which allows one to hear all the neurons in the population. One can then isolate a cell, by ear, and use that auditory isolation to help guide the electrical isolation of the trace."[25]

This depends on an extraordinary fact that Conway pointed out: because "neurons have distinctive waveforms, they each have a distinctive sound quality." Thus, *each neuron is a distinct individual identifiable by its "voice," its individual rhythm*. Therefore, listening to neurons "is akin to the way the auditory system can solve the 'cocktail party' problem: among dozens of speaking voices, one can pick out by ear one voice and track almost everything that voice says." As a result, listening "makes it actually much easier to identify and distinguish a neuron from the population noise than by oscilloscope." We return to this special kind of listening in the final chapter.

Conway notes that "the use of auditory information was for Hubel and Wiesel a prominent, and distinguishing feature, evident in how they disseminated their work," including Hubel's Nobel lecture (1981) and various videos they made documenting their experiments. We focus on the crux of their work: their discovery of the receptive fields of single neurons in the primary visual cortex (1959). Hubel and Wiesel set up their experiment so that an anesthetized cat or monkey would continuously view a visual image, composed of simple shapes and lines, that they could manipulate. In order to record the response of a single neuron in the visual cortex, Conway notes that "they used their own eyeballs to guide the placement of the electrodes. And then they recorded every cell that they encountered as they pushed the electrode ever deeper into the brain," noting their individual waveforms (rhythms). Their main problem was how to present the animal's eye

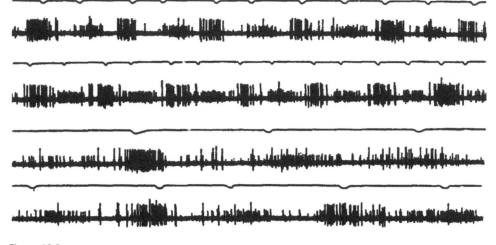

Figure 18.2
David Hubel's 1958 experiments on an unrestrained cat with a microelectrode in the striate cortex. There are two pairs of records, shown as four panels separated by horizontal lines: top to bottom: panels 1 and 2 show the first pair of records (faster hand movements), 3 and 4 the second pair (slower). In each pair, the upper line shows the stimulus: a hand movement interrupted a light beam falling on a photoelectric cell. The lower line shows the cat's neural response on an oscilloscope. Each line represents 4 seconds.

with various images that could be altered or changed while at the same time monitoring the neuron's response.

By the time of their experiments, cathode-ray oscilloscopes had become standard for observing neural impulses; nevertheless, Conway notes that "they rarely photographed action potentials on the oscilloscope, and when they did, they needed to dodge and burn the background of the photographic paper to make the traces visible." Hubel and Wiesel's challenge was to find a nimble and responsive way to use the neural signal to inform how the visual stimuli should most informatively be deployed or altered. This flexibility was especially necessary because initially it was not at all clear what (if any) structures or patterns of response were "hard-wired" in the visual cortex, as opposed to being learned or random behavior; as Hubel put it in his Nobel lecture, "It is hard, now, to think back and realize just how free we were from any idea of what cortical cells might be doing in an animal's daily life."[26]

The problem was how to make the neural signal as easily comprehensible as possible. For comparison, consider earlier experiments by Hubel in 1958 on an unrestrained cat, stimulated by small to-and-fro hand motions in front of the animal.[27] Figure 18.2 shows the stimulus and response as purely visual images, one above the other, the stimulus recorded via a photocell, the response via oscilloscope. The caption clarifies the details of the images, but the overall impression shows the difficulty of understanding what at first looks like a forest of lines; discerning their relation visually would take considerable practice.

Though the oscilloscope indeed showed the tracings of the neural response, it was far easier for Hubel and Wiesel just to listen to the amplified nerve sounds over a loudspeaker—an audio monitor, as Hubel called it—leaving their eyes free to watch the visual display they were presenting to the cat, whose neural output they were also hearing. This simultaneous hearing and seeing allowed them to mediate readily between the visual display and the hearing of the neural signal.

Hubel later recalled that "we set up our first experiments and they did not go well because at the beginning we could not make the cells fire at all. We'd shine light all over the screen and nothing seemed to work."[28] During his live Nobel lecture, Hubel described the monitor in detail for the benefit of nonspecialists:

In order to make a record, we take a video camera and direct it at the screen so that when we play back the tape, we can see what the animal would have seen, if it had been awake. And in order to make a record of the responses, we take the output of the microelectrode, which is connected to amplifiers and an oscilloscope, and we connect that to the soundtrack of the tape so that when we play back we can hear the cells firing. When one records from a single cell, one hears a single impulse as a click and if the cell is firing very vigorously, the clicks tend to merge and one hears a noise.[29]

He went on to play an experiment recorded on film; not hearing the soundtrack play, he told the projectionist, "We need sound," thus underlining its importance for what he wanted to present. During the film, he gave no spoken commentary, trusting that the audience would immediately understand what they were seeing and hearing: a spot of light on the screen seen by the cat would be registered by a hail of clicks, the evidence of a neuron firing in response.

In the text of his lecture, Hubel recounted that

our first real discovery came about as a surprise. We had been doing experiments for about a month. We were still using the Talbot-Kuffler ophthalmoscope and were not getting very far: the cells simply would not respond to our spots and annuli. One day we made an especially stable recording. . . . The cell [neuron] in question lasted nine hours, and by the end we had a very different feeling about what the cortex might be doing. For three or four hours we got absolutely nowhere. Then gradually we began to elicit some vague and inconsistent responses by stimulating somewhere in the midperiphery of the retina. We were inserting the glass slide with its black spot into the slot of the ophthalmoscope when suddenly over the audio monitor the cell went off like a machine gun. After some fussing and fiddling we found out what was happening. The response had nothing to do with the black dot. As the glass slide was inserted its edge was casting onto the retina a faint but sharp shadow, a straight dark line on a light background. That was what the cell wanted, and it wanted it, moreover, in just one narrow range of orientations. This was unheard of.[30]

Clearly, the audio monitor going "off like a machine gun" alerted the researchers that something significant had happened, though it took longer to clarify exactly what: not the black dot, as they first thought, but rather the edge of the slide, which they had not initially considered to be part of the intended "real" stimulus (as heard in ♪ video example 18.1). The rhythmic knowledge evoked by moving the slide revealed that it was the edge that really turned that neuron on. In Conway's description, "Listening to a neuron over the audio monitor is thrilling, much more like listening to a great pianist performing a beloved piece of music: a combination of delight at understanding something through hearing, and pleasure."

In his lecture, Hubel went on to show how the different kinds of cortical neurons (which he and Wiesel named simple, complex, and hypercomplex) were differentiated by the relation between the sound from the audio monitor and the visual stimulus, including the exact location, orientation, and direction sensitivity of each kind of neuron. This finding "was unheard of. . . . That the retinas mapped onto the visual cortex in a systematic way was of course well known, but it was far from clear what this apparently unimaginative remapping was good for. It seemed inconceivable that the information would enter the cortex and leave it unmodified," yet their correlations between sound and stimulus helped them find just such surprisingly simple mappings.[31] Their work showed that vision discerns outlines, edges, and thereby shapes. Yet that visual realization emerged through the thrill of hearing the neuron's "machine gun" firing. As with Adrian's encounter with the toad, the ear's ability to register critical changes in *time* enabled the neuroscientists to discern the activity of life through changing rhythms.

The power of such practices of listening indicates that they provide a kind of knowledge fundamentally different from the various forms of objectivity sought through visual means. In 1666, Hooke had already observed that whereas the eye could discover "Conveniences and Inconveniences at a great Distance as well as near at hand," the ear could receive "Warning or Information from Sound, where the Eye could not assist."[32] Indeed, research by Ernst Mach in 1865 showed that the ear responded more quickly to stimuli than the eye.[33]

In scientific atlases of images, whether of snowflakes, galaxies, or neurons, Daston and Galison delineated various kinds of objectivity they called *truth-to-nature* (idealized images of types), *mechanical objectivity* (seeking to avoid any such idealization or interpretation, for instance through photography), and *trained judgment*, which is "a capacity of both maker and user of atlas images to synthesize, highlight, and grasp relationships in ways that were not reducible to mechanical procedure."[34] One might surmise that practices like audio monitoring move immediately to the phase of trained judgment, not having modalities comparable to the idealizations of truth-to-nature or mechanical objectivity.

Yet none of these visually oriented terms really fit the sonic and rhythmic experiences we have been considering. A collection of sound recordings differs profoundly from a photographic atlas, which invites the different kinds of interpretation that Daston and Galison discuss. Where the viewer of an atlas must learn to interpret subtle visual differences, rhythmic knowledge provides the far more direct thrill of hearing a cat's neurons going "off like a machine gun." Because it operates directly on the felt sense of time, rhythmic knowledge has a force and felt immediacy notably different from visual scrutiny

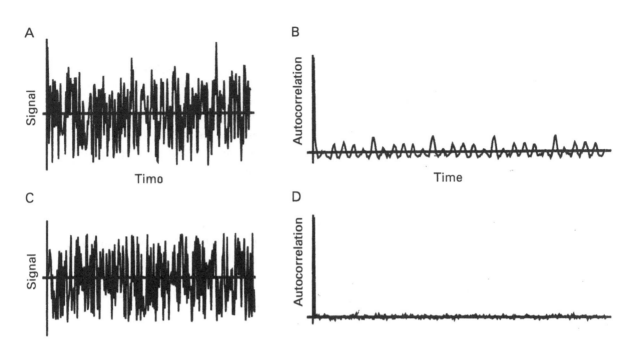

Figure 18.3
Signal and noise, shown with their calculated autocorrelation functions in Eric J. Heller's *Why You Hear What You Hear* (2012). Top: (a) a pure noise signal plus a "clean" signal (♪ sound example 18.2a); (b) the "clean" signal, shown by its computed autocorrelation (♪ sound example 18.2b). Bottom: (c) the noise by itself (♪ sound example 18.2c); (d) the autocorrelation verifying that this is pure noise. Visually, one cannot discriminate between pure noise (c) and noise plus clean signal (a), yet the ear can distinguish between them.

of phenomena in space, which often requires complex acts of interpretation. Summarizing their findings about scientific images, Daston and Galison proposed the provocative formulation "*seeing is being*"; by comparison, the examples we have been considering suggest that *hearing is living*.[35]

In due course, rhythmic knowledge led to the representation and manipulation of binary neural signals as digital data in computers, themselves modeled on artificial nervous systems.[36] Yet even the development of ever more powerful modes of data storage and analysis has not superseded the utility of hearing (or seeing) the result. Data analysis can take instrumental readings and give them a precise numerical form—bearing witness to the heritage of Herophilus and the Pythagoreans—but not with the immediacy of hearing or the graphic clarity of seeing. The very richness, density, and sheer volume of numerical data often require careful consideration how best to bin, organize, and analyze this material, each stage of which has many possibilities and pitfalls. The story of ultrasound has shown the power of combining acoustic data with visual representation; the final chapter presents other examples.

Whether through automated procedures or by direct inspection, the mind must distinguish signal from noise, a procedure in which hearing has advantages over sight. If some noise is added to a "clean" signal, the result is a ragged-looking wave (figure 18.3a). Calculating the autocorrelation function can extract the signal from the noise purely numerically (figure 18.3b). But the eye alone cannot really tell the difference between pure noise and signal plus noise (compare figures 18.3a and 18.3c), which both look equally ragged. On the other hand, Eric J. Heller points out that "we may be able to hear the original clean signal buried in the noise even if we cannot visually detect its presence" (♪ sound examples 18.2a, 18.2b, and 18.2c).[37] The ear can lock on to salient differences that are far less clear to the eye. To be sure, such statistical tools as autocorrelation can provide robust and quantitative measures important for careful analysis of any experiment. Yet such analysis is a heavily mediated process that depends on how it is done. Hearing gives a thrill that (in the case of Hubel and Wiesel) was important in the process of discovery, requiring (as Conway notes) subtle aural discrimination of the neural rhythms from the high background noise.

The Pythagorean program worked for hearing in ways it scarcely could have for sight. It is hard to imagine Pythagoras coming into an artist's studio and having an epiphany about the relation between colors and numbers that would have paralleled his legendary realization after hearing the hammers in the smithy. This stems from deep factors differentiating these senses: human hearing can span ten octaves of frequency, whereas our vision spans only about a major sixth in terms of color (a ratio of 400:700 in wavelength between violet and red).[38] That is, humans have never seen an octave in color, which would be the experience of two colors of wavelengths in a 2:1 ratio seen as "the same," in the way we hear two adjacent octave Cs on the piano as "the same note." We seem to have a diffuse awareness of ultraviolet light of wavelength 350 nanometers (though that would cause serious injury to our eyes); still, we do not register that as a "color" that could be compared with the red of light at 700 nanometers, an octave lower.[39] Thus, Pythagorean arguments connecting octaves with 2:1 ratios do not address our experience of colors.[40] All of these factors indicate why sonification—using hearing to access data—has proved an important tool in many fields and can enable insights into data that may go beyond what looking at graphs or what computers can provide.[41]

With this in mind, let us return briefly to Nagel's argument that we will never know what it is like to be a bat.[42] His conclusion that we will never penetrate the bat's subjectivity—its inner world of feeling—is persuasive, but we also cannot penetrate the subjectivity of other human beings, their unique inner world. Still, we extrapolate to other humans a considerable amount of sympathetic understanding based on our own inner experience and feelings. Nagel's assertion of the utter alienness of bats does not acknowledge that they and we share the same universal neural code. Though they use much higher frequencies for echolocation than we do for hearing, their neurons fire at comparable frequencies to ours. What, then, prevents us from entering into their world in the way Adrian (at least for a moment) saw through a toad's eye, both organisms sharing the same rhythms? The practice of audio monitoring is a kind of way-finding: like a bat flying in the dark, an experimenter listens for the individual voices of the neurons to navigate the darkness of the brain.

The Pythagorean project of understanding the body via rhythms expressed as numbers has come full circle. Herophilus began mathematizing (and musicalizing) the felt pulse; the responses to his idea eventually led to the actual hearing of the body's "analog" soundscape, eventually also its "ultrasoundscape." Going further into the nervous system, Adrian and others transduced single neural impulses into binary rhythms, first heard over loudspeakers, later recorded and processed as digital data—a stream of binary numbers. What began with numbers eventually returned to numbers. Yet the other side of the Pythagorean project remains important—not just numbers by themselves, but the connection between sounding bodies and sounding numbers as rhythm and music. As computers' growing power enables us to analyze physiological data, they also offer new powers of rendering that data visible and audible. Developments underway at the beginning of the twenty-first century show the growing reach and availability of those powers.

19

Echoes and Envoi

The discoveries we have discussed stretch over the past two thousand years yet continue to reverberate in the present. These final pages reflect on ways in which sounding bodies remain a vibrant part of the biomedical sciences, beginning with Adrian's technique of listening to neurons. Though emerging alongside other early uses of the telephone and loudspeaker, Adrian's sonic technique was not just a makeshift that disappeared when better technology came around. As noted in chapter 17, it led directly to the Matthews oscillograph—a loudspeaker capable of making visible records of its vibrations—that remained an important tool of neural electrophysiology through World War II. Though the oscillograph was supplanted by improved oscilloscopes, these did not replace the practice of listening to neurons.

Hubel and Wiesel treated the use of audio monitoring as a long-established and familiar technique, as indeed it already was. In an example, picked more or less randomly, a 1950 paper measuring spike discharges of single units (i.e., neurons) in the cerebellar cortex noted that "auditory monitoring of the potentials was carried out routinely."[1] Almost a century after Adrian first hooked up a loudspeaker to a toad neuron, a special practice of listening remains to this day a useful part of the tool kit of neuroscientists pursuing a wide range of investigations. Though to date there has been scarcely any discussion of this in the specialist literature, the technique of audio monitoring became craft knowledge, passed down from researcher to researcher through a kind of oral tradition that reaches back at least as far as Adrian, whose roots lie in the work of du Bois-Reymond and Helmholtz, as we have seen.[2]

This technique remains in practice throughout the world—for instance, in the laboratory of one prominent neuroscientist, Leslie M. Kay, a professor at the University of Chicago, whose research concerns the olfactory bulb. She explained that "really the only data worth recording is when you can actually hear the neurons. This is about localizing. Listening is the best way to home in" on a single neuron:

When your electrode is getting close to the [neuron] cell layer, it will be quiet or just static and then the static starts to sound structured. I'll say to students, do you hear that? If they don't, I'll say: keep listening. If I get a little further, then they start to hear the structure [♪ sound example 19.1].

What that means is that you are hearing thousands of neurons from far away. If it gets louder, you're going in the right direction. As you get close, you are now inside and can hear kind of a soft background but what you hear louder is a popping [♪ sound example 19.2]. When you isolate a cell, it gets very loud. It's a combination of frequency and volume that you are using with your ears and the speaker that we are using has an equalizer on it so that we can tune in on certain frequencies. When we are far away I might keep it [the frequency band] open and as we get closer I may narrow the band a little to where I can filter out the background and listen to the cell a little more. That's really a way of using your ears to get the electrode in the right place.[3]

Her commentary illustrates the kind of nuanced listening skills that her students need to acquire, as well as why those skills would be difficult to codify in print, being easier to convey in a dialogue that is more like a music lesson—a training in how to hear with sensitivity and discrimination. More than just technique, though, this learning calls for experiencing the thrill Conway describes in recognizing an individual neuron's rhythm, the delight in "understanding something through hearing."

Such listening skills are not restricted to research laboratories but find many uses in clinical settings. Listening to muscles remains a stethoscopic skill practiced by neurologists, who also use electromyography (EMG) to measure the electrical activity of muscles, amplifying those signals through loudspeakers as well as recording them on a computer.[4] Audio monitoring of nerve activity helps surgeons during delicate procedures; the vividly changing sound of nerve activity helps avoid damage to delicate facial nerves, for instance.[5] Likewise, audio monitoring helps position a fine needle implanted deep into the brain to treat Parkinsonian tremors.[6] For this and other such clinical procedures, audio monitoring is a valuable skill. Overall, modern operating rooms and intensive care units have become veritable sonic jungles pervaded by the telltale melodies and rhythms of heart monitors. So numerous and insistent are these multiple sonic alerts that the problem may be discriminating between more and less urgent warnings.[7]

Compared to learning to discriminate among stethoscopic sounds, neuroscientists find themselves probing into ever-changing sonic terrain as the electrodes move through different sections of the brain. Although all neurons have the same all-or-nothing response, heard as popping, different cortical areas have different sounds. Kay notes that "as you lower the electrode, if you know the cortical area you know what it sounds like and can hear what it sounds like when it's getting close [to a single neuron]. Anything connected to the olfactory system, you'll hear [from the audio monitor] kind of a dry *chhhh, chhhh* and you look at the animal and see if it matches the breathing. Because rats are obligate nasal breathers, don't breathe through the mouth at all, as they inhale, even if they are anesthetized, it activates the olfactory system." That is, the rhythm of an olfactory neuron echoes that of the respiration bringing odors past their receptors, though care must be taken to avoid any direct respiratory signal.

In contrast, to study the hippocampus researchers use a "tetrode," a bundle of four thin (15 micron) electrodes inside fused insulation: "If you had a big crowd and you used four microphones, you could use them to differentiate the signal. Later you can analyze the data from the mikes and deduce how many neurons you recorded." This reminds us that in neuroscience today, listening is only part of a larger process that then proceeds by computer analysis of the recorded signals: "Once you're in the right place, yes, you listen to it but now we record hundreds of neurons at the same time. We record the data and then it becomes just statistical, then we no longer listen." Here Kay has in mind the powerful analytical tools now readily available that allow sophisticated statistical analysis of the signals received from the neurons merely using the computing power of a laptop.

Indeed, even the power of a smartphone has given a new life to long-established sonic practices like auscultation. Several decades ago, medical students routinely worked on their listening skills so that they could acquire sufficient expertise with the stethoscope to meet the clinical standards of the time. Long-playing records of heart sounds as well as listening practica were standard features of medical education. Since then, the wider availability of electrocardiograms (ECG) of various kinds, as well as CT scanners, has led to more reliance on these tests and less on routine auscultation. To be sure, echocardiography is really a sonic practice that relies on ultrasound but that tends to be forgotten as the power of this and other ultrasound-based testing has tended to eclipse the humbler stethoscope. For example, echocardiography tends to be relegated to a group of specialized cardiologists. As a result, the motivation or even the need to listen for oneself has tended to diminish. Indeed, a whole genre of medical literature laments the passing of skillful listening through the stethoscope, which has become passé in medical training or relegated to nurses and ancillary medical professionals.[8] The stethoscope remains an iconic part of the physician's image, draped casually around the neck as a kind of badge of office, but it is used less and less because the needed skills are no longer emphasized or even taught.

Yet despite this, the stethoscope is entering into a kind of renaissance thanks to computer-assisted auscultation (CAA). The attachment of a computer interface to a digital stethoscope enables sophisticated analysis of the data, including machine learning techniques such as acoustic neural networking that can be "trained" on large data sets. As a result, this relatively simple and inexpensive instrument can be used to achieve much higher levels of accuracy than an ordinary stethoscope supplies. Thus, CAA permits diagnostic sensitivity far beyond the average skills of most practitioners, capable of becoming even better than a very skilled and experienced listener. Also, CAA can transmit data via smartphone to distant centers for specialized consultation when needed (figure 19.1).[9] Digital stethoscopes can also register cardiac data in the infrasonic range, which was not accessible by simple stethoscopes and promises to enhance the diagnostic power of CAA.[10]

These are important considerations even in wealthy countries because tests such as echocardiography are expensive and tend to be overprescribed. For instance, in one study of 3,460 adult referrals for "heart murmur," less than 50 percent actually had significant valvular disease.[11] The effective use of CAA could eventually eliminate many unnecessary

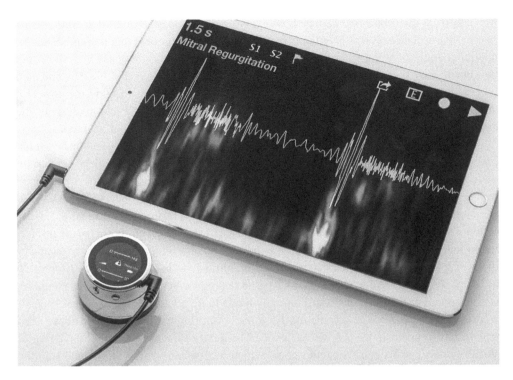

Figure 19.1
A digital stethoscope, Thinklabs One, linked to a tablet.

referrals and give much more useful screening for all kinds of cardiac problems. This problem of "innocent murmur" is especially important for pediatric patients, of whom up to 80 percent appear to have some kind of murmur but only 1 percent eventually have a pathological condition.[12] Also, certain cardiac problems of athletes that can cause sudden death could be spotted more frequently by CAA because they are elusive for ordinary auscultation.[13]

As important as these may be in wealthy countries, the promise of CAA is even more significant for the poorer countries to address the pressing lack of trained clinicians and large medical centers, not to speak of the surrounding economic and social challenges. Eventually CAA may be robust enough to be deployed even by minimally trained practitioners, who could learn to apply the electronic stethoscope and then use the computer analysis (as well as distant consultation) to give far better care to underserved populations. Because CAA is built around a simple stethoscope, it is basically an inexpensive and highly portable technology well suited to situations of great economic need or remoteness.

A similar development has already brought clinical ultrasound to much wider use. In place of relatively large and expensive equipment, a small and inexpensive transducer

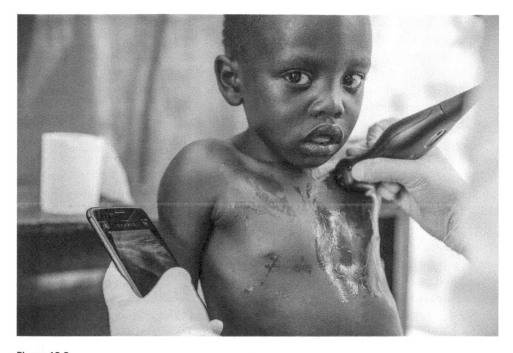

Figure 19.2
A child in Uganda being examined with a handheld ultrasound, which confirmed a diagnosis of pneumonia.
(Photo by Esther Ruth Mbabazi.)

can be attached to a smartphone to create a highly portable handheld ultrasound (HHU) device. For instance, the Butterfly iQ is about the size of an electric shaver, runs on batteries, and uses microchips rather than piezoelectric crystals so that it will not break if dropped; it gives a live image via a smartphone screen (figure 19.2). Its inventor, Jonathan Rothberg, has donated scanners to several medical charities in Africa, including Bridge to Health.[14] Teams from this Canadian charity have traveled across Uganda to deliver care using this technology. They check for pneumonia (a major killer of children that is often misdiagnosed) as well as examine organs ranging over the whole body. The Gates Foundation has supported the development of software intended to help untrained users operate the device. Through the smartphone, it can readily upload scans to the Internet so that they can be consulted remotely. In the case shown in figure 19.2, a physician back in Canada advised the user to hold the scanner at a sharper angle to project sound waves more deeply and reveal more about the situation inside a child's fluid-filled lungs. As a result, the child's pneumonia was securely diagnosed and treated. The handheld scanner can help diagnose problems ranging from scrofula to cancer, at least to give a preliminary scan that can guide further treatment; a briefcase-size scanner can perform fetal ultrasounds. This handheld version has also proved useful in diagnosing the peculiar lung conditions accompanying covid-19. Such new applications of sonic technology have great promise especially to extend the availability of modern medicine throughout the world.

Even more widespread is the use of otoacoustic emissions (OAE) to diagnose hearing loss in newborns. Already in 1683, Duverney, the great pioneer of otology, described "a Lady of *Picardy*, who upon the least violent Exercise perceiv'd so troublesome a pulsation, that it seemed to her that she had a *Pendulum* fix'd to her head, and this Pulsation was also heard by those who came near her," which he ascribed to a loudly pulsating vein.[15] In 1948, the physicist Thomas Gold argued from fundamental principles that the normal inner ear ought to produce sounds spontaneously, not just receive them.[16] Despite his compelling arguments, this startling suggestion was not given sufficient attention at the time, even by Georg von Békésy, the Nobel laureate who had made important contributions to the study of hearing. Discouraged by this lack of interest, Gold went on to become a prominent cosmologist. Finally, in 1977 David T. Kemp experimentally verified the existence of otoacoustic emissions as Gold had predicted them.[17] These emissions enable a powerful and simple test of the integrity of the inner ear. A small soft probe, placed in a newborn's outer ear, transmits a soft tone or click, which then will evoke a sound produced within a normal inner ear, whose spontaneous response does not require the baby to do anything and does not even disturb sleep. The whole process is inexpensive and noninvasive, readable through a handheld device (figure 19.3). As of 2010, the World Health Organization (WHO) noted that 94 percent of newborns in the United States received this screening, which is practiced in about 80 percent of countries worldwide. Still, many countries in Africa and Southeast Asia have "no serious organized effort to set up newborn and infant screening," despite the high prevalence of hearing loss (14 percent among school-age children in Nigeria, for instance).[18] The comparatively small expense required to provide these tests would drastically affect the lives of millions of children.

If you listen to the test tones used to evoke otoacoustic emissions in these tests (♪ sound example 19.3), you may realize that (unlike the lady of Picardy) you cannot hear the sounds being produced inside your own ears, sounds presumably filtered out by the brain because they are constantly present. Consequently, one must depend on a microphone to pick them up and record them for analysis. As with computer-assisted auscultation or ultrasound, one might infer that electronic data analysis has effectively superseded actual listening. Indeed, in the development of ultrasound, visual displays quickly replaced any attempt to "hear" the ultrasounds, even when shifted into the human hearing range (as in the early experiments of Griffin and Pierce listening to bats). If so, sound might seem merely a transitional stage in the process of generating a visual readout or data parameters. Nevertheless, ultrasound must distinguish between the echoes reverberating throughout the body, a complexity that tends to be hidden in visual displays.

The developments we have been studying in the past three chapters were among the earliest sonifications of data via telephone and loudspeaker. New ways of sonifying have been applied to data that had previously only been scrutinized visually or through statistical

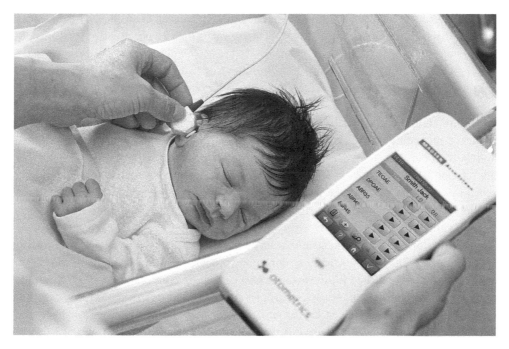

Figure 19.3
Using otoacoustic emissions (OAE) to check the inner ear of a newborn with a handheld Ohotoof Madsen Acuscreen.

indices. For instance, Mark D. Temple of Western Sydney University assigned sounds to DNA sequences by individual nucleotides (named A, T, G, or C), pairs, or groups of three (a codon).[19] He referred to Alan Turing's 1948 Manchester computer, which had a built-in "hooter" (loudspeaker) that could generate what Turing described as "something between a tap, a click, and a thump," which, if inserted in the program "at suitable points one is enabled to 'listen in' to the progress of the routine. Some indication of what is going on is given by the rhythm of the clicks that are heard."[20] With that in mind, Temple sonified the genetic codons for "start" (ATG) and "stop" (TGA, TAG, or TAA) distinctively, either by particular instruments (such as a snare drum for "start") or by using them actually to start and stop the sounds. The user can enter DNA sequences via a website and then choose various sonification options.[21] Comparing a score generated by the program with the actual sonification, for example of the first exon of the human *RAS* gene (figure 19.4; ♪ sound example 19.4), gives a vivid sense of the difference between seeing and hearing these data. This example of "DNA music" is rather minimalist in style, angular, and driving, like post-Stravinsky modern jazz. That, of course, is a result of the stylistic choices made in the algorithm (and its blues scale), opening interesting questions about the varying kinds of insight that might be enabled by different choices. Temple concluded that

DNA sonification "can help identify potential reading frames in complex DNA sequences and can identify mutations in repetitive DNA sequences that are obscure by visual inspection alone."[22]

Temple also sonified the genome of the coronavirus (SARS-CoV-2) only a few months after its first publication in early 2020.[23] Building on his earlier work, Temple now included real-time visual animations to accompany the sonification, "an important addition since with sonification data alone is it difficult to relate the auditory display to the underlying sequence information. The combination of the auditory and visual displays is more informative than either display in isolation." As Sterne has noted, we should not assume that the relation between different senses is "a zero-sum game, where the dominance of one sense by necessity leads to the decline of another."[24] Temple presented his coronavirus results on an interactive web page that allows the visitor to turn on and off many aspects of the sonification, in which "we layer up to twelve layers of audio, each relating to a RNA feature of interest."[25] The genome can be played either forward or backward, corresponding to its opposite directionality during the processes of transcription or translation (respectively) of the virus's life cycle in the host cell. This resembles the use of retrograde motion (playing melodic lines backward) one encounters in Bach's more recondite canons, as well as in twentieth-century serial music. On the website, one sees and hears the genome stream (forward or backward) through a "play-head" representing the locus of translation on the ribosome or of transcription on the RNA polymerase/replicase.

Temple remarks that in this version, he chose diatonic modes other than the ordinary major scale "since this is generally considered to be happy sounding and inappropriate for the data." Nevertheless, I found listening to this version of the coronavirus genome to be curiously cheerful and even rather soothing, despite the grim consequences it has wreaked on humanity. This doubtless reflected the ethos conveyed by his chosen diatonic modes (B-flat Aeolian or C Lydian for translation or transcription, respectively). In future iterations, he hopes that "it may be possible to choose the scale modes and key of choice" so that the listener might have further options.

Even in this initial format, though, Temple judges that "the auditory display is capable of detecting unusual features in the genome," particularly the virus's ability to undergo discontinuous transcription, switching templates during the process of transcription. To sonify this, Temple's play-head can skip to another point in the genome as indicated by the virus's relevant sequences. By both its name and function, the play-head reminds us that a ribosome "plays" the RNA passing over it in analogy to the way a digital play-head reads the tape passing through it. Both hark back to the older (analog) medium of a gramophone needle reading the groove of a vinyl record or a cassette tape passing across a physical play-head. These provocative analogies suggest that the perspectives of media and music theory may cast new light on genetic transcription.[26]

For instance, consider the question of tempo. Temple's initial version takes ninety-six minutes to play the whole coronavirus genome "forward" at five nucleotides per second,

Human RAS Coding DNA Sequence

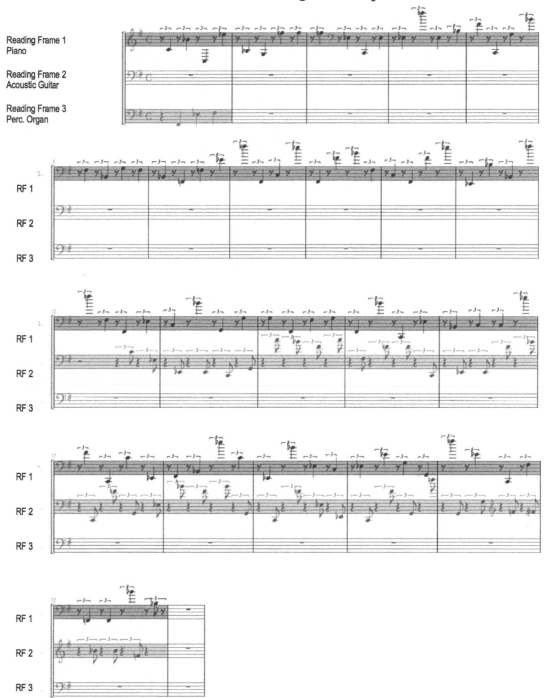

Figure 19.4
Mark Temple's 2017 score representing the DNA sequence of the first exon of the human RAS gene, as sonified in ♪ sound example 19.4. Comparing the score with sound shows that data that look similar on the page may sound noticeably different.

which is six times slower than its actual rate in vivo (thirty nucleotides per second). He chose that slower speed because "to play it any faster makes it difficult to interpret" and would require a different algorithm. Still, it would be intriguing to hear a "real-time" version lasting sixteen minutes, a duration much closer to human attention and comparable in length to many human musical works. By comparison, Temple's slowed-down version has a rather Wagnerian tempo and length, which comes across as "easy listening." As Alexander Rehding has discussed, radically slowed-down (or speeded-up) versions of Beethoven's Ninth convey startlingly different feelings through identical pitches and rhythmic patterns.[27] Yet tempo shapes rhythm on the largest scale. What might be learned by hearing the same sonification at different tempos? Those sixteen minutes may be the difference between life and death for the host who is "playing" the coronavirus genome involuntarily, as a result of infection.

To consider another approach to sonification, my final example considers work at Stanford by the neurologist Josef Parvizi and his collaborators, particularly Chris Chafe, a composer and cellist involved in computer music. They sought to diagnose seizures that are not convulsive, unlike the more familiar forms of grand mal epilepsy, but still pose important risks, even of death, in some cases. Indeed, as many as 90 percent of patients in an intensive care unit (ICU) may have such silent seizures, which, if sufficiently severe, should be treated as soon as possible, ideally within one hour.[28] Because the process of obtaining and reading an EEG can take many hours or even days if the needed staff are not available, Parvizi's group developed a handheld EEG (about the size of a smartphone) that interfaces with a headband worn by the patient to record the brain's electrical signals. The Ceribell EEG is simple enough to set up that it does not require a specialized technician needed to run an ordinary EEG; on average, a nurse or medical student without prior EEG experience took about six minutes to set up the "brain stethoscope" (figure 19.5).

This device gives a visual output in the usual form on its screen and transmits it wirelessly to a laptop for review, but its most significant innovation is sonifying the EEG data. To do this, Chafe devised algorithms that transpose the brainwaves from their actual range (0–100 Hertz, spanning infrasonic and low bass frequencies), which "mostly sounds like a low rumble," to a higher range the ear can interpret. Chafe notes that "there's no way to listen in real time without an algorithm such as ours."[29] As sonified and heard through the brain stethoscope, normal brainwaves have "a boring kind of background hum sound" (♪ sound example 19.5), as Chafe described it, whereas the non-convulsive seizures have a strikingly agitated sound, like an "irritated screaming thing" (♪ sound example 19.6).[30]

To test the diagnostic power of the sonification, Parvizi's group gave a few minutes of training to nurses and medical students not experienced in EEG techniques. One group heard a number of thirty-second examples that contrasted normal brain sounds with those experiencing nonconvulsive seizures; a control group received a similarly brief training in reading the visual display alone. Both groups were tested respectively against actual sounds or EEG graphs from patients who were monitored by expert epileptologists. Those

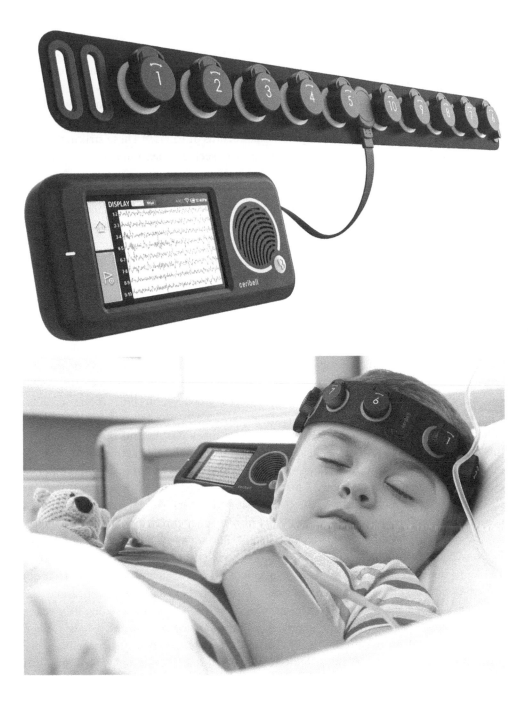

Figure 19.5
The Ceribell brain stethoscope, shown below in use. The headband attaches to a handheld read-out unit, which includes a speaker to play the real-time sonification of the EEG.

trained to listen for the sounds of silent seizures heard them with accuracy averaging 97 percent, compared to 76 percent for those purely examining the visual records and 88 percent for trained neurologists.[31] Further, those who responded to the sounds were able to make a more rapid determination, compared to those who had only the visual records, which required more time to read and interpret.[32] In comparison, one mid-twentieth-century textbook prescribed three months of practice for an average (scientific) person to reach 98 percent accuracy in reading the visual records of EEGs.[33] These striking results indicate that hearing has significant advantages over seeing, at least in contexts where an immediate response is important.

This accords with the immediacy that Adrian valued in sonic monitoring, as have later neuroscientists like Kay and many others. Yet the novel aspects of these advanced kinds of sonification have important implications. Whereas Adrian and Matthews struggled to make anything of the rumbling low frequencies of brain waves, Chafe's computer algorithm allow kinds of transposition and even orchestration that remain faithful to the temporal and rhythmic structure of the EEG data while also rendering it much easier for the ear to apprehend.[34] Such possibilities did not exist until the later twentieth century and emerged largely through musicians exploring the timbral and compositional possibilities opened up first by electronic media and then by the computer.

In 1963, electrical engineer Max Mathews prophesied that "with the aid of suitable output equipment, the numbers which a modern digital computer generates can be directly converted to sound waves. The process is completely general, and any perceivable sound can be so produced."[35] Responding to Mathews's prophecy, in 1967 composer John Chowning made "altogether a discovery of the 'ear,'" a new way to synthesize sounds he called frequency modulation (FM) synthesis.[36] An ingenious generalization of vibrato, this technique applied the basic principles behind Tartini tones (and hetrodyning) to audible frequencies using the emergent power of computers. Widely used in digital synthesizers, its commercial success enabled Chowning to found the Stanford Center for Computer Research in Music and Acoustics (CCRMA). Using such techniques, computers can free composers from all the ordinary constraints of instruments based on vibrating bodies, whose sounds are based on the octave (2:1) and the normal overtone series. Thus, Chowning was able to base his computer composition *Stria* (1977) on the "golden ratio" ($\varphi \approx 1.618 \ldots$) instead of the octave, "tones that cannot exist in the natural world."[37] His successor as director of CCRMA, Chafe brings to his compositions and sonifications the greater power and generality computer music has developed in the intervening fifty years. In that process, composition and sonification have increasingly merged; Chafe's *Earth Symphony* (2020) involves sonification of climate change data over the preceding fifty years; his *Tomato Music* (2011) was built from data and sounds collected from five vats of tomatoes ripening over ten days.

As Chafe points out, "Multiple time scales in music are crucial to its appreciation. Musical listening to datasets from outside of music can provide a way into understanding behaviors at multiple time scales."[38] Such multiform and simultaneous timescales characterize the body and living organisms in general. As I have discussed in *Polyphonic Minds*, this profoundly polyphonic quality seems essential for the functioning of the brain, for awareness and consciousness itself, no less than for the body and its many simultaneous processes.[39]

In the enormously difficult process of understanding the brain and nervous system, listening may offer a new kind of understanding, enabled by new kinds of sonification. Kay notes that olfactory neurons have a special "sound" connected with their link to respiration; other sensory systems have their own particular interconnections. Vision, in contrast, is not correlated to respiration; thus, the recorded sounds of neurons in the visual cortex are differentiable from those in the olfactory bulb, so that "hearing seeing" is different from "hearing smelling." This might be the salient test: Purely from a sound recording, could someone tell which sensory system or part of the brain they are hearing? As neuroscientists learn to differentiate the "chorus" of many neurons from the staccato popping of a single one, perhaps they could develop a stylistic and sonic language to describe how the various parts and regions of the brain *sound* different. The brain stethoscope brings forward features of the EEG that can be recognized purely visually or via data analysis, though less immediately and intuitively than by hearing. In the larger process of understanding the brain, hearing may be even more essential in disclosing larger patterns or features that might otherwise be lost in the deluge of data recorded visually or digitally.

At each stage in this long story, music, sound, and hearing have changed together. The rhythms of Herophilus reflected the musical practices he knew, blended with Pythagorean connections between number and sound. The brain stethoscope's computer music reflects the musical styles of its time, as did the Beethoven piano sonata (op. 90) that Helmholtz mentioned in his work on hearing or the Prokofiev piano sonata (op. 14) that Hubel played, with its driving rhythms.[40] Then too, as Adrian realized the toad was listening to him, his reaction was shaped by the sounds he heard from loudspeakers every day.

Modern medicine and biology have inherited from their Pythagorean origins both the orientation to ratios (and rationality) that links them to the physical sciences as well as their quest to hear the rhythms and frequencies of life. Ratios combined with hearing can discern the invisible interior of body and mind, thereby uniting thinking with feeling. This deep connection between number, music, and the body has grown over thousands of years, beginning with Pythagoras, who "with the eyes of the mind gazed upon those things that nature has denied to human sight."[41] Yet the mind also has ears, and we respond to rhythm through dance. As the life sciences continue to seek what lies beyond human sight, our senses of hearing and movement may continue to offer new ways to grasp the rhythms that are life itself.

Appendix: Two Papers on Muscle Sound by Hermann von Helmholtz

Experiments on Muscle Noise (1864)

1. One very clearly hears the well-known and often questioned muscle noise in certain circumstances in which the rubbing of the ear or the stethoscope on the skin covering the muscle are completely excluded, when you are in a quiet place, preferably at night, ears tightly plugged with sealing wax or wet paper, and then bring the muscles of the head (e.g. the masseters) into vigorous contraction. As long as the muscles remain in constant tension, one hears a muffled, roaring noise whose fundamental tone [*Grundton*] is not significantly altered by increased tension, while the intermixed roaring noise becomes louder and higher.

Not only the tension of the powerful masticators, the masseters, pterygoidei, and temporals, but also that of much weaker facial muscles, the orbiculares oris and palpebrarum, of the platysma myoides, the levator labii superioris alaeque nasi, of the tongue, etc, make audible noises, all of which are essentially of the same character, only louder, clearer, and purer, like the well-known noises one hears when one puts the stethoscope on the contracted muscles of the arm.

The pitch of the fundamental tone of the musical part of this noise is very difficult to determine because it lies below [925] the lower limit of perceptible pitches.[1] Mr. S. Haughton[2] recently had it determined by several people, for whom it sometimes corresponded to the C of 32 vibrations [per second], sometimes the D of 36; 35 to 36 was also the largest number that Wollaston found for it. I find the same for my masticators, though the pitch for the weaker facial muscles is a little bit lower.

2. I repeated these observations so that I would not produce the contraction of the muscles by my will but rather using an induction apparatus with a vibrating spring that, at the appropriate setting, could give up to 130 vibrations of the spring [per second] and the same number of opening beats [*Oeffnungsschläge*].[3] The induction apparatus stood in another room separated by two closed doors so nothing of its tone could be heard directly. As soon as I put the electrodes on my masseter and brought it into strong contraction, I

heard the tone [*Ton*] of the spring of the induction apparatus.[4] When I changed the adjustment of its [the spring's] screw, then I heard the change [in the muscle].

That the tone belongs to the contracted muscle and not to the direct effect of the electrical currents on the ear follows especially from the fact that the tone only became audible
when the current strength was increased sufficiently to give a contraction of the muscle.

3. Likewise, the tone was also heard, though less strongly, through a stethoscope on
a young man's arm muscles, which were brought into contraction by induction currents
flowing through them. In this case, the ear and auditory nerve of the observer were not
affected by electrical currents. Yet one should bear in mind the electric current directly
sets the contracted muscle to vibrating like a tensed wire. In order to exclude this possibility, I finally let the current flow through the median nerve of the upper arm [926] and
weakened its strength enough that, when applied directly to the muscle, it did not bring
it to contraction. As soon as the current affected the nerves strongly enough that strong
contractions of the forearm muscles occurred, I heard clearly the tone of the current-
interrupting spring. On the contrary, if I moved the electrodes on the upper arm a little to
the side so that the effect on the forearm muscles ceased, the tone also disappeared.

It follows that the periodic movement the wire delivered to the nerve in the form of
electrical shocks was led from the living nerve to the muscle with unchanged period and
finally set it into mechanical vibration, into sound vibrations. The number of vibrations
was about 130 per second.

First of all, these attempts seem to me to eliminate any doubt about the existence of a
specific contraction-dependent muscle noise and to set aside any explanation of it in terms
of friction of the muscle on the surrounding parts.

That an apparently evenly contracted muscle is in fact to be understood as rapidly
changing contrary molecular arrangements was inferred by Mr. E. du Bois-Reymond
from the appearance of the so-called secondary tetanus.[5] The speed of this change is one
of the most important reasons that one must attribute electrical muscle effects to the existence of very small electromotive molecules. But the evidence of such a change mostly
rested only on the fact that the muscle current of a tetanized muscle, passed through
another nerve, would likewise tetanize that muscle also. About ten changes per second
would be sufficient. If it now seems extremely likely that the number of internal changes
of a muscle tetanized by induction pulsations from a series of induction loops may be
about equal to the number of electric shocks, therefore I believe that direct proof for it
[i.e., du Bois-Reymond's theory about the change], as is supplied by the tone of the muscle, is of importance in these circumstances. [927]

I note that also in my investigations of the sensations of tone [*Tonempfindungen*] it was
also necessary for the auditory nerves to receive about 130 distinct excitations per second.

At the time, I had no equipment to produce more than 130 opening beats reliably with
regular periodicity, but I do not doubt that much higher tones are produced in the muscles.
When I let a tuning fork of 120 vibrations [per second] interrupt the current, I hear in the

muscle relatively strongly the tone of 240 vibrations, the higher octave of the pitch of the fork, which seemed to have been evoked through the 120 opening beats and the somewhat weaker 120 closing beats. The difference between the strengths of both types of beats was less in this case because the mercury flow was interrupted.

On the other hand, I used a tuning fork to induce tetanus in frog legs. The tuning fork was set between the poles of an electromagnet and, through the arc produced by its motion [the fork's oscillations], created an electric current in the form of steady sine waves in the coils of the electromagnet. I discovered that even 600 oscillations per second induced tetanus. However, I was unable to ascertain sound oscillations in the frog muscles.

On Muscle Tone (1866)

The author earlier discussed this subject and showed that when muscles of humans or rabbits have been set into tetanus by means of the currents of an induction apparatus whose spring carries out regular vibrations, instead of the normal muscle tone one hears a tone at the pitch of the vibrating spring of the induction apparatus. The usual devices of this type only give 40 to 60 vibrations per second; in the exposed sciatic nerve of a rabbit I had earlier induced tetanus through an induction apparatus in which a tuning fork of 120 vibrations interrupted the current and heard the corresponding tone of 120 vibrations from the animal as well as (though not quite so clearly) the first overtone of 240 vibrations. It is difficult to make the rapidly interrupted induction currents strong enough to affect human nerves through the skin because the mercury one must use at the interruption point quickly burns and turns to dust. Through careful adjustment of the appropriately attached secondary closures (partly metallic for the electromagnets, partly water decomposition [929] cells for the spark gap), using a tuning fork of 240 vibrations I managed to apply sufficiently strong impulses to attain tetanus in the median nerve of human forearm muscles, in which the tone of 240 vibrations was clearly audible, which indicated an extraordinarily high degree of mobility in the molecular apparatus of the muscle.

Since the muscle tone observed in this way is a phenomenon of low intensity, demanding quite remarkable attention from the observer, I have often tried to build resonance apparatuses to make it more clearly audible, especially because it was important to me to hear more clearly natural muscle tone that is at the limit of the deepest audible tones and thereby determine its nature. This was only very imperfectly possible acoustically, so that I thought it was possible rather to make vibrations of the muscles, especially in their deeper tones, visible to the eye.

To that end, I use steel springs (clock springs) long enough that their vibration period becomes equal to that of the sound to be perceived. These are inserted between four wire pins clamped at the ends of elastic boards partially separated by full-length cuts. If one places the board on the muscles so that one of its springy sections receives the muscular contractions, they are transferred to the clock spring, which comes into strong, easily visible resonant vibration. Using an apparatus giving 19.5 interruptions per second, human muscles came into strong resonant vibration when the spring was set at 19.5 vibrations [per second], weaker at 39 or 58.5, very weak at 78.

If one seeks the length of the spring that is best set into vibration through the natural contraction of the muscles, one finds it at 18 to 20 vibrations per second. The vibrations [930] in that range are not so regular, and therefore not as strong as they are with artificial tetanus. Because a steel spring resonates too long and therefore does not receive the transmitted mode of vibration fast enough, I found that a similar apparatus with tapered paper strips able to vibrate was better for observing natural muscle vibrations. Their oscillation period is best determined if one holds them against the vibrating spring of a suitably matched induction apparatus, and determines to which period of vibration they resonate most strongly.

These experiments thus show that the number of natural vibrations of human muscle is not 36 to 40, as Wollaston and Haughton believed they had observed, but only 18 to 20. What one hears as a muscle tone is therefore only the first overtone of the true muscle vibration, whose fundamental tone does not lie in the range of audible tones. Moreover, this natural muscle vibration is indeed approximately periodic, but not as exactly periodic as the movements of the vibrating tuning forks and steel springs.

In the hope of making the experiments significantly easier when experimenting with frogs, I have also experimented with their muscles. I succeeded in hearing traces of the tone of 120 vibrations when I hung a weight-bearing frog muscle to a [small] rod inserted into [my] auditory canal. On the other hand, one can very well observe the vibrations of the spring from 16 to 20 vibrations if one hangs the muscle on the above-described board holding the spring and, with an electrical tetanus of a corresponding number of beats, has it lift two ounces.

Spring vibrations of frequency 120 entirely failed to evoke isochrous electrical impulses from the nerve. In contrast, if I set the induction apparatus to 120 vibrations and the resonating spring to 16 vibrations, I saw weak vibrations of the spring that seemed to correspond to the natural vibration period of the frog spinal cord. [931] Moreover, it should be remarked that, as E. du Bois-Reymond first noticed and I myself confirmed, tetanus in the rabbit spinal cord caused by quickly vibrating currents also gave rise not to the sound of the current vibrations but to the natural muscle tone.

Currents of frequency 18 acting on the frog spinal cord, on the other hand, also induced strong isochronous vibrations in the spring. This frequency seems so close to the natural frequency of frog spinal cords that they will fully adapt themselves to currents at this frequency.

Notes

Introduction

1. For the place of music in the formation of physicians, see Kümmel 1977, 63–88.

2. Wellmann 2017 discusses the significance of rhythm in the biological sciences in ways that are complementary to my approach here.

3. Walker 1975, 1978, 1985; Gouk 1980, 1982, 1999, 2000a, 2000b; 2004, 2014, 2015; Kümmel 1977; Kassler 1995, 2001.

4. Kennaway 2012a, 2012b, 2014, 2015, 2016, 2019; Davies and Lockhart 2016; Lockhart 2017; Pottinger 2020; Raz 2014; Raz and Finger 2019; Steege 2011, 2012; Trippett 2018, 2019; Trippett and Walton 2019; Volmar 2012, 2013a, 2013b, n.d.; Wolf 2015.

5. Bijsterveld 2019; Pinch and Bijsterveld 2012; Brain and Wise 1994; Brain, Cohen, and Knudsen 2007; Brain 2015; Brittan 2011, 2017, 2019; Hadlock 2000; Hui, Kursell, and Jackson 2013; Hui 2013a, 2013b; Jackson 2004, 2006; Kursell 2011, 2013, 2015; 2018, 2019a, 2019b; Pantalony 2004, 2009; Prins 2012; Prins and Vanhaelen 2018; Rice 2011; Thompson 2002; Tkaczyk, Mills, and Hui 2020.

6. Kittler 1999, 2006, 2014; Sterne 2003; Erlmann 2010; I also found Schwartz 2011 a wonderful resource.

7. For an overview, see Davis 2017.

8. Longrigg 1993; Jouanna 1999, 2012: Nutton 2004; Siraisi 1990, 1975; Boner 2013; Cheung 2010; Finger 1994, 2000; Finger and Piccolino 2011; Duffin 1998; Nicolson and Fleming 2013.

9. For the history of otology, see Erlmann 2010; for bird and animal vocalization, see Catchpole 2008; Rothenberg 2008; Bruyninckx 2018.

10. For the history of chronobiology, see Daan 2010; for entrainment, see Pesic 2017, 267; for embryology, see Wellmann 2017.

11. See Harrington 1996, 34–71; Clements 2011; Gómez 2013. See also Trippett 2018, 208–216.

12. See Maslow 1966, 15–16.

Chapter 1

1. Longrigg 1993, 8–11.

2. *Iliad* 1:46–52.

3. Ibid., 1:472–474, as translated in Homer 1990, 15.

4. *Odyssey* 4:220–232. Translations not otherwise identified are my own.

5. Nutton 1995, 17. For the relation of medicine to pre-Socratic philosophy, see also Jouanna 1999, 259–269.

6. For a thoughtful treatment of the Pythagoreans' worldview and its influence, see Heller-Roazen 2011.

7. Pesic 2014b, 9–13.

8. Vogel 1966, 242. See also Schumacher 1963, 1965.

9. Provenza 2012, 93, argues that this was no invention of the Pythagoreans but "a manifestation of a very ancient practice, well attested in Greek culture, which resulted from the strong relation between religion and medicine" (95).

10. Iamblichus 1989, 49. See also West 1992, 32.

11. As cited in Provenza 2012, 100n43.

12. Ibid., 101, probably quoting Aristotle's lost work on the Pythagoreans.

13. Vogel 1966, 232–244, at 233. For the medical context in southern Italy, see also Burkert 1972, 292–294.

14. Riedweg 2005, 2–5, at 3; about the secret society, see 98–104.

15. See Burkert 1972.

16. *Metaphysics* 986a22ff, as translated in Longrigg 1993, 45, who discusses the relation between Alcmaeon and the Pythagoreans on 48–51.

17. Zhmud 2014, 98; Longrigg 1993, 47–63, at 48.

18. According to the Roman Aetius, cited in Longrigg 1993, 52.

19. According to Zhmud 2014, 99.

20. Ibid., 100.

21. Huffman 1993, 15.

22. As translated in ibid., 93.

23. Ibid., 75, 307.

24. Ibid. 84, discussing the concept of *archai* on 78–92.

25. In his celebrated treatment of the quadrature of lunes, see ibid., 80. This Hippocrates should be distinguished from the famous physician from Cos.

26. His student Archytas, who made important mathematical discoveries, also used language that paralleled that in Hippocratic texts. See Huffman 2005, 58–59, 89–90.

27. See Jouanna 1999, which gives an annotated list of this corpus (373–416) and discusses its authorship (56–71).

28. See ibid., 181–209.

29. See Lane Fox 2020, 112–118, who notes that these case histories were special to the Epidemic books in the Hippocratic corpus.

30. Jouanna 1999, 162–170, 344–346; Iamblichus 1989, 102.

31. *Regimen* 61, Hippocrates 1937, I.8.10–18 (243–245).

32. Jouanna 1999, 128–131.

33. *Phaedo* 115d, 117c (Plato 1997, 98–99). Subsequent citations from Plato generally include the reference for this translation, in parentheses.

34. *De natura hominis* I:10–12, II:1–2. I have used the generic "human being" (rather than "man") to translate *anthropos*, reflecting standard Greek usage.

35. Ibid., IV:4–10. For Hippocratic humorism, see Jouanna 1999, 314–317.

36. See Lane Fox 2020, 233. For the later history of the humors, see Jouanna 2012, 335–359.

37. See Lane Fox 2020, 57.

38. See Jouanna 1999, 335–341, and Lloyd 1987, 257–270.

39. See "Father Time" in Panofsky 1972, 69–93, on 71–75; gradually Chronos was conflated with Kronos or Saturn, oldest of the gods, who devoured his own children. In "Of Isis and Osiris," Plutarch noted that "the Greeks are used to allegorize Kronos (or Saturn) into chronos (time), and Hera (or Juno) into aer (air) and also to resolve the generation of Vulcan into the change of air into fire." For later uses of *kairos*, see Baert 2016.

40. *Precepts* 1.1–3.

41. As suggested by Lloyd 1970, 58.

42. "Pythagorici numerici" in his *De medicina* 3.4.15, cited in Longrigg 1993, 98.

43. See, for example, *Epidemics* 1.23–26, cited in Lloyd 1984, 100–101. For fevers, see Jouanna 1999, 150.

44. *Aphorisms* 1.20, cited in Lloyd 1984, 208.

45. *Epidemics* 1.11, cited in Lloyd 1984, 94.

46. Jouanna 1999, 341, finds the association with Pythagoreanism "tempting" but notes that "the period judged perfect by the Pythagoreans, the decade, plays no important role in the reckoning of critical days among the Hippocratic physicians" so that we should not conclude that "all arithmological analysis was Pythagorean."

47. Osler 1947, 49, as pointed out by Lloyd 1984, 32n.

48. Lloyd 1987, 260–261.

49. Hippocrates 1937, I.8.10–18 (243–245). See also Burkert 1972, 262–263.

50. Hippocrates 2010, 87–89, at 89.

51. Burkert 1972, 290–295. See also Lloyd 1970, 50–65 at 51.

52. Iamblichus 1989, 102; Vogel 1966, 235.

53. Regarding the Hippocratic Oath versus the oath sworn by the Asclepiads, see Jouanna 1999, 50–52, who discusses Hippocrates as a member of the Asclepiads on 10–12.

54. *Protagoras* 311c (749), *Phaedrus* 270b–d (547). See also Jouanna 1999, 5–7.

55. As asserted by Longrigg 1993, 108–148, at 147, who discusses his relation to Sicilian medicine on 108–148.

56. Pliny, quoted in ibid., 105.

57. See Huffman 2005, 3, 32–35, 41, 44.

58. *Phaedo* 61e–62e (53–54).

59. For these terms, see Lloyd 1970, 51, who does not connect them to the *Timaeus*.

60. *Timaeus* 69c (1270–1271).

61. Ibid., 30a (Plato 2016, 15).

62. Ibid., 53c–57d (Plato 1997, 1256–1260).

63. Ibid., 69c (1270–1271).

64. Ibid., 41e (1245).

65. Ibid., 90a (1289).

66. Ibid., 78a–81e (1278–1281).

67. Ibid., 88b (1287).

68. Ibid., 89a (1288).

69. *Republic* 424c (1056).

Chapter 2

1. *On the Heavens* 306a9–11 (Aristotle 1984, 1:500). Subsequent citations of Aristotle reference this edition in parentheses.

2. *Metaphysics* 985b24–26, 1090a24–25 (Aristotle, 2:1559, 1722).

3. Ibid., 1090a32–34, 1090b20 (Aristotle, 2:1723).

4. *Problems* 11.24, 42, as translated in Barker 2004, 2:93, 96

5. *Parts of Animals* 645a17–25 (Aristotle, 2:1004). For the effect of Aristotle's work in biology, see Longrigg 1993, 149–176.

6. Forster 2014, 22. The second quote is from von Staden 1989, 25.

7. Nutton 1995, 33–34.

8. Lewis 2017, 6–9.

9. However, Herophilus and his followers were critical of Hippocratic medicine; see Jouanna 1999, 63. Note the alternate spelling, "Herophilos"; the Latin version, Herophilus, is more commonly used.

10. Vesalius 2002, 3:45.

11. For his milieu and discoveries, see Harris 1973, 177–195; Nutton 1995, 34.

12. von Staden 1989, 38–39, citing Hyginus, *Fabula* 274.13. The alternate spelling "Agnodice" is also found.

13. According to ibid., 269–271.

14. Ibid., 278–279, 346. Wellmann 2017, 68, also comments on Herophilus's use of rhythm.

15. Galen, *Synopsis librorum suorum de pulsibus*, as cited in von Staden 1989, 355.

16. Ibid., 355–356.

17. Ibid., 356.

18. Ibid., 351.

19. Euclid, *Elements* VII, definition 2, and V, definitions 3, 4. Note also the careful clarification of "being in the same ratio" in definition 5.

20. According to Aristotle, fragment 519 R³ (Aristotle 1984, 2:2455). For its warlike character, see Plato, *Laws* 815a. Also called *pyrrihios*, it is still widely danced in Greece, including a mass performance staged for the opening of the 2004 Olympics in Athens.

21. Edgar Allan Poe, "The Rationale of Verse."

22. From his poem "The Garden."

23. The accent marks shown in Greek texts record a pitch, not a stress, accent.

24. The Hopkins line is from his "Pied Beauty," the Shakespeare from Sonnet 18. See von Staden 1989, 351.

25. Ibid., 349.

26. Ibid., 348.

27. Harris 1973, 397–431 at 410–418.

28. Von Staden 1989, 344–346.

29. According to Marcellinus, *De pulsibus* 11.463.260–267, cited in von Staden 1989, 354.

30. Lloyd 1987, 284.

31. Von Staden 1989, 288. For the work of Erasistratus, see Longrigg 1993, 205–219.

Chapter 3

1. Pliny, *Naturalis historia* 11.89.219, cited in von Staden 1989, 360.

2. Pliny, *Naturalis historia* 29.4.5–29.5.6, cited in von Staden 1989, 359.

3. See the discussion of the Pneumatic school of medicine in ibid., 285n158.

4. For Galen in relation to other medical sects of his time, see Sarton 1954, 30–38.

5. Galen, *De uteri dissectione*, cited in von Staden 1989, 143.

6. Galen, *On Medical Experience*, cited in von Staden 1989, 287–288.

7. Ibid., 288n166.

8. Thorndike 1923, 1:144.

9. Von Staden 1989, 346–347.

10. Ibid., 285–286.

11. Ibid., 358.

12. For Galen's pulse theory, see Harris 1973, 397–431.

13. See Siraisi 1975, 689–710, at 697–698.

14. Censorinus, *De die natali* 12,4–5, cited in von Staden 1989, 360. Censorinus also related musical numbers and astrology to embryology; see the excerpts in Godwin 1988, 17–19; 1993, 40–45.

15. Iamblichus 1891, 93 (sec. 32, lines 11–18) as translated in Lloyd 2013, 155–156.

16. For Aristides's life and the dating of his work, see Aristides Quintilianus 1983, 10–14.

17. *De musica* 106.9–20, as translated in Barker 2004, 2:506.

18. Ibid.

19. Ibid., 31.4–9 (2:433). Wellmann 2017, 21–28, comments on the development of this concept.

20. Barker 2004, 83.12–14 (2:434).

21. Ibid., 83.19–20 (2:434).

22. Ibid., 83.18–25 (2:486).

23. Ibid., 89.10–15 (2:491); for the breath, see 492n199.

24. Ibid., 89.15–25 (2:491–492).

25. Ibid., 82.26–28, 14–15 (2:485).

26. This was remarked by the music theorists Adrastus (first century CE) and Panaetius, whose own works were lost but whose ideas were recounted by Theon and Porphyry in the third century; see Barker 2004, 2:214, 238. One wonders whether they, unlike Aristotle's students, really observed this themselves; the octave is by far the loudest of sympathetic vibrations, though other concords will indeed selectively resound more faintly. In saying "concords," did they mean to include a specific list, or were they merely copying Aristotle's *Problems* inaccurately?

27. Ibid., Aristides *De musica* 86.22–23, 89.92–95 (2:489, 492), emphasis added.

28. Ibid., 89.85–95 (2:492).

29. Ibid., 106.18–20 (2:506).

30. Ibid., 30.20–24 (2:433).

31. Ibid., 55.29–56.4 (2:460).

32. Ibid., 58.11, 14–17 (2:462).

33. Ibid., 58.22–29 (2:462–463).

34. Ibid., 64.22–65.1 (2:468).

35. Ibid., 68:24–28 (2:471–472).

36. *On the Nature of Man* sec. 9, Lloyd 1984, 266.

37. Barker 2004, 40.13–15 (2:445). See also Palisca 2006, 75–77.

38. Henderson n.d., 454–455.

39. Strunk and Treitler 1998, 46.

40. Pesic 2017, 140–145, 197, 209.

41. The first five books were written in Milan (387) before his baptism, after which he returned to Africa and composed a sixth book in 391; see Augustinus 2002, x–xxviii.

42. For *numerus* as rhythm, see ibid., 7n1.

43. Ibid., 9 lines 25–27.

44. Ibid., 15 lines 9–14.

45. Ibid., 15 lines 15–18.

46. Ibid., 15 lines 23–25.

47. Ibid., 113 lines 1–7.

48. Ibid., 113 lines 18–22.

49. For the relation to Aristides, see Aristides Quintilianus 1983, 5.

50. Von Staden 1989, 361.

51. Boethius 1989.

52. Ibid., 7 [186], showing the standard page numbers from Boethius 1867 in square brackets.

53. Ibid., 10 [189].

54. Ibid., 3 [181], 5 [184].

55. Ibid., 5 [185].

56. Ibid., 6 [185].

57. Martianus Capella 1977, 1:55.

58. Von Staden 1989, 361.

59. Hrotsvit of Gandersheim 1989, 100. For Hrotsvit's use of music, see Chamberlain 1980.

60. Hrotsvit of Gandersheim 1989, 113.

61. Adamson 2006, 3.

62. Ibid., 162.

63. Ibid., 165.

64. Cited in ibid., 172.

65. Ibid., 173.

66. Ibid., 174, which only addresses the first two strings. Jouanna 2012, 335–359, notes the ambiguity of phlegm in Galen, who sometimes describes it as having "no effect on the character of the soul" (340), elsewhere as associated with despondency and forgetfulness (342).

67. See the excerpt from his "Maxims of the Philosophers" (*Nawadir al-falasifa*) in Godwin 1993, 91–98, at 97.

68. Qifṭī 1903, 361; Kümmel 1977, 61.

69. See the excerpt included in Godwin 1993, 112–122.

70. Sarton 1975, 2: part 1, 76.

71. Khodadoust et al. 2013, 291.

72. Ibid., 292.

73. Following the more recent translation of Holford-Strevens 1993, 475–476. The sections of the *Canon* on the pulse can be found in Abu-Asab, Amri, and Micozzi 2013, 200–221. Though not a scholarly edition, this translation was prepared from the original Arabic, unlike Gruner 1970, 283–322, which was based primarily on the medieval Latin translation.

74. See Holford-Strevens 1993, 476–477.

75. Ibid., 476.

Chapter 4

1. Paxton 1993. In this practice, the seven penitential psalms were emphasized.

2. Sweet 2006, 93–123, at 98. See also Glaze 1998.

3. Cited from Silvas 1999, 160–161. See Fassler 1998, 149–175, at 150.

4. *Causae et curae* 2:84:15–24, as translated in Sweet 2006, 99 (with some adjustments of the numerical nomenclature of parts).

5. Bingen 1999, 135. Cf. Cannon 1993, 84–86.

6. Ibid.

7. Bingen 1990, 532–533.

8. Bingen 1987, 358–359.

9. Ibid., 121.

10. For instance, see her description of the cosmos as a "vast instrument, round and shadowed, in the shape of an egg" in Bingen 1990, 93–98. For her description of the "harmony of the heavens," see Bingen 1994, 9–10.

11. Callahan 2000, 159.

12. Ibid.

13. Crombie 1995, 27. For his knowledge of medicine, see Crombie 1971, 75–76.

14. Baur 1912, 4–5.

15. Thorndike 1923, 2:436–453 at 445.

16. According to ibid., 158. For the development of medical education, see Siraisi 1990, 48–77.

17. For a survey of this period, see Kümmel 1977, 26–33.

18. For the school of Salerno, see Thorndike 1923, 1:731–741, and Siraisi 1990, 13–14, 57–58.

19. Capparoni 1936. On the Salernitan school, see Lawn 1963; Siraisi 1990, 13–14, 57–58; de Divitiis, Cappabianca, and de Divitiis 2004.

20. Ausécache 1998, 211. For Gilles's discussion of the dactylic meter in pulse, see Choulant 1826, 30.

21. Holsinger 2001, 173–175.

22. Palisca 1985, 51–66, at 54.

23. Ibid., 51, 54–55, 51.

24. Siraisi 1975, 691. For a survey of Peter's work, see Thorndike 1923, 2:874–947.

25. See Vecchi 1967.

26. See the discussion in Pesic 2017, 47–66. For rhythmic notation in this period, see Earp 2018; figure 4.2 only gives the barest outline for an evolving rhythmic notation that had many subtleties.

27. See Kibre 1978.

28. Vescovini 1987, 30–32.

29. Siraisi 1975, 692, thinks that Gentile treats the topic with "disdainful brevity," though still endorsing it.

30. Quoted from ibid., 694.

31. The spelling "Ramis" is also found; see the helpful discussion in Tomlinson 1993, 78–84.

32. Ramos de Pareja 1993, 110.

33. For al-Kindī's influence on Ficino, see Ficino 1989, 28, 48, 50, 51, 83, 86, and Prins 2012, 399.

34. See above, chapter 3.

35. Ficino 1989, 361; also available in Strunk and Treitler 1998, 385–389, at 388. Subsequent references show this latter reference in parentheses.

36. Ficino 1989, 357 (387).

37. Gouk 2004, 100.

38. First quote from Ficino letter 5, cited in Seta, Pirrotta, and Piperno 1989, 129.

39. Plotinus 1992, 369–370 (IV.4.40–41). See also Burnett 2018.

40. Ficino 1989, 361 (388).

41. Ibid., 361 (388–389).

42. See Farndell 2010, 68–69.

43. Ficino 1989, 361 (389).

44. Ibid., 359 (388).

45. Gaffurius 1969, 69, whose name is also given as Franchino Gafori. See also Bonge 1982, 168.

46. Ibid., 169–170, which discusses the problems of translation in these passages.

47. See Haar 1974, 81–82.

48. For Ficino, see Tomlinson 1993; for Campanella, see Walker 1975, 205–212.

49. See McDonald 2018.

50. Cited in ibid., 164.

51. For instance, the Zurich preachers Johann Ulrich Surgant (1502) and Heinrich Bullinger (1525), as well as the Nuremberg physician Ulrich Pinder (1510); see ibid., 168–169.

52. See Schleiner 1991.

53. McDonald 2018, 169.

54. Agrippa von Nettesheim 1992, 2.25, quoted from Agrippa 1993, 323–324, and cited in McDonald 2018, 169.

55. Agrippa von Nettesheim 1993, 345–354 at 351.

56. Ibid., 339.

57. Note that the four final notes correspond to the main division of the modes into *protus* (D), *deuterus* (E), *tritus* (F), and *tetradus* (G), each having two versions: "authentic" (in which the final is at the bottom of the range of the mode) and "plagal" (the final near the middle of the range).

58. Liban 1975, fol. F1r, as translated in McDonald 2018, 171.

59. *King Lear* IV:vii stage direction; *The Tempest* V.i.54–55; III.ii.129–130.

60. For the imagery in figure 4.6, see Mueller 1949.

61. Burton 1989, 2:3 (367).

62. See Pesic 2017, 137–145.

63. For the use of music as a stimulus, see Kümmel 1977, 324–344, which treats music as sedative on 344–362.

Chapter 5

1. Regier 2014, 1. All citations from Kepler's collected works, Kepler 1937, will be listed as KGW.

2. Caspar 1993, 109. Kepler was well enough informed about human pulse to make salient comparisons with astronomical time; see KGW 7:180, and Kümmel 1977, 48.

3. Jardine 1984, 228.

4. Rosen 2003, 44n62. See also Rothman 2017, 200.

5. Regier 2014, 1.

6. Rosen 2003, 43.

7. Kepler 2015, 47; KGW 3:19; as translated in Regier 2014, 9.

8. Boner 2013, 41, citing Kepler 1937, 4:177.18–25. See also Jardine 1984, 223–224, at 151, and Regier 2014, 8.

9. Boner 2013, 42; KGW 11, 2:48.23–28.

10. Boner 2013, 42; KGW 14, no. 130:640–651.

11. Escobar 2008, 32. See also the helpful discussion in Regier 2014.

12. Boner 2013, 169.

13. Kepler 1997, 387.

14. Ibid., 304.

15. Quoting Stephenson 1994, 2.

16. Kepler 2010, 33. See the helpful introductions by Owen Gingerich (3–11) and Guillermo Bleichmar (15–22); see also Pesic 2000, 87–90.

17. Kepler 2010, 85.

18. Ibid., 89.

19. Ibid., 93.

20. Ibid., 113.

21. KGW 1:268.26–38, as translated in Boner 2013, 31n113.

22. Ibid., 31.

23. Regier 2014, 1.

24. *On the Generation of Animals*, 736b29–737a1 Aristotle 1984, 1:1143.

25. Boner 2013, 32.

26. Ibid.

27. Ibid. 123; KGW 8:131.3–6.

28. Boner 2013, 113–114; KGW 8:216.37–39, 1:341.1–3. For helpful cautions about the meaning of "physiology" in this context, see Schechner 1997, 100, 143, 148–149, 259.

29. Boner 2013, 105; KGW 4:62.22–24, 19–22.

30. Caspar 1993, 184.

31. Boner 2013, 35, 107; KGW 15: no. 394, 74–78; 4:74.41–75.1.

32. Boner 2013, 159, 66; KGW 6:261.25–26, 16: no. 488, 42–44.

33. See Kepler 1997, 330–333.

34. Boner 2013, 66–67; KGW 16: no. 488, 93–95.

35. Boner 2013, 33n126; KGW 4:140.40–43. See also Juste 2010.

36. Boner 2013, 67.

37. Boner 2013, 154; KGW 6:406.32–37.

38. See Kepler 1997, 439–450, and Pesic 2014, 78–88; 2017, 145–147.

39. Boner 2013, 86–87; KGW 1:317.21–30.

40. Boner 2013, 87; KGW 1:317.37–318.2.

41. Boner 2013, 96–97; KGW 1:288.24–28.

42. See Rothman 2017, 188–197, which treats his relation to James on 200–203.

43. For James's visit to Tycho, see Thoren 1990, 334–335; Christianson 2002, 140–141. For Kepler's dedication, see Kepler 1997, 2–5, at 3.

44. Kepler 1997, 3–4.

45. Ibid., 4.

46. Ibid., 4–5.

47. Ibid., 5.

48. Ibid., 255–279.

49. See Rothman 2017, 204–216, for Kepler's relation to Bodin.

50. See Orr 2016.

51. Kepler 1997, 276.

52. Ibid., 275.

53. Ibid., 276.

54. Boethius 1989, 65–66 [212].

55. Kepler 1997, 446.

56. See Pesic 2000, 96–112.

57. See Pesic 2014, 82–86; 2017, 146.

58. For allusions to Ficino, see Kepler 1997, 322n41, 363n78.

59. For his relation to Wallenstein, see Caspar 1993, 338–345.

60. Kepler 1997, 308n28.

61. Ibid., 309n29; Caspar 1993, 339.

62. Kepler 1997, 373.

63. Caspar 1993, 339–340.

64. As translated in ibid., 341, 342.

65. Ibid., 342.

66. Ibid.

67. For instance, Kepler thought that the configuration of the heavens sixty-four minutes after his own birth foretold a "nearly fatal fever" he had undergone at age sixteen. See Boner 2013, 61–62; KGW 13: no. 117, 347–355.

Chapter 6

1. Morton 1969, 180–181, discussed in Daboo 2010, 1–2.

2. For instance, this was the view of Sigerist 1948, 113–114.

3. *Ion* 536c (Plato 1997, 945). See the insightful discussion of musical possession in Tomlinson 1993, 145–188.

4. As translated in De Martino 2005, 214. For the larger context, see Walker-Meikle 2014.

5. De Martino 2005, 217.

6. Ficino 2005, 363 (389).

7. Tomlinson 1993, 164–165.

8. Ibid., 125.

9. Ibid., 124–125.

10. Valletta 1706, as translated in De Martino 2005, 124–125.

11. For the relation of the Christian church to dance, see Daboo 2010, 93–99, at 95.

12. Acts 28:3.

13. Daboo 2010, 109–110.

14. Ibid., 110–111.

15. De Martino 2005, 221.

16. Ibid., 102.

17. Daboo 2010, 114.

18. Quoted from ibid., 113.

19. Ferdinandus 1621, 248; the modern edition by Arcuti 2002, 28, will subsequently be cited in parentheses. Translation mine.

20. Ferdinando specifies the left hypochondrium. In what follows, I shall use "tarantula" in this sense, noting that this term should not necessarily be understood as restricted to any particular species of arachnid.

21. Ibid., 248 (27).

22. Ibid., 254 (48).

23. Ibid., 253 (46).

24. Ibid., 264 (91).

25. Ibid., 264–266 (91–97).

26. Ibid., 266 (97).

27. Ibid., 254 (50).

28. Ibid., 261 (77–78).

29. Ibid., 262 (81).

30. Ibid., 262 (61–62).

31. Ibid., 254 (51).

32. Ibid., 258 (67).

33. Ibid., 254, 258 (51, 66).

34. Ibid., 258 (66).

35. Ibid., 259 (69).

36. Ibid., 258–259 (68).

37. Ibid., 250 (32–33).

38. Ibid., 266 (100).

39. Ibid., 267 (101).

40. Ibid., 261 (79).

41. Ibid., 267–268 (104).

42. Ibid., 258 (66–67).

43. Findlen 2004a, 1–48, at 1, which gives an excellent overview of his work and the controversies surrounding it.

44. Mayer-Deutsch 2004, 111.

45. *Ion* 55d–e (Plato 1997, 941).

46. Daboo 2010, 121. See also Cadenbach 2007.

47. As translated by De Martino 2005, 95.

48. Daboo 2010, 125.

49. Baldwin 2001, 90.

50. Daboo 2010, 127, translating from Kircher 1643, 872. The following citations indicate the original pagination in parentheses.

51. Ibid., 128 (873).

52. Ibid., 129 (768).

53. Findlen 2004b, 22.

54. As translated in Caminietzki 2004, 322–323.

55. Ibid., 321 (348).

56. Arcuti 2002, 251 (38); Baglivi 1696, 341; 1723, 370. All citations from this work follow this translation, cited in parentheses.

57. Baglivi 1696 288 (313).

58. Ibid., 287 (312).

59. Ibid., 309 (335–336).

60. Ibid., 294 (319).

61. Ibid., 321 (349).

62. Ibid., 322 (350).

63. Ibid., 322–323 (350–351).

64. Ibid., 332–333 (361–362).

65. Ibid., 333 (362).

66. Ibid., 336 (365).

67. Ibid., 310–311 (335–336).

68. Ibid., 16 (18–19).

69. Ibid., 152 (175).

70. Ibid., 274 (304).

71. Ibid., 210–211 (240).

72. For instance, Serao 1742, 198, cites an unnamed "valiant doctor" who became convinced of the reality of tarantism and used "crisis" to describe its climactic phase.

73. Baglivi 1696, 341 (370).

74. Ibid., 339 (368).

75. Ibid., 340 (368–369).

76. Ibid., 342 (371).

77. Ibid., 343 (372).
78. Ibid., 343–344 (372–373).
79. Ibid., 344 (373).
80. Ibid., 305 (331).
81. Ibid., 308–309 (334–335).
82. Ibid., 320–321 (348).
83. Ibid., 317 (344).
84. Daboo 2010, 141.
85. On Serao, see ibid., 141–143.
86. Ibid., 143; De Martino 2005, 239.
87. Daboo 2010, 145.
88. Serao 1742, 169.

Chapter 7

1. For a compendious survey, see Berg 1942. This history is carried forward in the more modern treatments by Grmek 1970 and Cheung 2010.
2. *Iliad* 4:122 (*neuron* as bowstring); 17:522 (*is* as sinew); Plato, *Timaeus* 82c–d, 84a (Plato 1997, 1282–1283).
3. Berg 1942, 337–338.
4. Von Staden 1989, 192–193.
5. Galen 2014, 260–319 (Book III, chapters 8–14). See Berg 1942, 346–347.
6. Galen 2014, 268–269.
7. As noted by Forrester in Fernel 2003, 1.
8. Forrester in Fernel 2003, 4.
9. Fernel 2003, 279 (79). Page numbers in parentheses refer to the original edition, Fernel 1567.
10. Ibid., 245 (68).
11. Fernel 2003, 321, 323; see Cheung 210, 70–73.
12. Fernel 2003, 329 (94).
13. Vesalius 2002, 45 (275 [375]). Page numbers in parentheses refer to the original edition, as reproduced in Vesalius 1964. Page numbers in brackets refer to the corrected page numbers of this edition.
14. Cicero, *De natura deorum* 2.133.13; cf. Garrison n.d. See also Kusukawa 2012, 216–218.
15. As noted by Lind in Vesalius 1969, xix–xx.
16. See Vesalius 2002, 3:2–3; 1998, 1:4–5.
17. On the originality of this mode of representation, see Singer 1926, 126.
18. Vesalius 1969, 1, 57.
19. Berg 1942, 325–355 at 352, 457.
20. See Krakeur and Krueger 1941, 219–232.
21. Ibid., 140.
22. Ibid., 168. See also Lindeboom 1979; Aucante 2006, 177–178, 193, 236, 421. For the larger context of Descartes's views, see Gaukroger 1995, 17–19, 394–395; 2002, 224–226.
23. As noted by Cheung 2010, 75.
24. Ibid., 73–74, 76 (Grew).
25. Hooke 1961, 1.
26. Ibid., 8.
27. Ibid.

28. Ibid., 113.

29. Ibid., 114.

30. Berg 1942, 373. Original entirely in italics.

31. Grmek 1970, 398.

32. Hooke 1961, 12.

33. Ibid., 16.

34. Ibid., 15.

35. Ibid.

36. Ibid., 15–16.

37. Ibid., 28, 30.

38. Ibid., 173.

39. Hooke 1678, 1.

40. Ibid., 2, 6.

41. Ibid., 6.

42. Ibid., 4.

43. Ibid., 5.

44. Ibid., 9.

45. Ibid.

46. Ibid., 11.

47. Ibid., 13–16.

48. Ibid., 16.

49. Willis 1681, 96. See the helpful discussion in Wallace 2003, 79–80.

50. Willis 1681, 126, 128.

51. Wallace 2003, 80.

52. As argued by ibid., 68–80.

53. Thoreau 1993, entry for September 5. Regarding metaphor in science, see Pesic 2017, 257–258.

54. Croone 2000, 75. In fact, this short treatise was appended to Willis's *Cerebri anatome*, causing some confusion about its authorship.

55. Croone 2000, 77. Regarding his theories, see Wallace 2003, 80–82.

56. Croone 2000, 77.

57. Ibid., 77–79.

58. Ibid., 79–81.

59. Borelli 1989, 8–9.

60. Ibid., 119, replacing this translation's "zither" by the original term, "kithara."

61. Ibid., 120.

62. Borelli 2015, 206.

63. Ibid., 147.

64. Ibid., 147, 168.

65. Ibid., 169.

66. Borelli 1989, 271–273.

67. Ibid.; on the spiral action, see also 257–260.

68. Ibid., 258.

69. Ibid., 318–319.

70. Borelli 233.

71. Ibid., 233–234.

72. Ibid. 234.

73. Ibid.

74. Ibid., 398–399.

75. Ibid., 354, refers to Glisson among other "distinguished authors." See Temkin 1964. For a general survey, see Berg 1942, 386–390.

76. Verworn 1913, 2–5, at 4.

77. Glisson 1677, 198.

78. Baglivi, *De praxi medica*, 322 (349).

79. Ibid., 249 (279).

80. Ibid., 227 (257).

81. See the excellent survey by Darrigol 2007.

82. See Berg 1942, 407–409, 420–426. Quote from Haller 1974, 7.

83. For these experiments, see Haller 1936; Steinke 2005; Eichberg 2009.

84. Bonnet 1770, 257.

85. Cheung 2010, 81–92, at 82n33.

86. See ibid., 83–84.

87. Ibid., 84–87, at 86.

88. Bonnet 1755, 13–14.

89. Bordeu 1818, 2:802.

90. Krakeur and Krueger 1941.

91. Diderot 1875, 2:30–35, at 30.

92. Ibid., 2:33.

93. As argued by Christensen 1993, 133–68, at 138n19. See also Christensen 1994.

94. More generally, see Lockhart 2017, 19–84, for a helpful discussion of animated statues. She emphasizes Étienne Bonnot de Condillac's *Traité des sensations* as giving a theoretical basis for such animation (19–43) in terms of sympathetic vibrations.

95. Diderot, *Œuvres complètes*, 5:477. My translation; see also Diderot 1964, 79. Further references cite their versions in parentheses, where applicable.

96. Ibid., 5:468 (71).

97. Ibid., 1:408.

98. Ibid., 9:279.

99. "Réfutation suivie de l'ouvrage d'Hélvetius intitulé De l'homme" (1774), ibid., 2:367–368.

100. "Salon de 1763," ibid., 10:210.

101. "La rêve de D'Alembert," ibid., 2:113 (100). Note that this work only appeared in 1830, though it was written in 1769.

102. Ibid.

103. Ibid., 114 (101).

104. Ibid., 114–115 (100–101).

105. Ibid., 2:123 (109).

106. Ibid., 2:141 (126).

107. Ibid. (126–127).

108. Ibid. 2:141–142 (127).

109. Ibid., 2:143 (128).

110. Darwin 1796, 1:367; Schwartz 2011, 157.

Chapter 8

1. This *regard médical* is the keynote of Foucault 1972. Even when he tries to include "the tactile and auditory dimensions," Foucault still asserts that they remain "under the dominant sign of the visible"; Foucault 1994, 165. On the history of anatomo-clinical medicine, see Keele 1963; Ackerknecht 1967; Maclean 2002. On the history of clinical listening, see Jackson 1992.

2. For references from Pietro through the mid-sixteenth century, see Kümmel 1977, 32–38.

3. *Hamlet* III.iv.140–142.

4. For a general description of his work, see Kümmel 1977, 38–44.

5. Rost 1953; Lyncker 1966.

6. See Kümmel 1977, 44–47.

7. For Montpellier, see Williams 2003.

8. For Marquet's career, see Costa 2008, 20–21.

9. Besides Herophilus, he placed his project in the lineage of "Avicenna, Savonarola, Saxon, and Fernel, and many other learned physicians of antiquity," who proposed this connection between music and cardiac rhythm, "though unable to execute it" (Marquet 1769, 4–5). All translations are by the author, unless otherwise noted. Among the few mentions of Marquet's work, see Kümmel 1977, 47–50, who emphasizes its relation to the strongly accented music of the time, and Kassler 1995, 220–221. There are brief descriptions in Pistacchio and Pistacchio 1996 and Grignon 1997. The most significant treatment of Marquet to date is in Sykes 2014, 60–63, which helpfully contextualizes different modalities of hearing.

10. Marquet 1769, 2–3.

11. Ibid., 1.

12. Ibid., 1–2.

13. Ibid., 3–4.

14. Here Marquet referenced "the great [Friedrich] Hoffmann [1660–1742]"; see Duchesneau 1982, 32–64.

15. Marquet 1769, 210, referring to Duverney 1683. Regarding Duverney, see Sykes 2014, 27–37; Gouk and Sykes 2011.

16. Marquet 1769, 199–200. Marquet also applied his mechanical understanding of the body to advocate the use of music in the treatment of melancholia in his "Oration on the Twenty-Third Aphorism of Hippocrates" (delivered in 1759, the year of his death), in Marquet 1770, 198–215 at 198.

17. Note that Marquet's first edition preceded the 1752 original publication of Quantz 1966. For discussion of issues regarding pulse measurement of tempo, see Pesic 2013. Wellmann 2017, 68, also comments on Marquet.

18. Marquet 1769, 37.

19. Ibid., 15–16.

20. Ibid., 36–37.

21. For instance, Quantz 1966, 283–291. For developments in the musical theory of rhythm, see Wellmann 2017, 61–75, commenting on Marquet at 68.

22. Marquet 1769, 32.

23. Ibid., 32–33.

24. Ibid., 33–34.

25. Vieussens 1715.

26. Cf. the assertion of Sykes 2014, 61, that Marquet's minuet overlaid "perceived pulse sounds over the form of the dance" so that "by visualizing the *sounds* of the pulse a better diagnosis could be made."

27. Marquet 1769, 40. I have corrected the evidently erroneous "devient rapide" to "becomes less rapid."

28. Regarding these critical reactions, see also Kümmel 1977, 51.

29. Bordeu 1756, 451. For Bordeu's relation to Montpellier, vitalism, and Diderot, see Williams 1994, 20–49; 2003, 147–162, 215–220. Bordeu's commentary on Marquet is also helpfully addressed in Sykes 2014, 58–70. See also the helpful discussion in Emch-Deriaz 2001.

30. Sykes 2014, 67.

31. Its editor, Marquet's son-in-law Pierre-Joseph Buchoz included a number of encomia (including his own), but these came preponderantly from physicians from Nancy, hence part of Marquet's professional circle, perhaps his friends or those who wished to honor his memory. See Pesic 2016, 12n36. Regarding Buchoz, see Erlmann 2010, 139.

32. Marquet 1769, "Jugement de Monsieur le Baron Duhaler sur cet ouvrage" (79–81), answered by "Réponse à l'auteur anonyme, sur sa satyre, Contré le Traité du Pouls" (82–83).

33. Buchoz also includes Ménuret's "Extrait du nouveau Traité du Pouls" (ibid., 86–95), which largely duplicates material in "Exposition en forme de critique, de la Doctrine du Pouls par la Musique, par les Auteur du Dictionnaire Encyclopédique, à l'article Pouls" (142–160). To Buchoz's credit, he includes the passages most critical of Marquet.

34. Much of this material also appeared soon thereafter in Ménuret's *Nouveau traité de la pouls* (1768).

35. See Williams 2003, 119–23, 226–230.

36. Ménuret de Chambaud 1751b, 13:220–221, also quoted in Marquet 1769, 144, 159.

37. Thus, for instance, Ozanam 1886, 111, after summarizing Marquet's ideas, finds that they have "no practical result and in no way advanced the semiotic sciences." I thank Martine Voiret for her expert advice on the eighteenth-century connotations of "détails pénibles."

38. Ménuret de Chambaud 1751b, 13:221–228

39. Ibid., 13:225.

40. Ménuret de Chambaud 1751a, 10:903.

41. Ibid., 10:904. a view he ascribes to one "dom Calmet."

42. Throughout, Ménuret cited his friend Roger 1758.

43. Ménuret de Chambaud 1751a, 907. For the connotations of "fibers" in Montequieu, see O'Neal 2002, 102–105.

44. Ménuret de Chambaud 1751a, 10:908, which mentions that his findings were "read at the Société Royal des Sciences" but does not give a date or further reference.

45. For instance, Marquet is not mentioned in Roger 1758 or in its later translation, Roger 1803.

46. Floyer 1707; Gibbs 1971, 1994. The minute hand had first appeared in watches about 1670. For the earlier history, see Levett and Agarwal 1979.

47. Porter 2001, 165–166.

48. Haller 1760, 2:171.

49. Ibid., 2:182.

50. For instance, Falconer 1796, 17, cites Marquet's value of 60 for adult resting pulse as coming from Haller.

51. This point is also emphasized by Fantini 2013, which discusses Haller on 259–260.

Chapter 9

1. Brechka 1970 discusses Van Swieten's reforms in the teaching and practice of medicine (132–142) and his interest in music (114). See also Erlmann 2010, 140–141.

2. Swieten 1754, 1:38.

3. These are discussed in ibid., 1:39–97.

4. Ibid., 11:478–479, sec. 1191.

5. Ibid., 8:241, sec. 826.

6. Duffin 2006, 269–70, esp. 269n78.

7. Lesky 1959.

8. For convenience, I will cite from the original text and translation given in Auenbrugger 1808, xxi. See the very helpful discussions of Auenbrugger in Sterne 2003, 119–122, and Volmar 2012, 22–36.

9. Neuburger 1922, 11.

10. His only mention of casks is Auenbrugger 1808, 46–47. I will bring forward evidence confirming the suggestion that "Auenbrugger's musical abilities may have contributed to his discovery of thoracic percussion" given in Jarcho 1961, 169. Dock 1935, 447, also thinks that Auenbrugger's musical training "must have influenced his choice of words."

11. This has been noted also by Schwartz 2011, 202–205.

12. For the score of the opera, see Salieri and Auenbrugger 1986. The editor's introduction provides useful commentary on the context of the work in the development of German Singspiel. For the printed libretto, see Schwan et al. 1986. Mozart's comment comes in his letter of December 10, 1783: Mozart 1997, 863.

13. Abert 2007, 258.

14. Glickman, n.d.

15. This work was rediscovered and included in Harbach et al. 1986.

16. Auenbrugger 1808, 2, 4.

17. Blades and Bowles, n.d.; Blades 1970, 263. Given his close association with the composer, Auenbrugger probably was aware of Salieri's innovative use of timpani with unusual tunings in operas written in 1772 and 1785, though these postdated Auenbrugger's clinical work of the 1750s and his ensuing 1761 publication detailing his discovery of diagnostic percussion.

18. For the relation between Corvisart and Auenbrugger's work, see O'Neal 1998, 2002, 162–185, and Volmar 2012, 37–40.

19. Auenbrugger 1808, 4.

20. Corvisart also emphasized the tactile aspects of percussion; see McCarthy 1999.

21. For details, see Pesic 2016, 19.

22. Auenbrugger 1808, 37–42, at 41.

23. Ibid., 132.

24. Ibid., 133.

25. Perhaps this is the ultimate source of the famous "dog that does not bark" invoked by Sherlock Holmes, influenced by his physician-creator, Arthur Conan Doyle.

26. Auenbrugger 1808, 171–176.

27. O'Sullivan 2012. See also Starobinski 1966; Roth 1991.

28. Auenbrugger 1808, 172.

29. Ibid., 178.

30. Ibid.

31. Among the most useful studies, see Reiser 1978, 23–44; Sterne 2003, 99–136; Duffin 1998; Volmar 2012, 38–46.

32. Duffin 1998, 17, 20, 80–82. Laennec's musical background is also discussed insightfully by Schwartz 2011, 206–211.

33. Duffin 1998, 252.

34. Ibid., 124. For Bayle, see Reiser 1978, 25.

35. Duffin 1998, 96–101, at 98.

36. Ibid., 42.

37. Ibid., 129.

38. Corvisart des Marets 1812, 19.

39. Jarcho 1961, 169.

40. For an insightful discussion of the growing significance of timbre in music, see Dolan 2013.

41. For the evocation of microsound in nineteenth-century music, see Brittan 2011.

42. For a detailed account of his work in cardiology, see Duffin 1989; Duffin 1998, chap. 8. For the changing practice of listening involved in auscultation, see Sterne 2003, 117–136.

43. Sterne 2003, 135–136, at 135. For a modern medical student's thoughtful account of learning to auscultate, see Rice 2011. See also the larger perspectives discussed in Bijsterveld 2019, 11–12 (on Laennec).

44. Grmek 1981; Duffin 1998, 134–138.

45. Laennec 1826, 64.

46. Ibid., 64ff.

47. Ibid., 436–437. More generally about this phenomenon, see Duffin 1998, 141–144.

48. Duffin 1998, 112–113.

49. Laennec 1826, 75.

50. Ibid., 743–744.

51. La Villemarqué 1867, viii (music), 120–122 (text).

52. Laennec is somewhat unclear on Erman, whom he also calls "Hermann" and identified only as secretary of the Academy in Berlin with whom he corresponded in 1820. The papers in question are Erman 1812; Wollaston 1812.

53. Laennec 1826, 746–752, at 752–753.

54. Ibid., 752–753.

55. Some of Laennec's editors saw fit to omit these coronary melodies as "mere curiosities." For example, the notated melody was omitted with this editorial footnote in Laennec 1830, 567. Laennec's musical description is described as "the most disarming passage in his entire opus" by Duffin 1998, 192, which also cites Cruveilhier's reaction and discusses the reception of Laennec's treatment of the heart on 174–206. For this and other reactions to stethoscopy, see Reiser 1978, 30–44, at 34–35.

Chapter 10

1. See Goldstein 2001, 155–156.

2. For an overview of the phenomenology of obsession, see Davis 2008, 3–30.

3. This work was identified as a precursor of Hector Berlioz's use of idée fixe by Brittan 2006, who acknowledged Ralph Locke for this information (214n14).

4. The most detailed treatment to date is Howe 2017. See also Georgallas 2016, whom I thank for kindly sending me a copy of this unpublished presentation and of the Piquer *Discorso*.

5. For a historical overview of these clinical categories, see Kümmel 1977, 277–287.

6. Hunter and Macalpine 1963, 27.

7. Ibid., 27–28.

8. Ibid., 28.

9. See, for instance, ibid., 377–378.

10. Ibid., 187. See Eadie 2003a, 2003b.

11. Hunter and Macalpine 1963, 187. See Willis 1674. "Mania" was translated as "madness" in Willis 1683.

12. Willis 1683, 193.

13. Ibid., 201.

14. Hunter and Macalpine 1963, 163. See the historical survey in Berrios 1996, 140–156, at 140–141.

15. Berrios 1996. 142.

16. Ibid., 141.

17. For an overview of these developments, see Davis 2008, 32–54.

18. "Yo fuí loco, y ya soy cuerdo"; Cervantes 1979, 668. See Cejador y Frauca 1905, 695.

19. For obsession in other literary works, see Davis 2008, 54–65.

20. Cervantes 1979, 147.

21. For the details of these changes, see Pesic 2019, 69–70.

22. Kamen 2001, 121–124, 159–160, 162–163, 178–179, 189–191. For the larger context of Felipe V's disordered predecessor, Carlos II, see Green 1993, 153–175.

23. For his frog delusion, see Kamen 2001, 165.

24. Coxe 1813, 3:89.

25. Kamen 2001, 200–202; Green 1993, 173; Kirkpatrick 1953, 94–106.

26. See Kirkpatrick 1953, 87–92, 107–136.

27. Piquer 1759, as translated in Pérez et al. 2011, which gives a helpful analysis of Piquer's work.

28. His later textbook (Piquer 1764, 94–95) cites Willis's *De febribus* (1659) but not Willis's work on mental conditions, which he does not seem to have known.

29. According to Pérez et al. 2011, 73, 74.

30. Petrie 1971, 3, 229.

31. For his melancholy, see Hargreaves-Mawdsley 1979, 100. For his remark to the ambassador, see Petrie 1971, 228, 164–165.

32. Petrie 1971, 164–165.

33. For the larger context, see Ullersperger 1871; Alvarez and Winston 1991.

34. Petrie 1971, 223–224. As king, Carlos IV was "good-hearted but weak and simple-minded . . . vacillating and confused," his "ineptitude" leading directly to the fall of the Bourbon dynasty and the Napoleonic takeover in 1808, according to Payne 1973, 415–416, 420.

35. Petrie 1971, 226–227.

36. Ibid., 224.

37. Belgray 1970, 20.

38. Ruiz Casaux y López de Carvajal 1959, 16, cited and translated in Belgray 1970, 26.

39. Ibid., 83.

40. Ibid., 87–88, 93.

41. Ibid., 320.

42. Lynch 1989, 294.

43. Ibid., 295.

44. Newell Jenkins rediscovered and recorded it, published as Brunetti 1960.

45. See Head 2014.

46. See Ramos 1997, 304–318.

47. The immediate cause seems to be the semitone E-flat–D in the first violins (m. 20), but the Largo had ended with an extended G–F# trill in the first oboe (mm. 17–19).

48. Iriarte 1779, 9, 40, as translated in Iriarte 1807, 48.

49. Piquer 1764.

50. As argued by Georgallas 2016.

51. Belgray 1970, xvi–xvii.

52. Ibid., xvii.

53. See Ivanovich 2014.

54. Though this reprise closely follows the original Andantino, it ends on the dominant of C rather than the original dominant of E-flat.

55. Ibid. This is the first time the symphony has moved away from E-flat and C minor, though only so far as to its parallel major.

56. Belgray 1970, 97–105, points out that several accounts of the enmity between Boccherini and Brunetti lack any documentary foundation so that "it is difficult to fathom the origin or determine the truth of the story."

57. Rothschild 1965, 39–40. The translation has been slightly adjusted to follow the orthography of the Spanish original, given in Solar-Quintes 1947.

58. Rothschild 1965, 40, however without giving further evidence for this assertion.

59. Belgray 1970, 103.

60. Rasch 2014, 291.

61. This is the candidate of Gérard 1969, 346, though Le Guin 2006, 69, judges that "the quintet referred to here cannot be identified."

62. Cf. Howe 2017; Georgallas 2016 reads the symphony in terms of mania-melancholia as bipolar disorder.

63. Cervantes 1979, 614.

64. See Durán 2006.

65. Esquirol 1838, 2:28–29.

66. For details of the evolving concepts of idée fixe and monomania, see Pesic 2019.

Chapter 11

1. See Gouk 2000a, 2004.

2. Kramer 2000; Gouk 2000b.

3. For a survey of Mesmer's career, see Gauld 1992, 1–22. For a helpful compilation of the vast literature on Mesmer, see Crabtree and Wozniak 1988.

4. Mesmer 1980, 5.

5. Cf. Pattie 1956.

6. Riskin 2002, 198.

7. See Pesic 2014a. For insightful treatment of the wider context of musical glasses, see Wolf 2015; Howard 2017, 31n32.

8. For the organological context of the harmonica, see Dolan 2013, 61–65.

9. Letter of July 13, 1762, to Giambatista Beccaria in Franklin 1959, 10:116–130.

10. Wolf 2015, 129, 114, 126.

11. Ibid., 118–119, quoting a poem by a fellow Philadelphian, Nathaniel Evans.

12. For two examples, see King 1945, 100, 118.

13. Chateaubriand 1932, 111.

14. Lipowski 1984.

15. Matthews 1975. For the female associations of this instrument, see Hoffmann 1991, 113–132, and Hadlock 2000. See also Zeitler 2013, 189–204.

16. For Nissen's view, see Abert 2007, 97n57, 58.

17. Walmsley 1967, 49.

18. According to Abert 2007, 1205–1206, the motet was written on June 17, 1791, the quintet K617 before May 23, 1791 (when Mozart entered it into his personal catalogue), and "probably also" the solo harmonica piece K356/617a (1205). See also Schmidt 2003.

19. Note, for instance the characteristic melodic gesture $\hat{5}$–($\hat{4}\#$)–$\hat{4}$–$\hat{3}$ in both works, as well as prominent use of subdominant (plagal) harmonies.

20. For his search for the voice of "his goddess" "dans les frémissement d'une harpe, dans les sons veloutés ou liquides d'un cor ou d'un harmonica," see Chateaubriand 1981, 1:95.

21. Rochlitz 1798, 97–98.

22. Pace 1958, 280. For connections between music and electricity, see Pesic 2014b, 181–215, and Smith 2016, 150–156.

23. Wolf 2015, 125.

24. Rochlitz 1798, 101–102; King, n.d. See also Zeitler 2013, 240–243.

25. Mesmer 1980, 18; Bonuzzi 2007, 46.

26. See Walmsley 1967, 51–59.

27. Walmsley 1967, 82–91; Fürst 2005; for Mesmer's account of the case, see Mesmer 1980, 58–65, 71–76. The spelling "Paradies" is also found.

28. King 1945, 110.

29. Ibid.

30. See Hadlock 2000, 528–532.

31. The image of the magnetic chain describing the transmission of poetic and musical inspiration goes back to Plato, *Ion* 533d–e, 535e–536c (Plato 1997, 941, 943–944).

32. Buranelli 1975, 125–126.

33. Pattie 1994, 73.

34. D'Eslon 1780, 90.

35. As judged by Hadlock 2000, 508.

36. Darnton 1968, 40.

37. Mesmer 1980, 67–68.

38. Franklin 1785, 24.

39. Darnton 1968, 151–152.

40. For "critical sleep," see Mesmer 1980, 112, 122–126.

41. Ibid., 103. His statement recalls the general form of Hooke's Law discussed in chapter 7.

42. See Gensini and Conti 2004; Wittern 1989.

43. See Mesmer 1980, 104–105.

44. D'Eslon 1780, 36–37.

45. Ratner 1985, 48–51.

46. Rosen 1998, 57.

47. See Hoyt 1996, which cites the use of *noeud* as term of dramaturgy on 150n32.

48. Darnton 1968, 143. Capitalization in the original.

Chapter 12

1. For instance, see Binet and Féré 1888; Braid 1899.

2. For an overview of Puységur's career, see Gauld 1992, 39–52.

3. Puységur 1820, 80.

4. Ibid., 50.

5. Ibid., 269.

6. Ibid., 165.

7. Ibid., 165–166. A *plateau électrique* was a small round piece of glass electrified by friction.

8. Richet 1884, 543.

9. Ibid., 200. For an overview of Richet and Charcot, see Gauld 1992, 297–317.

10. Harrington 1985.

11. Claretie 1881, 126. See Didi-Huberman 2003; Arnaud 2015.

12. For Donezetti's original version, see Donizetti 1941, 151r–165v. For a discussion of the change of instrumentation, see Hadlock 2000, 534–535, and Gossett 2006, 434–435. See also Sala 1994, 40.

13. Ashbrook 1965, 416–417.

14. Hadlock 2000, 534. For an insightful discussion of the reception history of this scene and its medical context, see Pottinger 2020.

15. This cadenza is generally interpolated at 4 bars before rehearsal number 32 in act III, number 14, according to Pugliese 2004. For the accompanied version, see Warner Classics 2014. For the unaccompanied cadenza, see Serban n.d.

16. Pugliese 2004, 35–37, at 37.

17. For Richet's influence, see Ellenberger 1970, 90; Bogousslavsky 2011, 200. For Charcot's study of magnetism-inducing metals, see Harrington 1988. For the roles of the women patients, see Borgstrom 2000.

18. Charcot 1882, 403, characterizes this type of hypnotism as "its type of perfect development," echoing the language of Puységur's *crise magnétique complète*.

19. For the use of the tam-tam in Charcot's clinics, see Richer 1881, 401, 600; Hustvedt 2011, 78, 196–199; Didi-Huberman 2003, 208–210.

20. Raz n.d.

21. Bourneville and Régnard 1879, 178–181.

22. Ibid., 179–180.

23. Claretie 1881, 129, who tells another such story of inadvertent catalepsy induced by cymbals on 128.

24. See Blades 1992, 382–385. For a comprehensive survey of the instrument's history, see Kreuzer 2018, 109–161.

25. Amiot 1827, 366. Amiot goes on to describe the forging of such a *lo* in detail, Because he describes tuning the newly forged *lo* against another an octave apart, one wonders whether he is describing a pitched gong. In a 1786 letter, Amiot mentions the "*yun-lo* commonly called 'tam-tam'"; see Hermans 2005, 56n212. For the science of fabricating tam-tams, see Kreuzer 2018, 135.

26. Ibid., 135.

27. Monelle 2006, 126.

28. Ibid.

29. Kreuzer 2018, 119.

30. See Tiersot 1908, 52–53, 56, 58–59, 212–213.

31. Martin 1868, 3:460. See also Johnson 1996, 139–140.

32. As given in Tiersot 1908, 52, 274; the *Moniteur* citation is from April 6, 1791.

33. As noted in ibid., 58. Here *timbre* may also refer to a "convex metal disc" or cymbal, from the Greek *tymbanon*, perhaps cognate with "clanging cymbal [κύμβαλον ἀλαλάζον]" in Paul's celebrated passage (1 Corinthians 13: 1). Chénier's poem "Du pouvoir de la musique" was dedicated to the operatic composer Étienne Méhul, at whose funeral Gossec's march was also played; see Monelle 2006, 129.

34. For the history of French orientalism, see McCabe 2008.

35. Fage 1844, 211. This was presumably the tam-tam Amiot sent to the duc de Chaulnes.

36. Berlioz 2002, 286.

37. Berlioz's *March funèbre pour la dernière scène d'Hamlet* Op. 18 no. 3, H103 (1848) used a tam-tam crescendo to fortissimo to underline the chorus's outcry "Ah!" (at rehearsal number 5), while soldiers in an off-stage platoon fire gun volleys in salute. To augment the shock from the sudden cutoff of this tremendous din, Berlioz directs the tam-tam to damp its sound abruptly.

38. See Kreuzer 2018, 135.

39. Hibberd 2004, 120, 119.

40. See Cardwell 1983.

41. See Devoto 1995; Bass 2001.

42. Cooper 1951, 56. For Richard Wagner's ambivalent relation to the tam-tam, see Kreuzer 2018, 142–151.

43. Charcot 1890, 17; Hustvedt 2011, 97.

44. See Ellenberger 1970, 96–101; Shorter 2008, 166–200.

45. Harrington 1985, 240.

46. Babinski 1901.

47. Hoffmann 2003, 164, which refers directly to the writings of Johann Ritter (see note 15 above). For Hoffmann's relation to mesmerism, see Ellenberger 1970, 159–162; Willis 2006, 28–82; Smith 2016.

48. Hoffmann 2014.

49. Radau 1869, 273; Kreuzer 2018, 142.

50. Freud 1966, 13.

51. Ellenberger 1970, 150, considers mesmeric crisis to be "a variety of what we call today the cathartic therapy."

52. Bernays 1880; Dalma 1963; Ellenberger 1970, 484.

53. Freud 1966, 28.

54. Freud 1963, 80–81.

Chapter 13

1. Mersenne 1963, 2. I thank Schwartz 2011, 96–97, for bringing this and the following citations of Hooke to my attention.

2. Hooke 1705, 39. For the dating of this work, see Hesse 1966, 68.

3. Hooke 1705, 39.

4. Ibid.

5. Ibid., 40.

6. Schwartz 2011, 97.

7. Hooke 1705, 135.

8. See Findlen 2013.

9. Spallanzani and Vassalli 1794.

10. The translation is included in Galambos 1941, here cited from an unpublished manuscript, Spallanzani 1941.

11. Ibid., 4.

12. Ibid., 11–14.

13. Ibid., 17.

14. Ibid., 23.

15. Jurine 1798, 139, a translation of Peschier 1798. Jurine's original memoir of 1794 was never published and seems lost, according to Dijkgraaf 1960, 9.

16. Jurine 1798, 140.

17. Galambos 1942, 135, quoting Jean Senebier's 1799 account of Spallanzani's later views.

18. Ibid., 136–137.

19. Cited in ibid., 137.

20. Whitaker 1906, 149, 147.

21. Hahn 1908, 191.

22. Maxim 1912, 148.

23. Ibid.

24. Ibid., 150.

25. For the early history of sonar, see Ainslie 2010, 6–15; for a comprehensive history through the 1960s, see Hackmann 1984. For Richardson, see S. A. Richardson 1957, 303. His patents are L. F. Richardson 1913a, 1913b.

26. See the fascinating account in Trippett 2018, 202–206.

27. L. F. Richardson 1913a, 3. Above water, he used a "bird call" sounding frequencies near 10,000 Hertz.

28. See Merklinger and Ellis 2017 and Hackmann 1984, 74–75.

29. As predicted by Gabriel Lippmann and observed by the Curies in 1881.

30. "Asdic" was the preferred British term (from "*Anti-Submarine Division*"), "sonar" the American coinage ("*Sound Navigation and Ranging*"). For the history of these terms, see Hackmann 1984, xxv–xxvi, which discusses Curie's work on 77–89.

31. Ibid., 81–82.

32. Hartridge 1920, 56. For his career, see "H. Hartridge MD" 1976.

33. I have followed the account of Page 1962, 20–25.

34. Ibid., 25–26.

35. Griffin 1958, 66.

36. Ibid., 67.

37. Ibid., 69.

38. As estimated by Belrose 2002.

39. As he described it in Tartini 1754, translated in Johnson 1985. For the history, see Jones 1935. For an insightful discussion in the context of Tartini's conception of nature, see Polzonetti 2001, 14–22.

40. See, for instance, Chung 2018.

41. For a careful discussion of the subtleties and controversies surrounding these phantoms, see Heller 2012, 493–504, addressing Tartini tones at 499–501.

42. Fessenden 1902, 2.

43. Hogan 1999, 1981–1982, a reprint of the 1913 original of that paper.

44. Helmholtz 1924, 2:66.

45. Nagel 1974, 438. See Chua and Rehding 2021, 173–174, for an interesting response.

Chapter 14

1. Kittler 1999, xli; see also Winthrop-Young 2002.

2. Nicolson and Fleming 2013, 15–16.

3. Ibid., 15–19, at 17.

4. Ibid., 18.

5. Ibid., 18, 28.

6. Ibid., 19–21.

7. Cited in ibid., 19.

8. Ibid., 15.

9. Ibid., 21.

10. Ibid., 21–22; White 1948.

11. White 1948, 45.

12. Nicolson and Fleming 2013, 32–35.

13. For PPI, see ibid., 23.

14. Ibid., 35–45, at 47.

15. Ibid., 39.

16. Cited in ibid., 49.

17. For the larger dimensions of this connection with childhood illness or injury, see Roe 1953.

18. Regarding this milieu, see Nicolson and Fleming 2013, 42–50.

19. See Donald 1974, 109.

20. Nicolson and Fleming 2013, 90–91.

21. Ibid., 109–110.

22. Donald 1974, 110.

23. For a detailed account of this reenactment, see Nicolson and Fleming 2013, 92–98, at 97–98.

24. Ibid., 98.

25. Ibid., 102.

26. Ibid., 111.

27. Donald 1974, 111.

28. Ibid.

29. Nicolson and Fleming 2013, 114.

30. Ibid., 115, For its early development, see Singh and Goyal 2007.

31. Donald, MacVicar, and Brown 1958, 1193–1194.

32. Cited in Nicolson and Fleming 2013, 117.

33. Ibid., 120.

34. Ibid., 124.

35. Ibid., 123.

36. Ibid., 124–125.

37. Ibid., 135.

38. Ibid., 206.

39. Ibid., 206–207, 220.

40. Ibid., 231.

41. Ibid., 194–195.

42. Ibid., 231.

43. Ibid., 233. See also Singh and Goyal 2007.

44. See Linebarger et al. 1999; Brannen and Bush 1984.

Chapter 15

1. See above, chapter 7, and Wallace 2003.

2. Gouk 2014, 44. Among more extensive studies, see, for instance, Rousseau 2004, Vila 1990; Lawrence 1979.

3. Gouk 2014, 46. Hartley calls this substance "isochronous," meaning that its vibrations occur in equal times like those of a string.

4. Ibid., 47.

5. As recounted in Wiltshire 2019; Eliot 1962, 20.

6. For this history, see McComas 2011, 11–20.

7. See Pesic 2014, 185–195.

8. Ibid., 190.

9. Müller 1838, 819–820, emphasis in the original.

10. Bell 1833, 21.

11. See Crary 1992, 89–90, and Jay 1993.

12. Sterne 2003, 55, giving a helpful overview of Müller on 60–62.

13. As rendered by Erlmann 2010, 205–206, who gives an insightful treatment of Müller on 202–216.

14. Ibid., 208.

15. As noted by ibid., 202–203.

16. Sterne 2003, 62.

17. Müller 1838, 673–674.

18. For this history, see McComas 2011, 11–27.

19. Finkelstein 2003, 2015.

20. For this phase of his work, see Olesko and Holmes 1994.

21. Koenigsberger 1965, 65.

22. For the larger context, see Canales 2009.

23. Piccolino and Bresadola 2013, 276. For a detailed account of this apparatus and Helmholtz's work with it, see Schmidgen and Schott 2014, 55–84, as well as Olesko and Holmes 1994, 68–69.

24. Helmholtz 1954, 122.

25. See Michelson 1880, 121–122.

26. Kleppner 2007, 8.

27. Ibid. and Michelson 1880, 120.

28. Harry 1987. See especially Wise 1997, 200–213.

29. See Helmholtz 1855 in Helmholtz 1882, 2:881–885, and the discussion in Olesko and Holmes 1994, 74–82.

30. Helmholtz 1954, 20.

31. For the larger context of these phonetic investigations, see Brain 2015, 64–92.

32. Koenigsberger 1965, 158.

33. Helmholtz 1954, 129.

34. Pantalony 2004.

35. See Pantalony 2009.

36. Cited from ibid., 428. See also Vogel 1993, 269.

37. Wollaston 1810, 9.

38. Erman 1812, 21–22; in fact, Wollaston's paper appeared in German translation in the same issue as Wollaston 1812.

39. Wollaston 1810, 2.

40. For the context of his contemporary work, see Cahan 2018, 319.

41. For the larger meanings of these words, see Erlmann 2010, 166–168.

42. Haughton 1863, 3.

43. Helmholtz 1864 in Helmholtz 1882, 2:924–927, at 925, translated in this book's appendix.

44. Ibid.

45. Helmholtz 1866 in Helmholtz 1882, 2:928–931, translated in this book's appendix. For the relation between *tonus* and tone, see Erlmann 2010, 128–129.

46. See Lenoir 1986, 28.

47. Ibid., 33. See Bernstein 1894, 131–132.

48. For a good overview of his work, see Lenoir 1986, 39–47.

Chapter 16

1. My account in this chapter relies on Volmar 2015, 60–80, quoted here on 85, as well as Dombois 2006. Because Volmar 2015 is (at this writing) available only in German, I recount a number of his detailed findings to make them more widely accessible.

2. du Bois-Reymond 1877, 575–576, as discussed in Volmar 2015, 62.

3. Ibid., 573, citing the work of R. Grossman he discussed in du Bois-Reymond 1875, 1:170.

4. Bernstein 1878, 123; Volmar 2015, 70.

5. Hermann 1878, 504.

6. Volmar 2015, 67.

7. d'Arsonval 1878, 832.

8. Ibid., 833; emphasis in the original.

9. Tarchanow 1878, 353, who subsequently described the greater sensitivity achievable with the Siemans and Halske telephones in Tarchanow 1879. Volmar uses the contemporary romanization Tarchanov.

10. Tarchanow 1878, 353–354.

11. Volmar 2015, 68.

12. For these developments of binaural listening, see Sterne 2003, 111–113, 155–156.

13. Bernstein 1881, 19.

14. Ibid., 22.

15. Cahan 2018, 576.

16. Stirling 1888, 156, who used tuning forks in many of his demonstrations; see 162, 171–173, 238, 241, 262, 301. See Volmar 2015, 62–63.

17. For these developments, see Volmar 2015, 66–80.

18. Sanderson 1895, 143.

19. Wedensky 1900, 139, emphasis original; for his work, see Volmar 2015, 73–76.

20. Wedensky 1900, 189, as noted by Volmar 2015, 80.

21. Wedensky 1883, 315.

22. Wedensky 1900, 144–145, emphasis original.

23. Snellen 1994.

24. Höber 1919, 305.

25. Ibid., 306.

26. Ibid., 309.

27. Ibid., 310–312, listing also work by Bernstein and others who applied their sonic findings to clinical studies.

28. Ibid., 312. For an insightful discussion of the microphonic amplification of "small sounds," see Abbate 2016, 806–820.

29. See, for instance, Cabot 1923 and the references in Scheminzky 1927, 500–501.

30. See the helpful commentary in Volmar 2015, 84.

31. Scheminzky 1927, 483.

32. Ibid., 498.

33. The term *soundscape* was introduced by Schafer 1993.

Chapter 17

1. For an excellent overview, see Finger 2000, 239–258.

2. Gotch and Horsley 1888, 19.

3. Ibid., 20.

4. See Lippmann 1875; Stock 2004.

5. McComas 2011, 63–64.

6. Lucas 1905, 125.

7. Lucas 1909, 8.

8. Ibid., 118–119.

9. Ibid., 133.

10. See the excellent treatment by Frank 1994.

11. Ibid., 128.

12. As noted by Erlanger and Gasser 1937, 2.

13. See Garten 1910, 553.

14. For the working conditions, see Hodgkin 1979, 13. McComas 2011, 64–67, discusses the early work of Adrian and Lucas.

15. Adrian 1932, 2.

16. Regarding the vibration problems, see Forbes and Thacher 1920, 458–463.

17. Gasser and Erlanger 1922, 501. For the estimate of the natural frequency, see Erlanger and Gasser 1937, 3.

18. Adrian 1921.

19. Ibid., 194.

20. Adrian and Lucas 1912.

21. Ibid., 218–224.

22. Cited in Frank 1994, 226.

23. Adrian and Forbes 1922, which included both afferent and efferent nerves and used a string galvanometer.

24. For this history, see McComas 2011, 98–100.

25. Here I am following the account of Frank 1994, 227.

26. Adrian et al. 1954, 17.

27. Adrian 1926, 49.

28. Ibid., 40; see Lucas 1912.

29. Between September 1924 and July 1925, according to Hodgkin 1979, 28, Adrian was led to this experiment by reports of electrical impulses reported by Forbes, Campbell, and Williams 1924.

30. According to his 1954 recollections in Adrian et al. 1954, 17. See OED s.v. "microphonic" A.2.b: "Designating, relating to, or vulnerable to unwanted noise or modulation produced in electrical equipment by vibration," giving examples dating from 1919.

31. Adrian et al. 1954, 18.

32. Adrian 1926.

33. Ibid., 64.

34. Adrian and Zotterman 1926, 156.

35. Hodgkin 1979, 30.

36. Adrian 1928, 118–119.

37. Adrian and Matthews 1927, 379.

38. See Hodgkin 1979, 31–33.

39. Ibid., 65n76.

40. Ibid., 31. For the summary of their work, see Adrian and Matthews 1927, 410–411.

41. See the summary in Adrian and Matthews 1927, 300–301.

42. On this point, Hodgkin 1979, 65n76, says only that Adrian and Rachel Matthews "had previously listened for impulses."

43. Cooper and Adrian 1924.

44. If they had been working with cephalopods, the optic nerve frequencies would have been higher—around 30–90 Hertz; for these frequencies. See Adrian and Matthews 1928, 276.

45. Ibid., 274.

46. Ibid., 297.

47. Ibid.

48. Hodgkin 1979, 5.

49. See Pollitt 1972, 54–60, at 58.

50. Hodgkin 1979, 33.

51. Adrian and Bronk 1929, 145.

52. Adrian and Bronk 1928, 84.

53. Ibid., 83.

54. Ibid., 84.

55. For the larger context of loudspeakers, see Schwartz 2011, 626–637.

56. See Thompson 2002, 239–241, at 240.

57. Ibid., 241.

58. Devlin 2018, 18–19.

59. Sterne 2003, 137.

60. Hodgkin 1979, 33.

61. B. H. Matthews 1928, 227.

62. See B. H. Matthews 1928; B. H. C. Matthews 1929; assessed by Hodgkin 1979, 35, and Gray 1990, 267. This device was sometimes also called an "oscilloscope" in contemporary sources, but to avoid confusion with the cathode ray device, I have used Matthews's original name.

63. Adrian and Bronk 1929.

64. Ibid., 120.

65. Ibid., 131, 141.

66. See the description and discussion in Hodgkin 1979, 132–134.

67. Ibid., 34, also notes Adrian's skill in roof climbing and fencing (8).

68. Adrian and Bronk 1929, 135n1.

69. Ibid., 135, cueing the note just cited after "loud speaker."

70. "The Nobel Prize in Physiology or Medicine 1932," n.d.a.

71. For the references to "gramophone records," see the note appended to ibid. I thank many people at the Nobel Foundation, Karolinska Insitutet, Trinity College Library, as well as Roger Keynes, Hugh Matthews, and Tilli Tansey for their efforts to find these recordings; I continue to hope that someday they will be found and heard again. For recording in medicine, see also Bijsterveld 2019, 98–99.

Chapter 18

1. Chua and Rehding 2021, 86–94, give a striking exploration of this idea.

2. Feld 2003, 224; see also Rice and Feld 2020.

3. See, for instance, Bijsterveld 2019, 6.

4. For instance, Helmholtz 1954, 149. See Otis 2001, 2002.

5. Adrian 1930, 339.

6. Garson 2015, 33; 2003.

7. Gamma (30–100 Hertz) and delta (around 4) waves were identified later. For Berger's discovery and its context, see Pesic 2017, 248–250.

8. For their statement clarifying Berger's priority, see Adrian and Matthews 1934. See also McComas 2011, 109–111.

9. Adrian and Matthews 1934a, 443.

10. Adrian and Matthews 1934b, 357.

11. Ibid., 382.

12. Adrian 1941, 160.

13. See ibid., 166, 171, 173–174, 182–183, 188.

14. Ibid., 173, 186–187, 189–191. See also Garson 2015, 32–33, 40, 42, 44, who connects signals with larger historical contexts, as does Schwartz 2011, 647–649, 836–837.

15. Quoted from Sterne 2003, 260; see also Volmar 2018.

16. Adrian and Umrath 1929, 142. The comment about Adrian's attitude to the oscillograph comes from Pfaffmann 2012, 334.

17. Adrian and Ludwig 1938, 444.

18. Adrian 1942, 460.

19. Adrian 1947, 23–24.

20. Ibid., 22–23.

21. Ibid., 23–24.

22. Nietzsche 1997, 143, aphorism 250.

23. Daston and Galison 2007, 372.

24. For instance, Hubel 1959; Hubel and Wiesel 1959, which was reissued as Robertson 2009. For a fuller account of their collaboration, see Hubel and Wiesel 2005, which mentions audio monitoring on 60–61; Hubel et al. 2015 revisits their early work and adds new information.

25. This and all other quotes from Conway come from email commentary he sent me in December 2020, for which I am most grateful.

26. Hubel 1982, 28.

27. Hubel 1958.

28. This comment is from a film commentary Hubel made for a reenactment of *Hubel and Wiesel Cat Experiment* n.d., 01″–26″.

29. "The Nobel Prize in Physiology or Medicine 1981," n.d.b. (video), 7′40″–8′23″. Note that his comments in the video version were sometimes more extensive and slightly different from the printed text.

30. Hubel 1982, 27–28. Conway notes that the time line for these experiments ("a month") is not accurate, based on his examination of the records.

31. Ibid., 28.

32. Hooke 1705, 8.

33. Debru 2001, 475; Schwartz 2011, 323–324.

34. Daston and Galison 2007, 314.

35. Ibid., 363. It should be added that they go on to discuss twenty-first-century experiments in which "seeing is action" because the acts of vision in those cases actually change or bring into existence the phenomena in question, such as in nanotechnology.

36. For the brain/computer analogy, see Pesic 2017, 251–268.

37. See Heller 2012, 87–88, at 87.

38. See Pesic 2014, 125–130, 156–157, 174, 222.

39. See above chapter 13 and figure 13.5.

40. Various proposals for "color harmony" connect 3:2 ratio in light frequency (for instance) with the "harmony" of those hues, but the sheer number and differences between such attempts indicate that they are far more problematic than the Pythagorean connections of ratio and consonance.

41. Regarding sonification, see Kramer 1994; Kramer et al. 1999; 2010; Hermann, Hunt, and Neuhoff 2011; Supper 2012; Sterne and Akiyama 2012; Grond and Hermann 2012; Schoon and Volmar 2012; Volmar 2013a 2013b; Dubus and Bresin 2013; Pesic and Volmar 2014; Morat and Volmar 2017; Bijsterveld 2019.

42. Nagel 1974, discussed above at the end of chapter 13.

Chapter 19

1. Brookhart, Moruzzi, and Snider 1950, 466.

2. McComas 2011, 100, does mention that "Adrian could also listen to" the nerve impulses via earphones or loudspeaker.

3. All quotations are from the author's interview with Leslie Kay on February 27, 2020.

4. See Pauletto and Hunt 2006; Crisp 2018.

5. See Kinney and Prass 1986.

6. See, for instance, Benazzouzz et al. 2002. My special thanks to Creig Hoyt and to Beate Diehl, who kindly directed me to the work of Alim-Louis Benabid, of which this is an example.

7. I thank Ari Winnick for pointing out these examples and their ubiquity.

8. See Tavel 1996; Rice 2011.

9. See Zühlke, Myer, and Mayosi 2012.

10. See Gordon and Lagerwerff 1976; Tavakolian et al. 2011; Shams, Zuckerwar, and Dimarcantonio 2016.

11. Movahed and Ebrahimi 2007.

12. Mclaren et al. 1975.

13. Mahnke 2009; Viviers et al. 2017.

14. See McNeil 2019.

15. Duverney 1683, 201–202, as translated in Duverney 1737, 141, brought to my attention by Schwartz 2011, 98–99.

16. Gold 1948. For his detailed personal account, see Gold 2012, 49–58.

17. Kemp 1978.

18. See "Newborn and Infant Hearing Screening" 2010, 15, 26.

19. Temple 2017.

20. Turing 1950, 24.

21. The website is Temple n.d.

22. Temple 2017, 9.

23. Temple 2020.

24. Sterne 2003, 16.

25. Temple 2020.

26. Regarding the gramophone, see Kittler 1999. For the relation between media theory and music theory, see Chua and Rehding 2021, 31–33.

27. See Rehding 2018.

28. According to Anonymous 2018.

29. Email communication, February 27, 2021.

30. Ibid.

31. Parvizi et al. 2018.

32. See Hobbs et al. 2018.

33. Gibbs and Gibbs 1951, 1:131, as discussed by Daston and Galison 2007, 328.

34. For the patents, see Chafe and Parvizi 2013a, 2013b, 2014.

35. Mathews 1963, 553.

36. See Chowning 1977; 2008, 3.

37. Chowning 2008, 5. For a detailed analysis, see Meneghini 2007.

38. "Sonification" n.d.

39. Pesic 2017.

40. Helmholtz 1954, 308. For Helmholtz, see Hui 2013b, 59, 87. I thank Carl Hubel for his recollections of his father's musical interests.

41. Ovid, *Metamorphoses* 15:63–64: "quae Natura negabat / visibus humanis, oculis ea pectoris hausit."

Appendix

[The originals of these two papers are Helmholtz 1864 and 1866, here following the reprinted text and pagination in Helmholtz 1882, 2:924–927, 928–931. Editorial comments are in brackets.]

1. [I have tried to render Helmholtz's *Ton* consistently as "tone," except where the context (such as here) clearly demanded "pitch" (*Tonhöhe*).]

2. "Outlines of a new theory of muscular action, being a thesis read for the degree of Doctor in Medicine etc." London 1863. [Haughton 1863.]

3. [The initial phase of vibration in which the spring (or fork) is beginning to move away from its equilibrium position; contrariwise, closing beats (*Schliessungsschläge*) return to that position.]

4. [Note Helmholtz's shift in terminology from "noise" (*Geräusch*) to "tone" (*Ton*).]

5. [By "tetanus," Helmholtz means not the disease but the condition of strong muscular contractions or spasms—in this case secondary because caused externally by the current.]

References

Abbate, Carolyn. 2018. "Sound Object Lessons." *Journal of the American Musicological Society* 69:793–829.

Abert, Hermann. 2007. *W. A. Mozart*. Edited by Cliff Eisen. Translated by Stewart Spencer. New Haven, CT: Yale University Press.

Abu-Asab, Mones, Hakima Amri, and Marc S. Micozzi, trans. 2013. *Avicenna's Medicine: A New Translation of the 11th-Century Canon with Practical Applications for Integrative Health Care*. Rochester, VT: Healing Arts Press.

Ackerknecht, Erwin Heinz. 1967. *Medicine at the Paris Hospital, 1794–1848*. Baltimore: Johns Hopkins University Press.

Adamson, Peter. 2006. *Al-Kindī: Life, Works, and Influence*. Oxford: Oxford University Press.

Adrian, E. D. 1921. "The Recovery Process of Excitable Tissues: Part II." *Journal of Physiology* 55 (3–4): 193–225.

Adrian, E. D. 1926. "The Impulses Produced by Sensory Nerve Endings. Part I." *Journal of Physiology* 61 (1): 49–72.

Adrian, E. D. 1928. *The Basis of Sensation: The Action of the Sense Organs*. London: Christophers.

Adrian, E. D. 1930. "The Mechanism of the Sense Organs." *Physiological Reviews* 10 (2): 336–347.

Adrian, E. D. 1932. *The Mechanism of Nervous Action: Electrical Studies of the Neurone*. Philadelphia: University of Pennsylvania Press.

Adrian, E. D. 1941. "Afferent Discharges to the Cerebral Cortex from Peripheral Sense Organs." *Journal of Physiology* 100 (2): 159–191.

Adrian, E. D. 1942. "Olfactory Reactions in the Brain of the Hedgehog." *Journal of Physiology* 100 (4): 459–473.

Adrian, E. D. 1947. *The Physical Background of Perception*. Oxford: Clarendon Press.

Adrian, E. D., and D. W. Bronk. 1928. "The Discharge of Impulses in Motor Nerve Fibres. Part I. Impulses in Single Fibres of the Phrenic Nerve." *Journal of Physiology* 66 (1): 81–101.

Adrian, E. D., and D. W. Bronk. 1929. "The Discharge of Impulses in Motor Nerve Fibres. Part II. The Frequency of Discharge in Reflex and Voluntary Contractions." *Journal of Physiology* 67 (2): 119–151.

Adrian, E. D., D. W. Bronk, B. Houssay, E. P. Joslin, W. Penfield, and L. Whitby. 1954. "Memorable Experiences in Research." *Diabetes* 3 (1): 17–27.

Adrian, E. D., and Alexander Forbes. 1922. "The All-or-Nothing Response of Sensory Nerve Fibres." *Journal of Physiology* 56 (5): 301–330.

Adrian, E. D., and Keith Lucas. 1912. "On the Summation of Propagated Disturbances in Nerve and Muscle." *Journal of Physiology* 44 (1–2): 68–124.

Adrian, E. D., and C. Ludwig. 1938. "Nervous Discharges from the Olfactory Organs of Fish." *Journal of Physiology* 94 (3): 441–460.

Adrian, E. D., and B. H. C. Matthews. 1934a. "The Berger Rhythm: Potential Changes from the Occipital Lobes in Man." *Brain* 57 (4): 355–385.

Adrian, E. D., and B. H. C. Matthews. 1934b. "The Interpretation of Potential Waves in the Cortex." *Journal of Physiology* 81 (4): 440–471.

Adrian, E. D., and B. H. C. Matthews. 1934c. "Discoverer of Cortical Rhythm." *British Medical Journal* 2 (3858): 1129.

Adrian, E. D., and Rachel Matthews. 1927a. "The Action of Light on the Eye. Part I. The Discharge of Impulses in the Optic Nerve and Its Relation to the Electric Changes in the Retina." *Journal of Physiology* 63 (4): 378–414.

Adrian, E. D., and Rachel Matthews. 1927b. "The Action of Light on the Eye. Part II. The Processes Involved in Retinal Excitation." *Journal of Physiology* 64 (3): 279–301.

Adrian, E. D., and Rachel Matthews. 1928. "The Action of Light on the Eye. Part III. The Interaction of Retinal Neurones." *Journal of Physiology* 65 (3): 273–298.

Adrian, E. D., and Karl Umrath. 1929. "The Impulse Discharge from the Pacinian Corpuscle." *Journal of Physiology* 68 (2): 139–154.

Adrian, E. D., and Yngve Zotterman. 1926. "The Impulses Produced by Sensory Nerve-Endings. Part II. The Response of a Single End-Organ." *Journal of Physiology* 61 (2): 151–171.

Agrippa von Nettesheim, Heinrich Cornelius. 1992. *De occulta philosophia libri tres*. Edited by V. Perrone Compagni. Leiden: Brill.

Agrippa von Nettesheim, Heinrich Cornelius. 1993. *Three Books of Occult Philosophy*. Edited by Donald Tyson. Translated by James Freake. St. Paul, MN: Llewellyn.

Ainslie, Michael. 2010. *Principles of Sonar Performance Modelling*. Berlin: Springer-Verlag.

Alvarez, Raquel, and C. M. Winston. 1991. "The History of Psychiatry in Spain." *History of Psychiatry* 2:303–313.

Amiot, Jean Joseph Marie. 1827. "Extraite d'une lettre inédite du Père Amiot, jésuite missionnaire à Péking, adressée à M. Bertin, ministre secrétair d'État, le 2 octobre 1784, sur le tam-tam et sur la musique chinoise." *Revue musicale*, no. 15:365–369.

Anonymous. 2018. "Scientists Detect Silent Seizures with Sound." *Chemical Engineering Progress* 114 (5): 9–10.

Arcuti, Silvana. 2002. *Epifanio Ferdinando e il morso della tarantola: appunti sul tarantismo*. Lecce: Pensa.

Aristides Quintilianus. 1983. *On Music, in Three Books*. Translated by Thomas J. Mathiesen. New Haven, CT: Yale University Press.

Aristotle. 1984. *The Complete Works of Aristotle*. Edited by Jonathan Barnes. Princeton, NJ: Princeton University Press.

Arnaud, Sabine. 2015. *On Hysteria: The Invention of a Medical Category between 1670 and 1820*. Chicago: University of Chicago Press.

Arsonval, Jacques-Arsène d'. 1878. "Téléphone employé comme galvanoscope." *Comptes rendus hebdomadaires des séances de l'Académie des sciences* 86:832–833.

Ashbrook, William. 1965. *Donizetti*. London: Cassell.

Aucante, Vincent. 2006. *La philosophie médicale de Descartes*. Paris: Press Universitaires de France.

Auenbrugger, Leopold. 1808. *Nouvelle méthode pour reconnaître les maladies internes de la poitrine par la percussion de cette cavité*. Translated by Jean Nicolas Corvisart des Marets. Paris: Chez Méquignon-Marvis.

Augustinus, Aurelius. 2002. *De musica liber VI*. Translated by Martin Jacobsson. Stockholm: Almqvist & Wiksell International.

Ausécache, Mireille. 1998. "Gilles de Corbeil ou le médecin pédagogue au tournant des XIIe et XIIIe siècles." *Early Science and Medicine* 3:187–214.

Babinski, Joseph. 1901. "Définition de l'hystérie." *Revue neurologique* 9:1074–1080.

Baert, Barbara. 2016. *Kairos or Occasion as Paradigm in the Visual Medium: "Nachleben," Iconography, Hermeneutics*. Louvain: Peeters.

Baglivi, Giorgio. 1696. *De praxi medica ad priscam observandi rationem revocanda*. Rome: Dominici Antonii Herculis.

Baglivi, Giorgio. 1723. *The Practice of Physick*. London: D. Midwinter.

Baldwin, Martha. 2001. "Matters Medical." In *The Great Art of Knowing: The Baroque Encyclopedia of Athanasius Kircher*, edited by Daniel Stolzenberg, 85–92. Stanford, CA: Stanford University Libraries.

Barker, Andrew. 2004. *Greek Musical Writings*. Cambridge: Cambridge University Press.

Bass, Richard. 2001. "Half-Diminished Functions and Transformations in Late Romantic Music." *Music Theory Spectrum* 23:41–60.

Baur, Ludwig. 1912. *Die Philosophischen Werke des Robert Grosseteste, Bischofs von Lincoln*. Münster: Aschendorff,.

Belgray, Alice Bunzl. 1970. "Gaetano Brunetti: An Exploratory Bio-Bibliographical Study." PhD diss., University of Michigan.

Bell, Charles. 1833. *The Nervous System of the Human Body; Embracing the Papers Delivered to the Royal Society on the Subject of the Nerves*. Washington, DC: Stereotyped by D. Green, for the Register and Library of Medicine and Chirurgical Science.

Belrose, J. S. 2002. "Reginald Aubrey Fessenden and the Birth of Wireless Telephony." *IEEE Antennas and Propagation Magazine* 44 (2): 38–47.

Benazzouzz, Abdelhamid, Sorin Breit, Adrian Koudsie, Pierre Pollak, Paul Krack, and Alim-Louis Benabid, 2002. "Intraoperative Microrecordings of the Subthalamic Nucleus in Parkinson's Disease." *Movement Disorders* 17 Suppl. 3:S145–S149.

Berg, Alexander. 1942. "Die Lehre von der Faser als Form- und Funktionselement des Organismus." *Virchows Archiv für pathologische Anatomie und Physiologie und für klinische Medizin* 309 (2): 333–460.

Berlioz. 2002. *Berlioz's Orchestration Treatise: A Translation and Commentary*. Edited by Hugh Macdonald. Cambridge: Cambridge University Press.

Bernays, Jacob. 1880. *Zwei Abhandlungen über die aristotelische Theorie des Drama*. Berlin: W. Hertz.

Bernstein, Julius. 1878. "Ueber Erzeugung von Tetanus und die Anwendung des akustischen Stromunterbrechers." *Archiv für die gesammte Physiologie des Menschen und der Thiere* 17:121–124.

Bernstein, Julius. 1881. "Telephonische Wahrnehmung der Schwankungen des Muskelstromes bei der Contraktion." *Bericht ueber die Sitzungen der Naturforschenden Gesellschaft zu Halle*, 18–27.

Bernstein, Julius. 1894. *Lehrbuch der Physiologie des thierischen Organismus, im speciellen des Menschen*. Stuttgart: Enke.

Berrios, G. E. 1996. *The History of Mental Symptoms: Descriptive Psychopathology since the Nineteenth Century*. Cambridge: Cambridge University Press.

Bijsterveld, Karin. 2019. *Sonic Skills: Listening for Knowledge in Science, Medicine and Engineering (1920s–Present)*. London: Palgrave Macmillan.

Binet, Alfred, and Charles Féré. 1888. *Animal Magnetism*. New York: D. Appleton.

Bingen, Hildegard of. 1987. *Hildegard of Bingen's Book of Divine Works: With Letters and Songs*. Edited by Matthew Fox. Santa Fe, NM: Bear & Company.

Bingen, Hildegard of. 1990. *Scivias*. Translated by Mother Columba Hart and Jane Bishop. New York: Paulist Press.

Bingen, Hildegard of. 1994. *Holistic Healing*. Translated by Patrick Madigan and Manfred Pawlik. Collegeville, MN: Liturgical Press.

Bingen, Hildegard of. 1999. *Hildegard of Bingen: On Natural Philosophy and Medicine*. Translated by Margret Berger. Cambridge: D. S. Brewer.

Blades, James. 1992. *Percussion Instruments and Their History*. Westport, CT: Bold Strummer.

Blades, James, and Edmund A. Bowles. N.d. "Timpani." *Grove Music Online*.

Boethius, Anicius Manlius Severinus. 1867. *Anicii Manlii Torquati Severini Boetii de Institutione arithmetica libri duo, De institutione musica libri quinque*. Edited by Gottfried Friedlein. Leipzig: Teubner.

Boethius, Anicius Manlius Severinus. 1989. *Fundamentals of Music*. Edited by Claude V. Palisca. Translated by Calvin M. Bower. New Haven, CT: Yale University Press.

Bogousslavsky, Julien. 2011. *Following Charcot: A Forgotten History of Neurology and Psychiatry*. Basel: Karger.

Bois-Reymond, Emil du. 1875. *Gesammelte Abhandlungen zur allgemeinen Muskel- und Nervenphysik*. Leipzig: Veit.

Bois-Reymond, Emil du. 1877. "Versuche am Telphon." *Archiv für Physiologie*, 573–576.

Boner, Patrick. 2013. *Kepler's Cosmological Synthesis: Astrology, Mechanism and the Soul*. Leiden: Brill.

Bonge, Dale. 1982. "Gaffurius on Pulse and Tempo: A Reinterpretation." *Musica Disciplina* 36:167–174.

Bonnet, Charles. 1755. *Essai de psychologie: ou, Considérations sur les opérations de l'âme, sur l'habitude et sur l'éducation*. London.

Bonnet, Charles. 1770. *La palingénésie philosophique, ou, Idées sur l'état passé et sur l'état futur des êtres vivans*. Geneva: Jean-Marie Bruyset.

Bonuzzi, Luciano. 2007. "Mozart e Mesmer: Da 'Bastiano e Bastiana' a 'Così fan tutte.'" In *Sig.r Amadeo Wolfgango Mozarte: da Verona con Mozart: personaggi, luoghi, accadimenti*, edited by Giuseppe Ferrari and Mario Ruffini, 45–58. Venice: Marsilio.

Bordeu, Théophile de. 1756. *Recherches sur le pouls, par rapport aux crises*. Paris: De Bure l'aîné.

Bordeu, Théophile de. 1818. *Oeuvres complètes de Bordeu*. Edited by Anthelme Richerand. Paris: Caille et Ravier.

Borelli, Giovanni Alfonso. 1989. *On the Movement of Animals*. Translated by Paul Maquet. Berlin: Springer-Verlag.

Borelli, Giovanni Alfonso. 2015. *Borelli's On the Movement of Animals—On the Force of Percussion*. Translated by Paul Maquet. Cham, Switzerland: Springer.

Borgstrom, Henrik. 2000. "Strike a Pose: Charcot's Women and the Performance of Hysteria at La Salpêtrière." *Theatre Annual* 53:1–14.

Bourneville, Désiré-Magloire, and Paul Régnard. 1879. *Nouvelle Iconographie de La Salpêtrière*. Paris: Aux Bureaux du progrès médical.

Braid, James. 1899. *Braid on Hypnotism: Neurypnology; Or, The Rationale of Nervous Sleep Considered in Relation to Animal Magnetism Or Mesmerism and Illustrated by Numerous Cases of Its Successful Application in the Relief and Cure of Disease*. Edited by Arthur Edward Waite. London: G. Redway.

Brain, Robert M. 2015. *The Pulse of Modernism: Physiological Aesthetics in Fin-de-Siècle Europe*. Seattle: University of Washington Press.

Brain, Robert M., R. S. Cohen, and Ole Knudsen, eds. 2007. *Hans Christian Ørsted and the Romantic Legacy in Science: Ideas, Disciplines, Practices*. Dordrecht: Springer.

Brain, Robert M., and M. Norton Wise. 1994. "Muscles and Engines: Indicator Diagrams and Helmholtz's Graphical Methods." In *Universalgenie Helmholtz: Rückblick nach 100 Jahren*, edited by Lorentz Krüger, 124–148. Berlin: Akademie Verlag.

Brannen, George E., and William H. Bush. 1984. "Ultrasonic Destruction of Kidney Stones." *Western Journal of Medicine* 140: 227–232.

Brechka, Frank T. 1970. *Gerard Van Swieten and His World, 1700–1772*. The Hague: M. Nijhoff.

Brittan, Francesca. 2006. "Berlioz and the Pathological Fantastic: Melancholy, Monomania, and Romantic Autobiography." *19th-Century Music* 29:211–239.

Brittan, Francesca. 2011. "On Microscopic Hearing: Fairy Magic, Natural Science, and the Scherzo Fantastique." *Journal of the American Musicological Society* 64:527–600.

Brittan, Francesca. 2017. *Music and Fantasy in the Age of Berlioz*. Cambridge: Cambridge University Press.

Brittan, Francesca. 2019. "The Electrician, the Magician, and the Nervous Conductor." *Nineteenth-Century Music Review*. https://doi.org/10.1017/S1479409820000099.

Brookhart, J. M., G. Moruzzi, and R. S. Snider. 1950. "Spike Discharges of Single Units in the Cerebellar Cortex." *Journal of Neurophysiology* 13 (6): 465–486.

Brunetti, Gaetano. 1960. *Il "maniático": Sinfonia no. 33 [e] Sinfonia in sol min., no. 22.* Edited by Newell Jenkins. Rome: L. del Turco.

Bruyninckx, Joeri. 2018. *Listening in the Field: Recording and the Science of Birdsong.* Cambridge, MA: MIT Press.

Buranelli, Vincent. 1975. *The Wizard from Vienna: Franz Anton Mesmer.* New York: Coward, McCann.

Burkert, Walter. 1972. *Lore and Science in Ancient Pythagoreanism.* Cambridge, MA: Harvard University Press.

Burnett, Charles. 2018. "Harmonic and Acoustic Theory: Latin and Arabic Ideas of Sympathetic Vibration as the Causes of Effects between Heaven and Earth." In *Sing Aloud Harmonious Spheres: Renaissance Conceptions of Cosmic Harmony,* edited by Jacomien Prins and Maude Vanhaelen, 32–43. New York: Routledge.

Burton, Robert. 1989. *The Anatomy of Melancholy.* Edited by Thomas C. Faulkner, Nicolas K. Kiessling, Rhonda L. Blair, J. B. Bamborough, and Martin Dodsworth. Oxford: Clarendon Press.

Cabot, Richard C. 1923. "A Multiple Electrical Stethoscope for Teaching Purposes: Preliminary Note." *Journal of the American Medical Association* 81 (4): 298–299.

Cadenbach, Ranier. 2007. "Einige apologetische Erwägungen zur musikgeschichlichen Relevanz von Athanasius Kirchers Phantasien zur Musiktherapie." In *Ars magna musices: Athanasius Kircher und die Universalität der Musik,* edited by Markus Engelhardt and Michael Heinemann, 227–252. Laaber: Laaber-Verlag.

Cahan, David. 2018. *Helmholtz: A Life in Science.* Chicago: University of Chicago Press.

Callahan, Christopher. 2000. "Music in Medieval Medical Practice: Speculations and Certainties." *College Music Symposium* 40:151–164.

Caminietzki, Carlos Ziller. 2004. "Baroque Science between the Old and the New World: Father Kircher and His Colleague Valentin Stansel (1621–1705)." In *Athanasius Kircher: The Last Man Who Knew Everything,* edited by Paula Findlen, 311–328. New York: Routledge.

Canales, Jimena. 2009. *A Tenth of a Second: A History.* Chicago: University of Chicago Press.

Cannon, Sue Spencer. 1993. *The Medicine of Hildegard of Bingen: Her Twelfth-Century Theories and Their Twentieth-Century Appeal as a Form of Alternative Medicine.* Los Angeles: University of California, Los Angeles.

Capparoni, P. 1936. *Il "Tractatus de pulsibus" di Alfano Io arcivescovo di Salerno.* Rome: Istituto nazionale medico farmacologico Serono.

Cardwell, Douglas. 1983. "The Well-Made Play of Eugène Scribe." *French Review* 56:876–884.

Caspar, Max. 1993. *Kepler.* Translated by Clarisse Doris Hellman. New York: Dover.

Catchpole, Clive. 2008. *Bird Song: Biological Themes and Variations.* 2nd ed. Cambridge: Cambridge University Press.

Cejador y Frauca, Julio, ed. 1905. *La lengua de Cervantes, gramatica y diccionario de la lengua castellana en el ingenioso hidalgo Don Quijote de La Mancha.* Madrid: Ratés.

Cervantes, Miguel de. 1979. *El ingenioso hidalgo Don Quijote de la Mancha.* 19th ed. Edited by Américo Castro. Mexico City: Editorial Porrúa.

Chafe, Christopher D., and Josef Parvizi. 2013a. Method of Sonifying Brain Electrical Activity. US Patent 2013324878 (A1), issued 2013.

Chafe, Christopher D., and Josef Parvizi. 2013b. Method of Sonifying Brain Electrical Activity. US Patent 10136862 (B2), issued 2013.

Chafe, Christopher D., and Josef Parvizi. 2014. Method of Sonifying Signals Obtained from a Living Subject. US Patent 2015150520 (A1), issued 2014.

Chamberlain, David. 1980. "Musical Learning and Dramatic Action in Hrotsvit's 'Pafnutius.'" *Studies in Philology* 77:319–343.

Charcot, Jean-Martin. 1882. "Sur les divers états nerveux déterminés par l'hypnotisation des hystériques." *Comptes rendus hebdomadaires des séances de l'Académie des sciences* 94:403–405.

Charcot, Jean-Martin. 1890. *Leçons sur les maladies du système nerveux.* Paris: Aux Bureaux du Progrès Médical.

Chateaubriand, François-René. 1932. *Les Natchez*. Edited by Gilbert Chinard. Baltimore, MD: Johns Hopkins University Press.

Chateaubriand, François-René. 1981. *Mémoires d'outre-tombe*. Edited by Georges Moulinier. Paris: Gallimard.

Cheung, Tobias. 2010. "Omnis Fibra Ex Fibra: Fibre OEconomies in Bonnet's and Diderot's Models of Organic Order." *Early Science and Medicine* 15 (1–2): 66–104.

Choulant, Ludwig, ed. 1826. *Aegidii Corboliensis carmina medica*. Leipzig: Leopold Voss.

Chowning, John M. 1977. "The Synthesis of Complex Audio Spectra by Means of Frequency Modulation." *Computer Music Journal* 1 (2): 46–54.

Chowning, John. 2008. "Fifty Years of Computer Music: Ideas of the Past Speak to the Future." In *Lecture Notes in Computer Science*, vol. 4969, 1–10. Berlin: Springer-Verlag. https://doi.org/10.1007/978-3-540-85035-9_1.

Christensen, Thomas. 1993. *Rameau and Musical Thought in the Enlightenment*. Cambridge: Cambridge University Press.

Christensen, Thomas. 1994. "Diderot, Rameau and Resonating Strings: New Evidence of an Early Collaboration." *Studies on Voltaire and the Eighteenth Century* 323:131.

Christianson, John Robert. 2002. *On Tycho's Island: Tycho Brahe, Science, and Culture in the Sixteenth Century*, abridged ed. Cambridge: Cambridge University Press.

Chua, Daniel K. L., and Alexander Rehding. 2021. *Alien Listening: Voyager's Golden Record and Music from Earth*. New York: Zone Books.

Chung, Minna Rose. 2018. "Tartini Tones: How to Use 'Difference Tones' to Perfect Your Intonation." *Strad* 129 (1536): 84–86.

Claretie, Jules. 1881. *La vie à Paris: 1880–1885. Année 2*. Paris: V. Havard.

Clements, Matthew. 2011. "Uexküll's Ecology: Biosemiotics and the Musical Imaginary." *Green Letters* 15 (1): 43–60.

Cooper, Martin. 1951. *French Music, from the Death of Berlioz to the Death of Fauré*. London: Oxford University Press.

Cooper, Sybil, and E. D. Adrian. 1924. "The Maximum Frequency of Reflex Response in the Spinal Cat." *Journal of Physiology* 59 (1): 61–81.

Corvisart des Marets, Jean Nicolas. 1812. *An Essay on the Organic Diseases and Lesions of the Heart and Great Vessels*. Translated by Carleton B. Chapman. Boston: Bradford & Read.

Costa, Nathalie Dos Santos. 2008. "François-Nicolas Marquet: sa vie, ses oeuvres et ses démêlés tardifs avec le Collège Royal de Médecine de Nancy." MD diss., Université Henri Poincaré, Nancy.

Coxe, William. 1813. *Memoirs of the Kings of Spain of the House of Bourbon, from the Accession of Philip V to the Death of Charles III: 1700 to 1788*, vol. 3. London: Longman, Hurst, Rees, Orme and Brown.

Crabtree, Adam, and Robert H. Wozniak. 1988. *Animal Magnetism, Early Hypnotism, and Psychical Research, 1766–1925: An Annotated Bibliography*. White Plains, NY: Kraus International Publications.

Crary, Jonathan. 1992. *Techniques of the Observer: On Vision and Modernity in the 19th Century*. Cambridge, MA: MIT Press.

Crisp, Kevin M. 2018. "Recording EMG Signals on a Computer Sound Card." *Journal of Undergraduate Neuroscience Education* 16:A210–A216.

Crombie, A. C. 1971. *Robert Grosseteste and the Origins of Experimental Science, 1100–1700*. Oxford: Clarendon Press.

Crombie, A. C. 1995. *The History of Science from Augustine to Galileo*. New York: Dover.

Croone, William. 2000. "On the Reason of the Movement of the Muscles." Edited by Margaret Nayler and August Ziggelaar. Translated by Paul Maquet. *Transactions of the American Philosophical Society* 90:1–130.

Daan, Serge. 2010. "A History of Chronobiological Concepts." In *The Circadian Clock*, edited by Albrecht Urs, 1–35. New York: Springer.

Daboo, Jerri. 2010. *Ritual, Rapture, and Remorse: A Study of Tarantism and Pizzica in Salento*. Oxford: Peter Lang.

Dalma, Juan. 1963. "La catarsis en Aristoteles, Bernays y Freud." *Revista de psíquiatría y psicología medical* 6:253–269.

Darnton, Robert. 1968. *Mesmerism and the End of the Enlightenment in France*. Cambridge, MA: Harvard University Press.

Darrigol, Olivier. 2007. "The Acoustic Origins of Harmonic Analysis." *Archive for History of Exact Sciences* 61:343–424.

Daston, Lorraine, and Peter Galison. 2007. *Objectivity*. Brooklyn: Zone Books.

Darwin, Erasmus. 1796. *Zoonomia; or, The Laws of Organic Life*. New York: T. & J. Swords.

Davies, J. Q., and Ellen Lockhart, eds. 2016. *Sound Knowledge: Music and Science in London, 1789–1851*. Chicago: University of Chicago Press.

Davis, Lennard J. 2008. *Obsession: A History*. Chicago: University of Chicago Press.

Davis, Lennard J. 2017. *The Disability Studies Reader*. 5th ed. New York: Routledge.

De Martino, Ernesto. 2005. *The Land of Remorse: A Study of Southern Italian Tarantism*. London: Free Association Books.

Debru, Claude. 2001. "Helmholtz and the Psychophysiology of Time." *Science in Context* 14:471–492.

D'Eslon, Charles. 1780. *Observations sur le magnétisme animal*. Paris: Didot.

Devlin, J. P. 2018. *From Analogue to Digital Radio: Competition and Cooperation in the UK Radio Industry*. London: Palgrave Macmillan.

Devoto, Mark. 1995. "The Strategic Half-Diminished Seventh Chord and the Emblematic Tristan Chord: A Survey from Beethoven to Berg." *International Journal of Musicology* 4:139–153.

Diderot, Denis. 1875. *Œuvres complètes de Diderot*. Edited by J. Assézat and M. Tourneux. Paris: Garnier.

Diderot, Denis. 1964. *Rameau's Nephew, and Other Works*. Translated by Jacques Barzun and Ralph H. Bowen. New York: Bobbs-Merrill.

Didi-Huberman, Georges. 2003. *Invention of Hysteria: Charcot and the Photographic Iconography of the Salpêtrière*. Cambridge, MA: MIT Press.

Dijkgraaf, Sven. 1960. "Spallanzani's Unpublished Experiments on the Sensory Basis of Object Perception in Bats." *Isis* 51 (1): 9–20.

Divitiis, Enrico de, Paolo Cappabianca, and Oreste de Divitiis. 2004. "The 'Schola Medica Salernitana': The Forerunner of the Modern University Medical Schools." *Neurosurgery* 55:722–745.

Dock, George. 1935. "Roziere de La Chassagne and the Early History of Percussion of the Thorax." *Annals of Medical History* 7:438–450.

Dolan, Emily I. 2013. *The Orchestral Revolution: Haydn and the Technologies of Timbre*. Cambridge: Cambridge University Press.

Dombois, Florian. 2006. "The 'Muscle Telephone': The Undiscovered Start of Audification in the 1870s, in Sounds of Science—Schall im Labor (1800–1930)." Preprint 346. Max Planck Institute for the History of Science. https://www.mpiwg-berlin.mpg.de/sites/default/files/Preprints/P346.pdf.

Donald, Ian. 1974. "Sonar—The Story of an Experiment." *Ultrasound in Medicine and Biology* 1 (2): 109–117.

Donald, I., J. MacVicar, and T. G. Brown. 1958. "Investigation of Abdominal Masses by Pulsed Ultrasound." *Lancet* 1 (7032): 1188–1195.

Donizetti, Gaetano. 1941. *Lucia Di Lammermoor*. Facsimile ed. of the autograph score. Milan: E. Bestetti.

Dubus, Gaël, and Roberto Bresin. 2013. "A Systematic Review of Mapping Strategies for the Sonification of Physical Quantities." *PLoS ONE* 9 (12): e82491.

Duchesneau, François. 1982. *La physiologie des lumières: empirisme, modèles et théories*. The Hague: M. Nijhoff.

Duffin, Jacalyn M. 1989. "The Cardiology of R. T. H. Laennec." *Medical History*, 42–71.

Duffin, Jacalyn. 1998. *To See with a Better Eye: A Life of R. T. H. Laennec*. Princeton, NJ: Princeton University Press.

Duffin, Jacalyn. 2006. "Jodocus Lommius's Little Golden Book and the History of Diagnostic Semeiology." *Journal of the History of Medicine and Allied Sciences* 61:249–287.

Durán, Manuel. 2006. *Fighting Windmills: Encounters with Don Quixote*. New Haven, CT: Yale University Press.

Duverney, Joseph-Guichard. 1683. *Traité de l'organe de l'ouie: contenant la structure, les usages & les maladies de toutes les parties de l'oreille*. Paris: chez Estienne Michallet.

Duverney, M. (Guichard Joseph). 1737. *A Treatise of the Organ of Hearing*. London: Samuel Baker.

Eadie, M. J. 2003a. "A Pathology of the Animal Spirits—the Clinical Neurology of Thomas Willis (1621–1675). Part I—Background, and Disorders of Intrinsically Normal Animal Spirits." *Journal of Clinical Neuroscience* 10:14–29.

Eadie, M. J. 2003b. "A Pathology of the Animal Spirits—the Clinical Neurology of Thomas Willis (1621–1675). Part II—Disorders of Intrinsically Abnormal Animal Spirits." *Journal of Clinical Neuroscience* 10:146–157.

Earp, Lawrence. 2018. "Notation II." In *The Cambridge History of Medieval Music*, edited by Mark Everist and Thomas Forrest Kelly, 674–717. Cambridge: Cambridge University Press.

Edelstein, Ludwig. 1967. *Ancient Medicine: Selected Papers of Ludwig Edelstein*, edited by Owsei Temkin and C. Lillian Temkin, translated by C. Lillian Temkin. Baltimore, MD: Johns Hopkins Press

Eichberg, Stephanie. 2009. "Constituting the Human via the Animal in Eighteenth-Century Experimental Neurophysiology: Albrecht von Haller's Sensibility Trials." *Medizinhistorisches Journal* 44:274–295.

Eliot, T. S. 1962. *The Complete Poems and Plays 1909–1950*. New York: Harcourt.

Ellenberger, Henri F. 1970. *The Discovery of the Unconscious: The History and Evolution of Dynamic Psychiatry*. New York: Basic Books.

Emch-Deriaz, Antoinette. 2001. "De l'importance de tater le pouls." *Canadian Bulletin of Medical History* 18:369–380.

Erlanger, Joseph, and Herbert S. Gasser. 1937. *Electrical Signs of Nervous Activity*. Philadelphia: University of Pennsylvania Press.

Erlmann, Veit. 2010. *Reason and Resonance: A History of Modern Aurality*. New York: Zone Books.

Erman, Paul. 1812. "Einige Bemerkungen über Muskular-Contraction." *Annalen der Physik* 40:1–30.

Escobar, Jorge M. 2008. "Kepler's Theory of the Soul: A Study on Epistemology." *Studies in History and Philosophy of Science* 39:15–41.

Esquirol, Étienne. 1838. *Des maladies mentales: considérées sous les rapports médical, hygiénique et médico-legal*. Paris: J. B. Bailliére.

Fage, Adrien de la. 1844. *Histoire générale de la musique et de la danse*. Paris: Comptoir des Imprimeurs Unis.

Falconer, W. 1796. *Observations Respecting the Pulse; Intended to Point out with Greater Certainty, the Indications Which It Signifies; Especially in Feverish Complaints*. London.

Fantini, Bernardino. 2013. "Forms of Thought between Music and Science." In *The Emotional Power of Music*, edited by Tom Cochrane and Bernardino Fantini, 257–270. Oxford: Oxford University Press.

Farndell, Arthur, trans. 2010. *All Things Natural: Ficino on Plato's Timaeus*. London: Shepheard-Walwyn.

Fassler, Margot. 1998. "Composer and Dramatist: 'Melodious Singing and the Freshness of Remorse.'" In *Voice of the Living Light: Hildegard of Bingen and Her World*, edited by Barbara Newman, 149–175. Berkeley: University of California Press.

Feld, Steven. 2003. "A Rainforest Acoustemology." In *The Auditory Culture Reader*, edited by Micahel Bull and Les Black, 223–239. Oxford: Berg.

Ferdinandus, Epiphanius. 1621. *Centum Historiae seu observationes, & casus medici*. Venice: Ballionus.

Fernel, Jean. 1567. *Universa medicina*. Paris: Andreas Wechel.

Fernel, Jean. 2003. *The Physiologia of Jean Fernel (1567)*. Translated by J. M. Forrester. Philadelphia: American Philosophical Society.

Fessenden, Reginald A. 1902. Wireless signaling. US Patent 706740A, filed September 28, 1901, and issued August 12, 1902.

Ficino, Marsilio. 1989. *Three Books on Life*. Translated by Carol V. Kaske and John R. Clark. Binghamton, NY: Medieval & Renaissance Texts & Studies.

Findlen, Paula, ed. 2004a. *Athanasius Kircher: The Last Man Who Knew Everything*. New York: Routledge.

Findlen, Paula. 2004b. "The Last Man Who Knew Everything . . . or Did He?" In *Athanasius Kircher: The Last Man Who Knew Everything*, edited by Paula Findlen, 1–48. New York: Routledge.

Findlen, Paula. 2013. "Laura Bassi and the City of Learning." *Physics World* 26 (9): 30–34.

Finger, Stanley. 1994. *Origins of Neuroscience: A History of Explorations into Brain Function*. New York: Oxford University Press.

Finger, Stanley. 2000. *Minds behind the Brain: A History of the Pioneers and Their Discoveries*. Oxford: Oxford University Press.

Finger, Stanley, and Marco Piccolino, 2011. *The Shocking History of Electric Fishes: From Ancient Epochs to the Birth of Modern Neurophysiology*. Oxford: Oxford University Press.

Finger, Stanley, and William Zeitler. 2015. "Pathological Connections: Benjamin Franklin and His Glass Armonica: From Music as Therapeutic to Pathological." In *Music, Neurology, and Neuroscience: Historical Connections and Perspectives*, edited by Eckart Altenmüller, Stanley Finger, and François Boller, 98–125. Amsterdam: Elsevier.

Finkelstein, Gabriel. 2003. "M. Du Bois-Reymond Goes to Paris." *British Journal for the History of Science* 36 (3): 261–300.

Finkelstein, Gabriel. 2015. "Mechanical Neuroscience: Emil Du Bois-Reymond's Innovations in Theory and Practice." *Frontiers in Systems Neuroscience* 9 (133).

Floyer, Sir John. 1707. *The Physician's Pulse-Watch*. London: Samuel Smith & Benj. Walford.

Forbes, A., C. J. Campbell, and H. B. Williams. 1924. "Electrical Records of Afferent Nerve Impulses from Muscular Receptors." *American Journal of Physiology* 69 (2): 283–303.

Forbes, Alexander, and Catharine Thacher. 1920. "Amplification of Action Currents with the Electron Tube in Recording with the String Galvanometer." *American Journal of Physiology* 52 (3): 409–471.

Forster, E. M. 2014. *Alexandria: A History and Guide*. London: Tauris Parke Paperbacks.

Foucault, Michel. 1972. *Naissance de la clinique: une archéologie du regard médical*, 2nd ed. Paris: Presses Universitaires de France.

Foucault, Michel. 1994. *The Birth of the Clinic: An Archaeology of Medical Perception*. New York: Vintage.

Frank. 1994. "Instruments, Nerve Action, and the All-or-None Principle." *Osiris* 9:208–235.

Franklin, Benjamin. 1785. *Report of Dr. Benjamin Franklin: And Other Commissioners Charged by the King of France, with the Examination of the Animal Magnetism, as now Practised at Paris*. London: J. Johnson.

Franklin, Benjamin. 1959. *The Papers of Benjamin Franklin*. Edited by Leonard W. Labaree. New Haven, CT: Yale University Press.

Freud, Sigmund. 1963. "The Moses of Michelangelo (1914)." In *Character and Culture*, 80–106. New York: Scribner.

Freud, Sigmund. 1966. "An Autobiographical Study (1914)." In *The Standard Edition of the Complete Psychological Works of Sigmund Freud*, translated by James Strachey, 20:7–74. London: Hogarth Press.

Fürst, Marion. 2005. *Maria Theresia Paradis: Mozarts berühmte Zeitgenossin*. Cologne: Böhlau.

Gaffurius, Franchinus. 1969. *The* Practica Musicae *of Franchinus Gafurius*. Translated by Irwin Young. Madison: University of Wisconsin Press.

Galambos, Robert. 1941. "The Production and Reception of Supersonic Sounds by Flying Bats." PhD diss., Harvard University.

Galen. 2014. *On the Natural Faculties*. Translated by Arthur John Brock. Cambridge, MA: Harvard University Press.

Garrison, Daniel H. N.d. "The Name 'Fabrica' | Vesalius." Accessed August 27, 2019, at http://www.vesalius-fabrica.com/en/original-fabrica/inside-the-fabrica/the-name-fabrica.html.

Garson, Justin. 2003. "The Introduction of Information into Neurobiology." *Philosophy of Science* 70 (5): 926–936.

Garson, Justin. 2015. "The Birth of Information in the Brain: Edgar Adrian and the Vacuum Tube." *Science in Context* 28 (1): 31–52.

Garten, Siegfried. 1910. "Ein Beitrag zur Kenntnis der positiven Nachschwankung des Nervenstromes nach elektrischer Reizung." *Pflüger's Archiv für die gesamte Physiologie des Menschen und der Tiere* 136 (1): 545–563.

Gasser, H. S., and Joseph Erlanger. 1922. "A Study of the Action Currents of Nerve with the Cathode Ray Oscillograph." *American Journal of Physiology* 62 (3): 496–524.

Gaukroger, Stephen. 1995. *Descartes: An Intellectual Biography*. Oxford: Clarendon Press.

Gaukroger, Stephen. 2002. *Descartes' System of Natural Philosophy*. Cambridge: Cambridge University Press.

Gauld, Alan. 1992. *A History of Hypnotism*. Cambridge: Cambridge University Press.

Gensini, Gian Franco, and Andrea A. Conti. 2004. "The Evolution of the Concept of 'Fever' in the History of Medicine: From Pathological Picture per se to Clinical Epiphenomenon (and Vice Versa)." *Journal of Infection* 49 (2): 85–87.

Georgallas, Virginia. 2016. "The Maniac's Affliction: Music, Madness, and Caprice in Late Eighteenth-Century Spain." Unpublished conference presentation, American Musicological Society, Vancouver, BC.

Gérard, Yves. 1969. *Thematic, Bibliographical, and Critical Catalogue of the Works of Luigi Boccherini*. London: Oxford University Press.

Gibbs, D. D. 1971. "The Physician's Pulse Watch." *Medical History* 15:187–190.

Gibbs, Denis. 1994. "The Almshouses of Lichfield: Cradles of Pulse-Timing." *Journal of Medical Biography* 2 (2): 89–93.

Gibbs, Frederic A., and Erna L. Gibbs, eds. 1951. *Atlas of Electroencephalography*. Reading, MA: Addison-Wesley.

Glaze, Florence Eliza. 1998. "Medical Writer: 'Behold the Human Creature.'" In *Voice of the Living Light: Hildegard of Bingen and Her World*, edited by Barbara Newman, 125–148. Berkeley: University of California Press.

Glickman, Sylvia. N.d. "Auenbrugger, Marianna von." *Grove Music Online*.

Glisson, Francis. 1677. *Tractatus de ventriculo et intestinis*. Amsterdam: Jacobum Juniorem.

Godwin, Joscelyn, ed. 1988. *Music, Mysticism and Magic: A Sourcebook*. London: Arkana.

Godwin, Joscelyn, ed. 1993. *The Harmony of the Spheres: A Sourcebook of the Pythagorean Tradition in Music*. Rochester, VT: Inner Traditions International.

Gold, T. 1948. "Hearing. II. The Physical Basis of the Action of the Cochlea." *Proceedings of the Royal Society of London. Series B, Biological Sciences (1934–1990)* 135 (881): 492–498.

Gold, Thomas. 2012. *Taking the Back off the Watch: A Personal Memoir*. Edited by Simon Mitton. Berlin: Springer-Verlag.

Goldstein, Jan. 2001. *Console and Classify: The French Psychiatric Profession in the Nineteenth Century*. Chicago: University of Chicago Press.

Gómez, Javier González-Velandia. 2013. "Sinfonía de la vida: Aspectos musicales en la obra de Jakob von Uexküll." *Eikasia*, no. 51:237–258.

Gordon, E. S., and J. M. Lagerwerff. 1976. "Electronic Stethoscope with Frequency Shaping and Infrasonic Recording Capabilities." *Aviation, Space, and Environmental Medicine* 47 (3): 312–316.

Gossett, Philip. 2006. *Divas and Scholars: Performing Italian Opera*. Chicago: University of Chicago Press.

Gotch, Francis, and Victor Horsley. 1888. "Observations upon the Electromotive Changes in the Mammalian Spinal Cord following Electrical Excitation of the Cortex Cerebri. Preliminary Notice." *Proceedings of the Royal Society of London* 45:18–26.

Gouk, Penelope. 1980. "The Role of Acoustics and Music Theory in the Scientific Work of Robert Hooke." *Annals of Science* 37 (5): 573.

Gouk, Penelope. 1982. "Music in the Natural Philosophy of the Early Royal Society." PhD diss., Imperial College London.

Gouk, Penelope. 1999. *Music, Science, and Natural Magic in Seventeenth-Century England*. New Haven: Yale University Press.

Gouk, Penelope. 2000a. "Music, Melancholy, and Medical Spirits in Early Modern Thought." In *Music as Medicine: The History of Music Therapy since Antiquity*, edited by Peregrine Horden, 173–194. Aldershot: Ashgate.

Gouk, Penelope. 2000b. "Sister Disciplines? Music and Medicine in Historical Perspective." In *Musical Healing in Cultural Contexts*, edited by Penelope Gouk, 171–196. Florence: Taylor and Francis.

Gouk, Penelope. 2004. "Raising Spirits and Restoring Souls: Early Modern Medical Explanations for Music's Effects." In *Hearing Cultures: Essays on Sound, Listening, and Modernity*, edited by Veit Erlmann, 87–105. Oxford: Berg.

Gouk, Penelope. 2014. "Music and the Nervous System in Eighteenth-Century British Medical Thought." In *Music and the Nerves, 1700–1900*, edited by James Kennaway, 44–71. London: Palgrave Macmillan.

Gouk, Penelope. 2015. "An Enlightenment Proposal for Music Therapy: Richard Brocklesby on Music, Spirit, and the Passions." In *Music, Neurology, and Neuroscience: Evolution, the Musical Brain, Medical Conditions, and Therapies*, edited by Eckart Altenmüller, Stanley Finger, and François Boller, 159–185. Amsterdam: Elsevier. https://doi.org/10.1016/bs.pbr.2014.11.026.

Gouk, Penelope, and Ingrid Sykes. 2011. "Hearing Science in Mid-Eighteenth-Century Britain and France." *Journal of the History of Medicine and Allied Sciences* 66:507–545.

Gouk, Penelope, Jacomien Prins, James Kennaway, and Wiebke Thormahlen, eds. 2019. *The Routledge Companion to Music, Mind and Well-Being*. New York: Routledge.

Gray, John Archibald Browne. 1990. "Bryan Harold Cabot Matthews, 14 June 1906–23 July 1986." *Biographical Memoirs of Fellows of the Royal Society* 35:263–279.

Green, Vivian. 1993. *The Madness of Kings: Personal Trauma and the Fate of Nations*. New York: St. Martin's Press.

Griffin, Donald R. 1958. *Listening in the Dark: The Acoustic Orientation of Bats and Men*. New Haven, CT: Yale University Press.

Grignon, Georges. 1997. "François-Nicolas Marquet, Le pouls et le menuet (1682–1759)." *La lettre du musée*, no. 2 (1998): 6–7.

Grmek, M. 1970. "La notion de fibre vivante chez les médecins de l'école iatrophysique." *Clio Medica* 5 (4): 297.

Grmek, Mirko D. 1981. "L'invention de l'auscultation médiate, retouches à un cliché historique." *Revue du Palais de ta Découverte* 22:107–116.

Grond, Florian, and Thomas Hermann. 2012. "Aesthetic Strategies in Sonification." *AI & Society* 27 (2): 213–222. https://doi.org/10.1007/s00146-011-0341-7.

Gruner, Oskar Cameron, trans. 1970. *A Treatise on the Canon of Medicine of Avicenna, Incorporating a Translation of the First Book*. New York: Kelley.

"H. Hartridge MD." 1976. *British Medical Journal* 1 (6011): 716.

Haar, James. 1974. "The Frontispiece of Gafori's *Practica Musicae* (1496)." *Renaissance Quarterly* 27 (1): 7–22. http://www.jstor.org/stable/2859295.

Hackmann, Willem Dirk. 1984. *Seek and Strike: Sonar, Anti-Submarine Warfare, and the Royal Navy, 1914–54*. London: HMSO.

Hadlock, Heather. 2000. "Sonorous Bodies: Women and the Glass Harmonica." *Journal of the American Musicological Society* 53: 507–42.

Hahn, Walter Louis. 1908. "Some Habits and Sensory Adaptations of Cave-Inhabiting bats. II." *Biological Bulletin* 15 (4):165–193.

Haller, Albrecht von. 1760. *Elementa physiologiae corporis humani*, vol. 2. Venice.

Haller, Albrecht von. 1936. "A Dissertation on the Sensible and Irritable Parts of Animals." *Bulletin of the Institute of the History of Medicine* 4:651–699.

Haller, Albrecht von. 1974. *Primae lineae physiologiae*. Hildesheim: Georg Olms.

Harbach, Barbara, Elisabetta De Gambarini, Marianne Martinez, and Marianne Auenbrugger. 1986. *Women Composers for the Harpsichord*. Bryn Mawr, PA: Elkan-Vogel.

Hargreaves-Mawdsley, W. N. 1979. *Eighteenth-Century Spain, 1700–1788: A Political, Diplomatic and Institutional History*. Totowa, NJ: Rowman and Littlefield.

Harrington, Anne. 1985. "Hysteria, Hypnosis, and the Lure of the Invisible: The Rise of Neo-Mesmerism in Fin-de-Siècle French Psychiatry." In *The Anatomy of Madness: Essays in the History of Psychiatry*, edited by W. F. Bynum, Roy Porter, and Michael Shepherd, 3:226–246. London: Tavistock.

Harrington, Anne. 1988. "Metals and Magnets in Medicine: Hysteria, Hypnosis and Medical Culture in Fin-de-Siècle Paris." *Psychological Medicine* 18:21–38.

Harrington, Anne. 1996. *Reenchanted Science: Holism in German Culture from Wilhelm II to Hitler*. Princeton, NJ: Princeton University Press.

Harris, C. R. S. 1973. *The Heart and the Vascular System in Ancient Greek Medicine, from Alcmaeon to Galen*. Oxford: Clarendon Press.

Harry, J. D. 1987. "Early Designs of the Myograph." *Medical Instrumentation* 21 (5): 278–282.

Hartridge, H. 1920. "The Avoidance of Objects by Bats in Their Flight." *Journal of Physiology* 54 (1–2): 54–57.

Haughton, Samuel. 1863. *Outlines of a New Theory of Muscular Action*. London: Williams and Norgate.

Head, Matthew. 2014. "Fantasia and Sensibility." In *The Oxford Handbook of Topic Theory*, edited by Danuta Mirka, 259–278. New York: Oxford University Press.

Heller, Eric J. 2012. *Why You Hear What You Hear: An Experiential Approach to Sound, Music, and Psychoacoustics*. Princeton, NJ: Princeton University Press.

Heller-Roazen, Daniel. 2011. *The Fifth Hammer: Pythagoras and the Disharmony of the World*. New York: Zone Books.

Helmholtz, Hermann von. 1855. "Ueber die Geschwindigkeit einiger Vergänge in Muskeln und Nerven (1855)." In *Wissenschaftliche Abhandlungen*, 2:881–885. Leipzig: J. A. Barth.

Helmholtz, Hermann von. 1864. "Versuche über das Muskelgeräusch." *Verhandlungen des naturhistorisch-medicinischen Vereins zu Heidelberg*, 155–157.

Helmholtz, Hermann von. 1866. "Ueber den Muskelton." *Verhandlungen des naturhistorisch-medicinischen Vereins zu Heidelberg* 4:88–90.

Helmholtz, Hermann von. 1882. *Wissenschaftliche Abhandlungen*. Leipzig: J. A. Barth.

Helmholtz, Hermann von. 1924. *Helmholtz's Treatise on Physiological Optics*. Edited by James P. C. Southall. Translated by J. von Kries. New York: Optical Society of America.

Helmholtz, Hermann von. 1954. *On the Sensations of Tone as a Physiological Basis for the Theory of Music*. Edited by Alexander John Ellis. New York: Dover.

Henderson, Jeffrey. N.d. "Lucian, Philosophies for Sale." Loeb Classical Library. Accessed September 12, 2018, at https://www.loebclassics.com/view/lucian-philosophies_sale/1915/pb_LCL054.455.xml.

Hermann, Ludimar. 1878. "Ueber electrophysiologische Verwendungen des Telephons: Anhang zu den Untersuchungen über die Actionsströme." *Archiv für die gesammte Physiologie des Menschen und der Thiere* 16:504–509.

Hermann, Thomas, Andy Hunt, and John G. Neuhoff, eds. 2011. *The Sonification Handbook*. Berlin: Logos.

Hermans, Michel. 2005. "Joseph-Marie Amiot: une figure de la rencontre de 'l'autre' au temps des Lumières." In *Les danses rituelles chinoises d'après Joseph-Marie Amiot: aux sources de l'ethnochorégraphie*, edited by Yves Lenoir and N. Standaert, 11–78. Namur, Belgium: Presses universitaires de Namur.

Hesse, Mary B. 1966. "Hooke's Philosophical Algebra." *Isis* 57 (1): 67–83.

Hibberd, Sarah. 2004. "'Dormez donc, mes chers amours': Hérold's *La Somnambule* (1827) and Dream Phenomena on the Parisian Lyric Stage." *Cambridge Opera Journal* 16:107–132.

Hippocrates. 1937. *Works*, vol. 4. Translated by W. H. S. Jones. Cambridge, MA: Harvard University Press.

Hippocrates. 2010. *Works*, vol. 4. Translated by Paul Potter. Cambridge, MA: Harvard University Press.

Hobbs, Kyle, Prashanth Krishnamohan, Catherine Legault, Steve Goodman, Josef Parvizi, Kapil Gururangan, and Michael Mlynash. 2018. "Rapid Bedside Evaluation of Seizures in the ICU by Listening to the Sound of Brainwaves: A Prospective Observational Clinical Trial of Ceribell's Brain Stethoscope Function." *Neurocritical Care* 29 (2): 302–312.

Höber, Rudolf. 1919. "Ein Verfahren zur Demonstration der Aktionsströme." *Pflügers Archiv für die gesamte Physiologie des Menschen und der Tiere* 177:305–312.

Hodgkin, Alan. 1979. "Edgar Douglas Adrian, Baron Adrian of Cambridge, 30 November 1889–4 August 1977." *Biographical Memoirs of Fellows of the Royal Society* 25:1–73.

Hoffmann, E. T. A. 2003. *E. T. A. Hoffmann's Musical Writings*. Edited by David Charleton. Translated by Martyn Clarke. Cambridge: Cambridge University Press.

Hoffmann, E. T. A. 2014. *Kreisler; Berganza; Magnetiseur: textkritische Edition: Autographe der Bibliotheca Bodmeriana*. Edited by Katerina Latifi. Frankfurt am Main: Stroemfeld.

Hoffmann, Freia. 1991. *Instrument und Körper: Die musizierende Frau in der bürgerlichen Kultur*. Frankfurt am Main: Insel Verlag.

Hogan, J. L. 1999. "The Heterodyne Receiving System, and Notes on the Recent Arlington-Salem Tests." *Proceedings of the IEEE* 87 (11): 1979–1990.

Holford-Strevens, Leofranc. 1993. "The Harmonious Pulse." *Classical Quarterly* 43:475–479.

Holsinger, Bruce W. 2001. *Music, Body, and Desire in Medieval Culture: Hildegard of Bingen to Chaucer*. Stanford, CA: Stanford University Press.

Homer. 1990. *The Iliad*. Translated by Robert Fagles. New York: Viking.

Hooke, Robert. 1678. *Lectures de Potentia Restitutiva, or Of Spring Explaining the Power of Springing Bodies*. London: John Martyn.

Hooke, Robert. 1705. *The Posthumous Works*. London: Richard Waller.

Hooke, Robert. 1961. *Micrographia; or, Some Physiological Descriptions of Minute Bodies Made by Magnifying Glasses, with Observations and Inquiries Thereupon*. New York: Dover.

Howard, Patricia. 2017. *Gluck*. London: Routledge.

Howe, Blake. 2017. "Music and the Agents of Obsession." *Music Theory Spectrum* 38:218–240.

Hoyt, Peter A. 1996. "The Concept of Développement in the Early Nineteenth Century." In *Music Theory in the Age of Romanticism*, edited by Ian Bent, 140–162. Cambridge: Cambridge University Press.

Hrotsvit of Gandersheim. 1989. *The Plays of Hrotsvit of Gandersheim*. Translated by Katharina Wilson. New York: Garland.

Hubel and Wiesel Cat Experiment. N.d. Accessed May 20, 2020. at https://www.youtube.com/watch?v=IOHayh06LJ4.

Hubel, David H. 1958. "Cortical Unit Responses to Visual Stimuli in Nonanesthetized Cats." *American Journal of Ophthalmology* 46 (3): 110–122.

Hubel, D. H. 1959. "Single Unit Activity in Striate Cortex of Unrestrained Cats." *Journal of Physiology* 147:226–238.

Hubel, David H. 1982. "Evolution of Ideas on the Primary Visual Cortex, 1955–1978: A Biased Historical Account." *Bioscience Reports* 2 (7): 435–469.

Hubel, David H., and Torsten N. Wiesel. 1959. "Receptive Fields of Single Neurones in the Cat's Striate Cortex." *Journal of Physiology* 148: 574–591.

Hubel, David H., and Torsten N. Wiesel. 2005. *Brain and Visual Perception: The Story of a 25-Year Collaboration*. New York: Oxford University Press.

Hubel, David H., Torsten N. Wiesel, Erin M. Yeagle, Rosa Lafer-Sousa, Bevil R. Conway. 2015. "Binocular Stereoscopy in Visual Areas V-2, V-3, and V-3A of the Macaque Monkey." *Cerebral Cortex* 25 (4): 959–971.

Huffman, Carl A. 1993. *Philolaus of Croton: Pythagorean and Presocratic*. Cambridge: Cambridge University Press.

Huffman, Carl A. 2005. *Archytas of Tarentum: Pythagorean, Philosopher, and Mathematician King*. Cambridge: Cambridge University Press.

Hui, Alexandra. 2013a. "Changeable Ears: Ernst Mach's and Max Planck's Studies of Accommodation in Hearing." *Osiris* 28:119–145.

Hui, Alexandra. 2013b. *The Psychophysical Ear: Musical Experiments, Experimental Sounds, 1840–1910*. Cambridge, MA: MIT Press.

Hui, Alexandra, Julia Kursell, and Myles W. Jackson, eds. 2013. *Music, Sound, and the Laboratory from 1750–1980*. Chicago: University of Chicago Press.

Hunter, Richard, and Ida Macalpine, eds. 1963. *Three Hundred Years of Psychiatry, 1535–1860: A History Presented in Selected English Texts*. London: Oxford University Press.

Hustvedt, Asti. 2011. *Medical Muses: Hysteria in Nineteenth-Century Paris*. New York: Norton.

Iamblichus. 1891. *Iamblichi De communi mathematica scientia liber*. Edited by Nicolaus Festa. Leipzig: B. G. Teubner.

Iamblichus. 1989. *On the Pythagorean Life*. Translated by Gillian Clark. Liverpool: Liverpool University Press.

Iriarte, Tomás de. 1779. *La música*. Madrid: En la imprenta real de la Gazeta.

Iriarte, Tomás de. 1807. *Music: A Didactic Poem, in Five Cantos*. Translated by John Belfour. London: W. Miller.

Ivanovich, Roman. 2014. "The Brilliant Style." In *The Oxford Handbook of Topic Theory*, edited by Danuta Mirka, 330–354. New York: Oxford University Press.

Jackson, Myles. 2004. "Physics, Machines, and Musical Pedagogy." *History of Science* 42: 371–418.

Jackson, Myles W. 2006. *Harmonious Triads: Physicists, Musicians, and Instrument Makers in Nineteenth-Century Germany*. Cambridge, MA: MIT Press.

Jackson, Stanley W. 1992. "The Listening Healer in the History of Psychological Healing." *American Journal of Psychiatry* 149:1623–1632.

Jarcho, Saul. 1961. "Auenbrugger, Laennec, and John Keats." *Medical History* 5:167–172.

Jardine, Nicholas. 1984. *The Birth of History and Philosophy of Science: Kepler's A Defence of Tycho against Ursus, with Essays on Its Provenance and Significance*. Cambridge: Cambridge University Press.

Jay, Martin. 1993. *Downcast Eyes: The Denigration of Vision in Twentieth-Century French Thought*. Berkeley: University of California Press.

Johnson, Frederic. 1985. "Tartini's '*Trattato di musica seconda la vera scienza dell'armonia*': An Annotated Translation with Commentary." PhD diss., Indiana University.

Johnson, James H. 1996. *Listening in Paris: A Cultural History*. Berkeley: University of California Press.

Jones, Arthur Taber. 1935. "The Discovery of Difference Tones." *American Physics Teacher* 3 (2): 49–51.

Jouanna, Jacques. 1999. *Hippocrates*. Translated by M. B. DeBevoise. Baltimore, MD: John Hopkins University Press.

Jouanna, Jaques. 2012. *Greek Medicine from Hippocrates to Galen*. Leiden: Brill.

Jurine, M. de. 1798. "Experiments on Bats Deprived of Sight." *Philosophical Magazine* 1 (2): 136–140.

Juste, David. 2010. "Musical Theory and Astrological Foundations in Kepler: The Making of the New Aspects." In *Music and Esotericism*, edited by Laurence Wuidar, 177–195. Leiden: Brill.

Kamen, Henry. 2001. *Philip V: The King Who Reigned Twice*. New Haven, CT: Yale University Press.

Kassler, Jamie Croy. 1995. *Inner Music: Hobbes, Hooke, and North on Internal Character*. Madison, NJ: Fairleigh Dickinson University Press.

Kassler, Jamie Croy. 2001. *Music, Science, Philosophy: Models in the Universe of Thought*. Aldershot: Ashgate.

Keele, Kenneth D. 1963. *The Evolution of Clinical Methods in Medicine*. London: Pitman Medical.

Kemp, D. T. 1978. "Stimulated Acoustic Emissions from within the Human Auditory System." *Journal of the Acoustical Society of America* 64 (5): 1386–1391. https://doi.org/10.1121/1.382104.

Kennaway, James. 2012a. *Bad Vibrations: The History of the Idea of Music as Cause of Disease*. Farnham, Surrey: Ashgate.

Kennaway, James. 2012b. "Musical Hypnosis: Sound and Selfhood from Mesmerism to Brainwashing." *Social History of Medicine* 25:271–289.

Kennaway, James. 2014. "Introduction: The Long History of Neurology and Music." In *Music and the Nerves, 1700–1900*, edited by James Kennaway, 1–17. London: Palgrave Macmillan UK.

Kennaway, James. 2015. "Historical Perspectives on Music as a Cause of Disease." In *Progress in Brain Research*, edited by Eckart Altenmüller, Stanley Finger, and François Boller, 216:127–145. Amsterdam: Elsevier.

Kennaway, James. 2016. "Lebenskraft, the Body and Will Power: The Life Force in German Musical Aesthetics." In *The Early History of Embodied Cognition, 1740–1920*, edited by John A. McCarthy, Stephanie M. Hilger, Heather I. Sullivan, and Nicholas Saul, 124–144. New York: Brill.

Kennaway, James. 2019. "Music and the Body in the History of Medicine." In *The Oxford Handbook of Music and the Body*, edited by Youn Kim and Sander L. Gilman, 333–348. New York: Oxford University Press.

Kepler, Johannes. 1937. *Gesammelte Werke*. Edited by Walther von Dyck, Max Caspar, and Franz Hammer. Munich: C. H. Beck.

Kepler, Johannes. 1997. *The Harmony of the World*. Translated by E. J. Aiton, A. M. Duncan, and J. V. Field. Philadelphia: American Philosophical Society.

Kepler, Johannes. 2010. *The Six-Cornered Snowflake*. Translated by Jacques Bromberg. Philadelphia: Paul Dry Books

Kepler, Johannes. 2015. *Astronomia Nova*. New rev. ed. Translated by William H. Donahue. Santa Fe, NM: Green Lion Press.

Khodadoust, Kazem, Mohammadreza Ardalan, Kamyar Ghabili, Samad E. J. Golzari, and Garabed Eknoyan. 2013. "Discourse on Pulse in Medieval Persia—the *Hidayat* of Al-Akhawayni (?–983 AD)." *International Journal of Cardiology* 166:289–293.

Kibre, Pearl. 1978. "'Astronomia' or 'Astrologia Ypocratis.'" In *Science and History: Studies in Honor of Edward Rosen*, edited by Erna Hilfstein, Pawel Czartoryski, and Frank D. Grande, 135–156. Wroclaw: Ossolineum.

King, A. Hyatt. 1945. "The Musical Glasses and Glass Harmonica." *Proceedings of the Royal Musical Association* 72:97–122.

King, Alec Hyatt. N.d. "Musical Glasses." *Grove Music Online*.

Kircher, Athanasius. 1643. *Magnes, siue, De arte magnetica opus tripartitum*. Cologne: Apud Iodocvm Kalcoven.

Kinney, Sam E., and Richard Prass. 1986. "Facial Nerve Dissection by use of Acoustic (Loudspeaker) Facial EMG Monitoring." *Otolaryngology—Head and Neck Surgery* 95:458–463.

Kirkpatrick, Ralph. 1953. *Domenico Scarlatti*. Princeton, NJ: Princeton University Press.

Kittler, Friedrich A. 1999. *Gramophone, Film, Typewriter*. Stanford, CA: Stanford University Press.

Kittler, Friedrich A. 2006. *Musik und Mathematik*. Munich: Wilhelm Fink Verlag.

Kittler, Friedrich A. 2014. *The Truth of the Technological World: Essays on the Genealogy of Presence*. Stanford, CA: Stanford University Press.

Kleppner, Daniel. 2007. "Master Michelson's Measurement." *Physics Today* 60 (8): 8–9.

Koenigsberger, Leo. 1965. *Hermann von Helmholtz*. New York: Dover.

Krakeur, Lester G., and Raymond L. Krueger. 1941. "The Mathematical Writings of Diderot." *Isis* 33:219–232.

Kramer, Cheryce. 2000. "Music as Cause and Cure of Illness in Nineteenth-Century Europe." In *Music as Medicine: The History of Music Therapy since Antiquity*, edited by Peregrine Horden, 338–352. Aldershot: Ashgate.

Kramer, Gregory, ed. 1994. *Auditory Display: Sonification, Audification, and Auditory Interfaces*. Reading, MA: Addison-Wesley.

Kramer, Gregory, Bruce Walker, Terri Bonebright, Perry Cook, John H. Flowers, Nadine Miner, and John Neuhoff. 1999. "Sonification Report: Status of the Field and Research Agenda." Santa Fe, NM: International Community for Auditory Display (ICAD).

Kramer, Gregory, Bruce Walker, Perry Cook, and Nadine Miner, eds. 2010. *Sonification Report: Status of the Field and Research Agenda*. DigitalCommons@University of Nebraska—Lincoln.

Kreuzer, Gundula. 2018. *Curtain, Gong, Steam: Wagnerian Technologies of Nineteenth-Century Opera*. Oakland, CA: University of California Press.

Kümmel, Werner Friedrich. 1977. *Musik und Medizin: Ihre Wechselbeziehung in Theorie und Praxis von 800 bis 1800*. Freiburg: Alber.

Kursell, Julia. 2011. "A Gray Box: The Phonograph in Laboratory Experiments and Fieldwork, 1900–1920." In *The Oxford Handbook of Sound Studies*, edited by Trevor Pinch and Karin Bijsterveld. Oxford: Oxford University Press..

Kursell, Julia. 2013. "Experiments on Sound Color in Music and Acoustics: Helmholtz, Schoenberg, and Klangfarbenmelodie." *Osiris* 28:191–211.

Kursell, Julia. 2015. "A Third Note: Helmholtz, Palestrina, and the Early History of Musicology." *Isis* 106 (2): 353–366.

Kursell, Julia. 2018. *Epistemologie des Hörens: Helmholtz' physiologische Grundlegung der Musiktheorie*. Munich: Wilhelm Fink Verlag.

Kursell, Julia. 2019a. "Hearing in the Music of Hector Berlioz." In *Nineteenth-Century Opera and the Scientific Imaginstion*, edited by David Trippett, 109–133. Cambridge: Cambridge University Press.

Kursell, J. 2019b. "Listening to More Than Sounds: Carl Stumpf and the Experimental Recordings of the Berliner Phonogramm-Archiv." *Technology and Culture* 60 (2, Supplement): S39–S63.

Kusukawa, Sachiko. 2012. *The Canon of the Human Body: Vesalius's De humani corporis fabrica*. Chicago: University of Chicago Press.

Laennec, R. T. H. 1826. *Traité de l'auscultation médiate et des maladies des poumons et du coeur*. Paris: Chaudé.

Laennec, R. T. H. 1830. *A Treatise on the Diseases of the Chest and on Mediate Auscultation*. Edited by John Forbes. New York: Samuel Wood & Sons.

Lane Fox, Robin. 2020. *The Invention of Medicine: From Homer to Hippocrates*. New York: Basic Books.

La Villemarqué, Théodore Hersart. 1867. *Barzaz Breiz; chants populaires de la Bretagne*. 6th ed. Paris: Didier.

Lawn, Brian. 1963. *The Salernitan Questions: An Introduction to the History of Medieval and Renaissance Problem Literature*. Oxford: Clarendon Press.

Lawrence, Christopher. 1979. "The Nervous System and Society in the Scottish Enlightenment." In *Natural Order: Historical Studies of Scientific Culture*, edited by Barry Barnes and Steven Shapin, 19–40. Beverly Hills, CA: Sage.

Le Guin, Elisabeth. 2006. *Boccherini's Body: An Essay in Carnal Musicology*. Berkeley: University of California Press.

Lenoir, T. 1986. "Models and Instruments in the Development of Electrophysiology, 1845–1912." *Historical Studies in the Physical and Biological Sciences* 17:1–54.

Lesky, Erna. 1959. "Leopold Auenbrugger—Schüler van Swietens." *Deutsche Medizinische Wochenschrift* 84:717–725.

Levett, J., and G. Agarwal. 1979. "The First Man/Machine Interaction in Medicine: The Pulsilogium of Sanctorius." *Medical Instrumentation* 13:61–63.

Lewis, Orly. 2017. *Praxagoras of Cos on Arteries, Pulse and Pneuma: Fragments and Interpretation*. Leiden: Brill.

Liban, Jerzy. 1975. *De musicae laudibus oratio*. Cracow: Polskie Wydawnictwo Muzyczne.

Lindeboom, Gerrit Arie. 1979. *Descartes and Medicine*. Amsterdam: Rodopi.

Linebarger, Eric J., David R. Hardten, Gaurav K. Shah, and Richard L. Lindstrom. 1999. "Phacoemulsification and Modern Cataract Surgery." *Survey of Ophthalmology* 44:123–147.

Lipowski, Z. J. 1984. "Benjamin Franklin and Princess Czartoryska: An Unknown Therapeutic Encounter." *Pennsylvania History* 51 (2): 167–171.

Lippmann, Gabriel. 1875. "Relations entre les phénomènes électriques et capillaires." *Annales de Chimie et de Physique*, series 5. 5:494–549.

Lloyd, G. E. R. 1970. *Early Greek Science: Thales to Aristotle*. New York: Norton.

Lloyd, G. E. R., ed. 1984. *Hippocratic Writings*. Translated by J. Chadwick. Harmondsworth: Penguin Classics.

Lloyd, G. E. R. 1987. *The Revolutions of Wisdom: Studies in the Claims and Practice of Ancient Greek Science.* Berkeley: University of California Press.

Lloyd, G. E. R. 2013. *Greek Science after Aristotle.* New York: Random House.

Lockhart, Ellen. 2017. *Animation, Plasticity, and Music in Italy, 1770–1830.* Oakland: University of California Press.

Longrigg, James. 1993. *Greek Rational Medicine: Philosophy and Medicine from Alcmaeon to the Alexandrians.* London: Routledge.

Lucas, K. 1912. "On a Mechanical Method of Correcting Photographic Records Obtained from the Capillary Electrometer." *Journal of Physiology* 44 (3): 225–242.

Lucas, Keith. 1905. "On the Gradation of Activity in a Skeletal Muscle-Fibre." *Journal of Physiology* 33 (2): 125–137.

Lucas, Keith. 1909. "The 'All or None' Contraction of the Amphibian Skeletal Muscle Fibre." *Journal of Physiology* 38 (2–3): 113–133.

Lynch, John. 1989. *Bourbon Spain 1700–1808.* Oxford: Blackwell.

Lyncker, Peter. 1966. "Samuel Hafenreffer, 1587–1660: Leben, Werk, seine Bedeutung für die Dermatologie." MD diss., Tübingen.

Maclean, Ian. 2002. *Logic, Signs, and Nature in the Renaissance: The Case of Learned Medicine.* Cambridge: Cambridge University Press.

Mahnke, C. Becket. 2009. "Automated Heartsound Analysis/Computer-Aided Auscultation: A Cardiologist's Perspective and Suggestions for Future Development." In *Proceedings of the 2009 Annual International Conference of the IEEE Engineering in Medicine and Biology Society,* 3115–3118. Piscataway, NJ: IEEE.

Marquet, François Nicolas. 1769. *Nouvelle méthode facile et curieuse, pour connoitre le pouls par les notes de la musique.* 2nd ed. Paris: Didot.

Marquet, François Nicolas. 1770. *Traité de l'apoplexie, paralysie, et autres affections soporeuses développées par l'expérience.* Paris: Costard.

Martianus Capella. 1977. *The Marriage of Philology and Mercury.* Translated by William Harris Stahl, Richard Johnson, and E. L. Burge. New York: Columbia University Press.

Martin, Henri. 1868. *Histoire de France populaire, depuis les temps les plus reculés jusqu'à nos jours,* vol. 3. Paris: Furne, Jouvet et Cie.

Maslow, Abraham H. 1966. *The Psychology of Science: A Reconnaissance.* New York: Harper.

Mathews, M. V. 1963. "The Digital Computer as a Musical Instrument." *Science* 142 (3592): 553–557.

Matthews, B. H. 1928. "A New Electrical Recording System for Physiological Work." *Journal of Physiology* 65 (3): 225–242.

Matthews, Betty. 1975. "The Davies Sisters, J. C. Bach and the Glass Harmonica." *Music and Letters,* no. 2: 150–169.

Matthews, Bryan H. C. 1929. "A New Electrical Recording System." *Journal of Scientific Instruments* 6 (7): 220–226.

Maxim, Hiram. 1912. "The Sixth Sense of the Bat." *Scientific American Supplement* 74:148–150.

Mayer-Deutsch, Angela. 2004. "'Quasi-Optical Palingenesis': The Circulation of Portraits and the Image of Kircher." In *Athanasius Kircher: The Last Man Who Knew Everything,* edited by Paula Findlen, 105–129. New York: Routledge.

McCabe, Ina Baghdiantz. 2008. *Orientalism in Early Modern France: Eurasian Trade, Exoticism and the Ancien Regime.* Oxford: Berg.

McCarthy, O. R. 1999. "Getting a Feel for Percussion." *Vesalius* 5:3–10.

McComas, Alan J. 2011. *Galvani's Spark: The Story of the Nerve Impulse.* Oxford: Oxford University Press.

McDonald, Granley. 2018. "The Reception of Ficino's Theory of World Harmony in Germany." In *Sing Aloud Harmonious Spheres: Renaissance Conceptions of Cosmic Harmony,* edited by Jacomien Prins and Maude Vanhaelen, 160–182. New York: Routledge.

Mclaren, M. J., D. M. Hawkins, H. J. Koornhof, K. R. Bloom, D. M. Bramwell-Jones, E. Cohen, G. E. Gale, et al. 1975. "Epidemiology of Rheumatic Heart Disease in Black Schoolchildren of Soweto, Johannesburg." *British Medical Journal* 3 (5981): 474–478.

McNeil Jr., Donald G. 2019. "In African Villages, These Phones Become Ultrasound Scanners." *New York Times*, April 15, 2019. https://www.nytimes.com/2019/04/15/health/medical-scans-butterfly-iq.html.

Meneghini, Matteo. 2007. "An Analysis of the Compositional Techniques in John Chowning's Stria." *Computer Music Journal* 31 (3): 26–37.

Ménuret de Chambaud. 1751a. "Musique, effets de la." In *Encyclopédie, ou Dictionnaire raisonné des sciences, des arts et des métiers*, edited by Denis Diderot and Jean le Rond d'Alembert, 10:903. Paris: Briasson.

Ménuret de Chambaud. 1751b. "Pouls." In *Encyclopédie, ou Dictionnaire raisonné des sciences, des arts et des métiers*, edited by Denis Diderot and Jean le Rond d'Alembert, 13:205. Paris: Briasson.

Merklinger, Harold M., and Dale D. Ellis. 2017. "Fessenden and Boyle: Two Canadian Sonar Pioneers." *Proceedings of Meetings on Acoustics* 30:1–20.

Mersenne, Marin. 1963. *Harmonie Universelle, contenant la théorie et la pratique de la musique*. Paris: Centre national de la recherche scientifique.

Mesmer, Franz Anton. 1980. *Mesmerism: A Translation of the Original Scientific and Medical Writings of F. A. Mesmer*. Translated by George Bloch. Los Altos, CA: W. Kaufman.

Michelson, Albert A. 1880. *Experimental Determination of the Velocity of Light: Made at the U. S. Naval Academy, Annapolis*. Washington, DC: U.S. Nautical Almanac Office.

Monelle, Raymond. 2006. *The Musical Topic: Hunt, Military and Pastoral*. Bloomington: Indiana University Press.

Morat, Daniel, and Axel Volmar, eds. 2017. *Netzwerk Hör-Wissen im Wandel: Wissensgeschichte des Hörens in der Moderne*. Berlin: De Gruyter.

Morton, H. V. 1969. *A Traveller in Southern Italy*. New York: Dodd, Mead.

Movahed, Mohammad-Reza, and Ramin Ebrahimi. 2007. "The Prevalence of Valvular Abnormalities in Patients Who Were Referred for Echocardiographic Examination with a Primary Diagnosis of 'Heart Murmur.'" *Echocardiography* 24 (5): 447–451.

Mozart, Wolfgang Amadeus. 1997. *The Letters of Mozart and His Family*. Edited by Stanley Sadie and Fiona Smart. Translated by Emily Anderson. London: Palgrave Macmillan.

Mueller, William R. 1949. "Robert Burton's Frontispiece." *PMLA* 64:1074–1088.

Müller, Johannes. 1838. *Elements of Physiology*. Translated by William Baly. London: Taylor & Walton.

Neuburger, Max. 1922. *Leopold Auenbrugger und sein Inventum Novum: eine historische Skizze*. Vienna: M. Salzer.

"Newborn and Infrant Hearing Screening." 2010. Geneva: World Health Organization. https://www.who.int/blindness/publications/Newborn_and_Infant_Hearing_Screening_Report.pdf?ua=1.

Nicolson, Malcolm, and John E. E. Fleming. 2013. *Imaging and Imagining the Fetus: The Development of Obstetric Ultrasound*. Baltimore, MD: Johns Hopkins University Press.

Nietzsche, Friedrich. 1997. *Daybreak: Thoughts on the Prejudices of Morality*. Translated by M. Clark and B. Leiter. Cambridge: Cambridge University Press.

Nutton, Vivian. 1995. "Medicine in the Greek World, 800–50 BC." In *The Western Medical Tradition: 800 B.C.–1800 A.D.*, by Lawrence I. Conrad, Michael Neve, Vivian Nutton, Roy Porter, and Andrew Wear, 11–38. Cambridge: Cambridge University Press.

Nutton, Vivian. 2004. *Ancient Medicine*. London: Routledge.

Olesko, Kathryn M., and Frederic L. Holmes. 1994. "Experiment, Quantification, and Discovery: Helmholtz's Early Physiological Researches, 1843–50." In *Hermann von Helmholtz and the Foundations of Nineteenth-Century Science*, edited by David Cahan, 50–108. Berkeley: University of California Press.

O'Neal, John C. 1998. "Auenbrugger, Corvisart, and the Perception of Disease." *Eighteenth-Century Studies* 31:473–489. http://www.jstor.org/stable/30053888.

O'Neal, John C. 2002. *Changing Minds: The Shifting Perception of Culture in Eighteenth-Century France*. Newark: University of Delaware Press.

Orr, D. Alan. 2016. "'God's Hangman': James VI, the Divine Right of Kings, and the Devil." *Reformation and Renaissance Review* 18:137–154.

Osler, William. 1947. *The Principles and Practice of Medicine.* 16th ed. Edited by Henry A. Christian. New York: Appleton-Century.

O'Sullivan, Lisa. 2012. "The Time and Place of Nostalgia: Re-Situating a French Disease." *Journal of the History of Medicine and Allied Sciences* 67:626–649. http://jhmas.oxfordjournals.org/content/early/2011/12/08/jhmas.jrr058.

Otis, Laura. 2001. "The Other End of the Wire: Uncertainties of Organic and Telegraphic Communication." *Configurations* 9 (2): 181–206.

Otis, Laura. 2002. "The Metaphoric Circuit: Organic and Technological Communication in the Nineteenth Century." *Journal of the History of Ideas* 63:105–128.

Ozanam, Charles. 1886. *La circulation et le pouls: Histoire, physiologie, sémeiotique, indications therapeutiques.* Paris: J. B. Baillière et fils.

Pace, Antonio. 1958. *Benjamin Franklin and Italy.* Philadelphia: American Philosophical Society.

Page, Robert Morris. 1962. *The Origin of Radar.* Garden City, NY: Anchor Books.

Palisca, Claude V. 1985. *Humanism in Italian Renaissance Musical Thought.* New Haven, CT: Yale University Press.

Palisca, Claude V. 2006. *Music and Ideas in the Sixteenth and Seventeenth Centuries.* Chicago: University of Illinois Press.

Panofsky, Erwin. 1972. *Studies in Iconology: Humanistic Themes in the Art of the Renaissance.* New York: Harper & Row.

Pantalony, David. 2004. "Seeing a Voice: Rudolph Koenig's Instruments for Studying Vowel Sounds." *American Journal of Psychology* 117 (3): 425–442.

Pantalony, David. 2009. *Altered Sensations: Rudolph Koenig's Acoustical Workshop in Nineteenth-Century Paris.* Dordrecht: Springer.

Parvizi, Josef, Kapil Gururangan, Babak Razavi, and Chris Chafe. 2018. "Detecting Silent Seizures by Their Sound." *Epilepsia* 59 (4): 877–884.

Pattie, Frank A. 1956. "Mesmer's Medical Dissertation and Its Debt to Mead's *De imperio solis ac lunae.*" *Journal of the History of Medicine and Allied Sciences* 11:275–287.

Pattie, Frank A. 1994. *Mesmer and Animal Magnetism: A Chapter in the History of Medicine.* Hamilton, NY: Edmonston Publishing.

Pauletto, Sandra and Andy Hunt. 2006. "The Sonification of EMG Data." *Proceedings of the 12th International Conference on Auditory Display, London, UK, June 20–23, 2006,* 152–157.

Paxton, Frederick S. 1993. "Liturgy and Anthropology: A Monastic Death Ritual of the Eleventh Century." *Studies in Music Thanatology* 2:1–20.

Payne, Stanley G. 1973. *A History of Spain and Portugal.* Madison: University of Wisconsin Press.

Pérez, Jesus, Ross J. Baldessarini, Núria Cruz, Paola Salvatore, and Eduard Vieta. 2011. "Andres Piquer-Arrufat (1711–1772): Contributions of an Eighteenth-Century Spanish Physician to the Concept of Manic-Depressive Illness." *Harvard Review of Psychiatry* 19:68–77.

Peschier, J. 1798. "Extrait des expériences de Jurine sur les chauve-souris qu'on a privé de la vue." *Journal de physique, de chimie et d'histoire naturelle* 3:145–148.

Pesic, Peter. 2000. *Labyrinth: A Search for the Hidden Meaning of Science.* Cambridge, MA: MIT Press.

Pesic, Peter. 2013. "Thomas Young and Eighteenth Century Tempi." *Performance Practice Review* 18 (1). http://scholarship.claremont.edu/ppr/vol18/iss1/2.

Pesic, Peter. 2014a. "Francis Bacon, Violence, and the Motion of Liberty: The Aristotelian Background." *Journal of the History of Ideas* 75:69–90.

Pesic, Peter. 2014b. *Music and the Making of Modern Science.* Cambridge, MA: MIT Press.

Pesic, Peter. 2016. "Music, Mechanism, and the 'Sonic Turn' in Physical Diagnosis." *Journal of the History of Medicine and Allied Sciences* 71:144–172.

Pesic, Peter. 2017. *Polyphonic Minds: Music of the Hemispheres.* Cambridge, MA: MIT Press.

Pesic, Peter. 2019. "Music, Melancholia, and Mania: Gaetano Brunetti's Obsessional Symphony." *19th-Century Music* 43:67–85.

Pesic, Peter. 2020. "Composing the Crisis: From Mesmer's Harmonica to Charcot's Tam-Tam." *Nineteenth-Century Music Review.* https://doi.org/10.1017/S1479409820000087.

Pesic, Peter, and Axel Volmar. 2014. "Pythagorean Longings, the Rhetoric of String Theory, and the Sonification of High-Energy Physics." *Journal of Sonic Studies* 6. https://www.researchcatalogue.net/view/109371/109372/0/53.

Petrie, Charles. 1971. *King Charles III of Spain: An Enlightened Despot.* New York: J. Day Co.

Pfaffmann, C. 2012. "Taste Neurophysiology, Sensory Coding, and Behavior." In *Foundations of Sensory Science*, by H. Autrum, L. M. Beidler, H. Davis, and H. Engström, 325–349. Berlin: Springer Science & Business Media.

Piccolino, Marco, and Marco Bresadola. 2013. *Shocking Frogs: Galvani, Volta, and the Electric Origins of Neuroscience.* Oxford: Oxford University Press.

Pinch, T. J., and Karin Bijsterveld. 2012. *The Oxford Handbook of Sound Studies.* Oxford: Oxford University Press.

Piquer, Andrés. 1759. "Discurso sobre la enfermedad del rey nuestro Señor Don Fernando VI, que Dios guarde." Madrid: Biblioteca Nacional de España.

Piquer, Andrés. 1764. *Praxis medica: ad usum Scholae Valantinae.* Madrid: Ediciones Joachim Ibarra.

Pistacchio, Bonifacio, and Eleonora Pistacchio. 1996. "François Nicolas Marquet (1687–1759) e il suo metodo per annotare con le note musicali il polso dell'uomo." *Rivista di storia della medicina* 27:189–204.

Plato. 1997. *Complete Works.* Edited by John M. Cooper and D. S. Hutchinson. Indianapolis, IN: Hackett.

Plato. 2016. *Timaeus.* 2nd ed. Translated by Peter Kalkavage. Indianapolis: Hackett.

Plotinus. 1992. *The Enneads.* Translated by Stephen Mackenna. Burdett, NY: Larson Publications.

Pollitt, J. J. 1972. *Art and Experience in Classical Greece.* Cambridge: Cambridge University Press.

Polzonetti, Pierpaolo. 2001. *Tartini e la musica secondo natura.* Lucca: LIM.

Porter, Roy. 2001. *The Cambridge Illustrated History of Medicine.* Cambridge: Cambridge University Press.

Pottinger, Mark. 2020. "*Lucia* and the Auscultation of Disease in Mid-Nineteenth-Century France." *Nineteenth-Century Music Review.* https://doi.org/10.1017/S1479409820000075.

Prins, Jacomien. 2012. "The Music of the Pulse in Marsilio Ficino's *Timaeus* Commentary." In *Blood, Sweat, and Tears: The Changing Concepts of Physiology from Antiquity into Early Modern Europe*, edited by Manfred Horstmanshoff, Helen King, and Claus Zittel, 393–413. Leiden: Brill.

Prins, Jacomien, and Maude Vanhaelen, eds. 2018. *Sing Aloud Harmonious Spheres: Renaissance Conceptions of Cosmic Harmony.* New York: Routledge.

Provenza, Antonietta. 2012. "Aristoxenus and Music Therapy: Fr. 26 Wehrli within the Tradition on Music and Catharsis." In *Aristoxenus of Tarentum: Discussion*, edited by Carl A. Huffman, 91–128. New Brunswick, NJ: Transaction.

Pugliese, Romana Margherita. 2004. "The Origins of *Lucia di Lammermoor*'s Cadenza." *Cambridge Opera Journal* 16 (1): 23–42.

Puységur, Amand Marc Jacques de Chastenet, marquis de. 1820. *Mémoires pour servir à l'histoire et à l'établissement du magnétisme animal.* Paris: Dentu.

Qifṭī, ʿAlī ibn Yūsuf. 1903. *Tʾarīḫ al-Ḥukamāʾ.* Edited by August Müller and Julius Lippert. Leipzig: Dieterich Verlagsbuchhandlung.

Quantz, Johann Joachim. 1966. *On Playing the Flute.* Translated by Edward R. Reilly. New York: Schirmer Books.

Radau, R[odolphe]. 1869. *Die Lehre vom Schall: gemeinfassliche Darstellung der Akustik.* Munich: R. A. Oldenbourg.

Ramos de Pareja, Bartolomeo. 1993. *Musica Practica.* Translated by Clement A. Miller. Neuhausen-Stuttgart: Hänssler-Verlag.

Ramos, Rene M. 1997. "The Symphonies of Gaetano Brunetti (ca. 1744–1798)." PhD diss., Indiana University.

Rasch, Rudolf. 2014. "The Art of Repetition as Practiced by Luigi Boccherini in His Sonatas for Keyboard and Violin Opus 5." In *Boccherini Studies*, edited by Christian Speck, 4:291–296. Bologna: Ut Orpheus Edizioni.

Ratner, Leonard G. 1985. *Classic Music: Expression, Form and Style*. New York: Schirmer Books.

Raz, Carmel. 2014. "'The Expressive Organ within Us': Ether, Ethereality, and Early Romantic Ideas about Music and the Nerves." *19th-Century Music* 38 (2): 115–144. https://doi.org/10.1525/ncm.2014.38.2.115.

Raz, Carmel. N.d. "Of Sound Minds and Tuning Forks: Charcot's Acoustic Experiments at the Salpêtrière." Accessed December 9, 2017, at http://musicologynow.ams-net.org/2015/10/of-sound-minds-and-tuning-forks .html.

Raz, Carmel and Stanley Finger. 2019. "Musical Glasses, Metal Reeds and Broken Hearts: Two Cases of Melancholia Treated by New Musical Instruments." In T*he Routledge Companion to Music, Mind and Well-Being*, edited by Penelope Gouk, Jacomien Prins, James Kennaway, and Wiebke Thormahlen, 1:77–92. London: Taylor and Francis.

Regier, Jonathan. 2014. "Kepler's Theory of Force and His Medical Sources." *Early Science and Medicine* 19 (1): 1–27.

Rehding, Alexander. 2018. *Beethoven's Symphony No. 9*. New York: Oxford University Press.

Reiser, Stanley Joel. 1978. *Medicine and the Reign of Technology*. Cambridge: Cambridge University Press.

Rice, Tom. 2011. "Sounding Bodies: Medical Students and the Acquisition of Stethoscopic Perspectives." In *The Oxford Handbook of Sound Studies*, edited by Trevor Pinch and Karin Bijsterveld, 298–319. Oxford: Oxford University Press.

Rice, Tom and Steven Feld. 2020. "Questioning Acoustemology: An Interview with Steven Feld." *Sound Studies*, https://www.tandfonline.com/doi/full/10.1080/20551940.2020.1831154

Richardson, Lewis Fry. 1913a. Apparatus for Warning a Ship of Its Approach to Large Objects in a Fog. UK Patent GB191209423 (A), filed April 20, 1912, and issued March 6, 1913.

Richardson, Lewis Fry. 1913b. Apparatus for Warning a Ship at Sea of Its Nearness to Large Objects Wholly or Partly under Water. UK Patent GB191211125 (A), filed May 10, 1912, and issued March 27, 1913.

Richardson, Stephen A. 1957. "Lewis Fry Richardson (1881–1953): A Personal Biography." *Conflict Resolution* 1: 300–304.

Richer, Paul. 1881. *Etudes cliniques sur l'hystéro-épilepsie ou grande hystérie*. Paris: Delahaye et Lecrosnier.

Richet, Charles. 1884. *L'homme et l'intelligence: fragments de physiologie et de psychologie*. Paris: Alcan.

Riedweg, Christoph. 2005. *Pythagoras: His Life, Teaching, and Influence*. Translated by Steven Rendell. Ithaca, NY: Cornell University Press.

Riskin, Jessica. 2002. *Science in the Age of Sensibility: The Sentimental Empiricists of the French Enlightenment*. Chicago: University of Chicago Press.

Robertson, Brian. 2009. "A Celebration of the 50th Anniversary of David Hubel and Torsten Wiesel's Receptive Fields of Single Neurones in the Cat's Striate Cortex." *Journal of Physiology* 587 (12): 2721–2732.

Rochlitz, Friedrich. 1798. "Ueber die vermeynte Schädlichkeit des Harmonikaspiels." *Allgemeine musikalische Zeitung*, November 14, 1798.

Roe, Anne. 1953. *The Making of a Scientist*. New York: Dodd, Mead.

Roger, Joseph-Louis. 1758. "Tentamen de vi soni et musices in corporum humanum." MD diss., Montpellier.

Roger, Joseph-Louis. 1803. *Traité des effets de la musique sur le corps humain*. Translated by Etienne-Jules Sainte-Marie. Paris: Brunot.

Rosen, Charles. 1998. *The Classical Style: Haydn, Mozart, Beethoven*. Exp. ed. New York: Norton.

Rosen, Edward, trans. 2003. *Kepler's* Somnium: *The Dream, or Posthumous Work on Lunar Astronomy*. Mineola, NY: Dover.

Rost, G. A. 1953. "[Samuel Hafenreffer, author of the first textbood on dermatology in German speaking countries]." *Zeitschrift Fur Haut- Und Geschlechtskrankheiten* 14 (7): 227–230.

Roth, Michael S. 1991. "Dying of the Past: Medical Studies of Nostalgia in Nineteenth-Century France." *History and Memory* 3:5–29.

Rothenberg, David. 2008. *Thousand Mile Song: Whale Music in a Sea of Sound*. New York: Basic Books.

Rothman, Aviva. 2017. *The Pursuit of Harmony: Kepler on Cosmos, Confession, and Community*. Chicago: University of Chicago Press.

Rothschild, Germaine de. 1965. *Luigi Boccherini: His Life and Work*. Translated by Andreas Mayor. London: Oxford University Press.

Rousseau, G. S. 2004. *Nervous Acts: Essays on Literature, Culture, and Sensibility*. Houndmills, Basingstoke, Hampshire: Palgrave Macmillan.

Ruiz Casaux y López de Carvajal, Juan Antonio. 1959. *La musica en la corte de don Carlos IV y su influencia en la vida musical española*. Madrid: Real Academia de Bellas Artes de San Fernando.

Sala, Emilio. 1994. "Women Crazed by Love: An Aspect of Romantic Opera." Translated by William Ashbrook. *Opera Quarterly* 10 (3): 19–41.

Salieri, Antonio, and Leopold Auenbrugger. 1986. *Der Rauchfangkehrer*. New York: Garland.

Sanderson, J. B. 1895. "The Electrical Response to Stimulation of Muscle, and Its Relation to the Mechanical Response." *Journal of Physiology* 18 (1–2): 117–160.

Sarton, George. 1954. *Galen of Pergamon*. Lawrence: University of Kansas Press.

Sarton, George. 1975. *Introduction to the History of Science*. Huntington, NY: R. E. Krieger Pub. Co.

Scheminzky, Ferdinand. 1927. "Untersuchungen über die Verstärkung und graphische Registrierung von Schallerscheinungen über Herz und Lunge mittels Elektronenröhren; Konstruktion eines Elektrostethoskops." *Zeitschrift für die gesamte experimentelle Medizin* 57 (1): 470–501.

Schleiner, Winfried. 1991. *Melancholy, Genius, and Utopia in the Renaissance*. Wiesbaden: Otto Harrassowitz.

Schmidgen, Henning. 2014. *The Helmholtz Curves: Tracing Lost Time*. Translated by Nils F. Schott. New York: Fordham University Press.

Schmidt, Matthias. 2003. "Das Andere der Aufklärung: Zur Kompositionästhetik von Mozarts Glasharmonika-Quintett KV 617." *Archiv für Musikwissenschaft* 60:279–302.

Schechner, Sara. 1997. *Comets, Popular Culture, and the Birth of Modern Cosmology*. Princeton, NJ: Princeton University Press.

Schoon, Andi, and Axel Volmar, eds. 2012. *Das geschulte Ohr: Eine Kulturgeschichte der Sonifikation*. Bielefeld: transcript-verlag.

Schumacher, Joseph. 1963. *Antike Medizin; die naturphilosophischen Grundlagen der Medizin in der griechischen Antike*. 2nd ed. Berlin: De Gruyter.

Schumacher, Joseph. 1965. *Die Anfänge abendländischer Medizin in der griechischen Antike*. Stuttgart: W. Kohlhammer.

Schwan, Christian Friedrich, Leopold Auenbrugger, Christian Lichtenberg, Josef Franz von Göz, and Johann Gottlieb Stephanie der Jüngere. 1986. *Librettos IV*. New York: Garland.

Schwartz, Hillel. 2011. *Making Noise: From Babel to the Big Bang and Beyond*. Brooklyn, NY: Zone Books.

Serao, Francesco. 1742. *Della tarantola o sia falangio di Puglia lezioni accademiche*. Naples.

Serban, Andrei. N.d. *Natalie Dessay: Lucia in Paris*. Accessed July 21, 2019, at http://video.alexanderstreet.com/watch/natalie-dessay-lucia-in-paris.

Seta, Fabrizio della, Nino Pirrotta, and Franco Piperno. 1989. *In cantu et in sermone*. Florence: L. S. Olschki.

Shams, Qamar A., Allan J. Zuckerwar, and Albert L. Dimarcantonio. 2016. Infrasonic Stethoscope for Monitoring Physiological Processes. US Patent 20160095571A1, filed March 16, 2015, and issued April 7, 2016.

Shorter, Edward. 2008. *From Paralysis to Fatigue: A History of Psychosomatic Illness in the Modern Era*. New York: Simon and Schuster.

Sigerist, Henry E. 1948. "The Story of Tarantism." In *Music and Medicine*, edited by Dorothy M. Schullian and Max Schoen, 96–116. New York: H. Schuman.

Silvas, Anna. 1999. *Jutta and Hildegard: The Biographical Sources*. University Park: Penn State Press.

Singer, Charles. 1926. *The Evolution of Anatomy*. New York: Knopf.

Singh, Siddharth, and Abha Goyal. 2007. "The Origin of Echocardiography." *Texas Heart Institute Journal* 34 (4): 431–438.

Siraisi, Nancy G. 1975. "The Music of Pulse in the Writings of Italian Academic Physicians (Fourteenth and Fifteenth Centuries)." *Speculum* 50:689–710.

Siraisi, Nancy G. 1990. *Medieval and Early Renaissance Medicine: An Introduction to Knowledge and Practice*. Chicago: University of Chicago Press.

Smith, Alexis B. 2016. "Ritter's Musical Blood Flow through Hoffmann's Kreisler." In *The Early History of Embodied Cognition 1740–1920: The Lebenskraft-Debate and Radical Reality in German Science, Music, and Literature*, edited by John McCarthy, Stephanie M. Hilger, Heather I. Sullivan, and Nicholas Saul, 145–162. Leiden: Brill.

Snellen, H. A. 1994. *Willem Einthoven (1860–1927) Father of Electrocardiography: Life and Work, Ancestors and Contemporaries*. Dordrecht: Springer Netherlands.

Solar-Quintes, Nicolás. 1947. "Nuevos documentos sobre Luigi Boccherini." *Anuario Musical* 2:81.

"Sonification." N.d. Chris Chafe. Accessed June 7, 2020, at http://chrischafe.net/portfolio/sonification-2/.

Spallanzani, Lazzaro. 1941. "Letters on a Suspected New Sense in Bats." Translated by Dale McAdoo. MCZ Archives, Harvard.

Spallanzani, Lazzaro, and Anton Maria Vassalli. 1794. *Lettere sopra il sospetto di un nuovo senso nei pipistrelli dell'abate Lazzaro Spallanzani . . . con le risposte dell'abate Antonmaria Vassalli*. Turin: Stamperia Reale.

Staden, Heinrich von. 1989. *Herophilus: The Art of Medicine in Early Alexandria*. Cambridge: Cambridge University Press.

Starobinski, Jean. 1966. "The Idea of Nostalgia." *Diogenes* 53:81–103.

Steege, Benjamin. 2011. "Janáček's Chronoscope." *Journal of the American Musicological Society* 64 (3): 647–687.

Steege, Benjamin. 2012. *Helmholtz and the Modern Listener*. Cambridge: Cambridge University Press.

Steinke, Hubert. 2005. *Irritating Experiments: Haller's Concept and the European Controversy on Irritability and Sensibility, 1750–90*. Amsterdam: Rodopi.

Stephenson, Bruce. 1994. *The Music of the Heavens: Kepler's Harmonic Astronomy*. Princeton, NJ: Princeton University Press.

Sterne, Jonathan. 2003. *The Audible Past: Cultural Origins of Sound Reproduction*. Durham, NC: Duke University Press.

Sterne, Jonathan, and Mitchell Akiyama. 2012. "The Recording That Never Wanted to Be Heard and Other Stories of Sonification." In *The Oxford Handbook of Sound Studies*, edited by Trevor Pinch and Karin Bijsterveld, 544–560. Oxford: Oxford University Press.

Stirling, William. 1888. *Outlines of Practical Physiology*. London: Charles Griffin & Co.

Stock, John T. 2004. "Gabriel Lippmann and the Capillary Electrometer." *Bulletin of the History of Chemistry* 29 (1): 16–20.

Strunk, W. Oliver, and Leo Treitler, eds. 1998. *Source Readings in Music History*. New York: Norton.

Supper, Alexandra. 2012. *Lobbying for the Ear: The Public Fascination with and Academic Legitimacy of the Sonification of Scientific Data*. Maastricht: Universitaire Pers Maastricht.

Sweet, Victoria. 2006. *Rooted in the Earth, Rooted in the Sky: Hildegard of Bingen and Premodern Medicine*. New York: Routledge.

Swieten, Gerard van. 1754. *The Commentaries upon the Aphorisms of Dr. Herman Boërhaave*. London.

Sykes, Ingrid. 2014. *Society, Culture and the Auditory Imagination in Modern France: The Humanity of Hearing*. Houndmills, Basingstoke, Hampshire: Palgrave Macmillan.

Tarchanow, J. 1878. "Das Telephon als Anzeiger der Nerven- und Muskelströme beim Menschen und den Thieren." *St. Petersburger medicinische Wochenschrift* 3:353–354.

Tarchanow, J. 1879. "Das Telephon im Gebiete der thierischen Electricität." *St. Petersburger medicinische Wochenschrift* 4:93–95.

Tartini, Giuseppe. 1754. *Trattato di musica seconda la vera scienza dell'armonia*. Padova.

Tavakolian, Kouhyar, Brandon Ngai, Andrew P. Blaber, and Bozena Kaminska. 2011. "Infrasonic Cardiac Signals: Complementary Windows to Cardiovascular Dynamics." In *Conference Proceedings . . . of the Annual*

International Conference of the IEEE Engineering in Medicine and Biology Society. IEEE Engineering in Medicine and Biology Society. Annual Conference, 4275–4278. Piscataway, NJ: IEEE.

Tavel, Morton E. 1996. "Cardiac Auscultation: A Glorious Past—But Does It Have a Future?" *Circulation* 93 (6): 1250–1253.

Temkin, Owsei. 1964. "The Classical Roots of Glisson's Doctrine of Irritation." *Bulletin of the History of Medicine* 38:297–328.

Temple, Mark D. 2017. "An Auditory Display Tool for DNA Sequence Analysis." *BMC Bioinformatics* 18 (1):221. https://doi.org/10.1186/s12859-017-1632-x.

Temple, Mark D. N.d. "DNA Sonification." Accessed June 13, 2020, at http://dnasonification.org.

Temple, Mark D. 2020. "Real-Time Audio and Visual Display of the Coronavirus Genome." *BMC Bioinformatics* 29 (1): 431.

"The Nobel Prize in Physiology or Medicine 1932." N.d.a. NobelPrize.Org. Accessed March 1, 2020. https://www.nobelprize.org/prizes/medicine/1932/adrian/lecture/.

"The Nobel Prize in Physiology or Medicine 1981." N.d.b. NobelPrize.Org. Accessed May 21, 2020. https://www.nobelprize.org/prizes/medicine/1981/hubel/lecture/.

Thompson, Emily Ann. 2002. *The Soundscape of Modernity: Architectural Acoustics and the Culture of Listening in America, 1900–1933*. Cambridge, MA: MIT Press.

Thoreau, Henry David. 1993. *A Year in Thoreau's Journal, 1851*. Edited by H. Daniel Peck. New York: Penguin Books.

Thoren, Victor E. 1990. *The Lord of Uraniborg: A Biography of Tycho Brahe*. Cambridge: Cambridge University Press.

Thorndike, Lynn. 1923. *A History of Magic and Experimental Science*. New York: Columbia University Press.

Tiersot, Julien. 1908. *Les fêtes et les chants de la révolution française*. Paris: Hachette.

Tkaczyk, Viktoria, Mara Mills, and Alexandra Hui, eds. 2020. *Testing Hearing: The Making of Modern Aurality*. Oxford: Oxford University Press.

Tomlinson, Gary. 1993. *Music in Renaissance Magic: Toward a Historiography of Others*. Chicago: University of Chicago Press.

Trippett, David. 2018. "Music and the Transhuman Ear: Ultrasonics, Material Bodies, and the Limits of Sensation." *Musical Quarterly* 100 (2): 199–261.

Trippett, David. 2019. "Sound as Hermeneutic, or Helmholtz and the Quest for Objective Perception." *19th-Century Music* 43 (2): 99–120. https://doi.org/10.1525/ncm.2019.43.2.99.

Trippett, David, and Benjamin Walton, eds. 2019. *Nineteenth-Century Opera and the Scientific Imagination*. Cambridge: Cambridge University Press.

Turing, A. M. 1950. "Programmers' Handbook for Manchester Electronic Computer Mark II. Computing Machine Laboratory. University of Manchester." 1950. http://www.alanturing.net/programmers_handbook/.

Ullersperger, Johann Baptist. 1871. *Die Geschichte der Psychologie und der Psychiatrik in Spanien von den ältesten Zeiten bis zur Gegenwart*. Würzburg: Stuber.

Valletta, Ludovico. 1706. *De phalangio apulo opusculum*. Naples: de Bonis.

Vecchi, G. 1967. "Medicina e musica, voci e strumenti nel 'Conciliator' (1303) di Pietro da Abano." *Quadrivium* 8:5–22.

Verworn, Max. 1913. *Irritability*. New Haven, CT: Yale University Press.

Vesalius, Andreas. 1964. *De humani corporis fabrica libri septem*. Brussels: Culture et Civilisation.

Vesalius, Andreas. 1969. *The Epitome of Andreas Vesalius*. Translated by L. R. Lind. Cambridge, MA: MIT Press.

Vesalius, Andreas. 1998. *On the Fabric of the Human Body. Book I*. Translated by William Frank Richardson and John Burd Carman. San Francisco: Norman Pub.

Vesalius, Andreas. 2002. *On the Fabric of the Human Body Books III and IV*. Translated by William Frank Richardson and John Burd Carman. Novato, CA: Norman Pub.

Vescovini, Graziella Federici. 1987. "Peter of Abano and Astrology." In *Astrology, Science, and Society: Historical Essays*, edited by Patrick Curry, 19–39. Woodbridge, Suffolk: Boydell Press.

Vieussens, Raymond. 1715. *Traité nouveau de la structure et des causes du mouvement naturel du coeur.* Avignon.

Vila, Anne C. 1998. *Enlightenment and Pathology: Sensibility in the Literature and Medicine of Eighteenth-Century France.* Baltimore, MD: Johns Hopkins University Press.

Viviers, Pierre L., Jo-Anne H. Kirby, Jeandré T. Viljoen, and Wayne Derman. 2017. "The Diagnostic Utility of Computer-Assisted Auscultation for the Early Detection of Cardiac Murmurs of Structural Origin in the Periodic Health Evaluation." *Sports Health* 9 (4): 341–345.

Vogel, C. J. de. 1966. *Pythagoras and Early Pythagoreanism: An Interpretation of Neglected Evidence on the Philosopher Pythagoras.* Assen, The Netherlands: Van Gorcum.

Vogel, Stephan. 1993. "Sensations of Tone, Perception of Sound, and Empiricism." In *Hermann von Helmholtz and the Foundations of Nineteenth-Century Science*, edited by David Cahan, 259–287. Berkeley: University of California Press.

Volmar, Axel. 2012. "Klang-Experimente: eine Geschichte der auditiven Kultur der Naturwissenschaften seit 1800." PhD diss., University of Siegen.

Volmar, Axel. 2013. "Listening to the Cold War: The Nuclear Test Ban Negotiations, Seismology, and Psychoacoustics, 1958–1963." *Osiris* 28:80–102.

Volmar, Axel. 2013b. "Sonic Facts for Sound Arguments: Medicine, Experimental Physiology, and the Auditory Construction of Knowledge in the 19th Century." *Journal of Sonic Studies* 4 (1). http://journal.sonicstudies.org/vol04/nr01/a13.

Volmar, Axel. 2015. *Klang-Experimente: Die auditive Kultur der Naturwissenschaften, 1761–1961.* Frankfurt am Main: Campus.

Volmar, Axel. 2018. "Experiencing High Fidelity: Sound Reproduction and the Politics of Music Listening in the Twentieth Century." In *The Oxford Handbook of Music Listening in the 19th and 20th Centuries*, edited by Christian Thoraeu and Hansjakob Ziemer, 395–412. Oxford: Oxford University Press.

Volmar, Axel. N.d. "Listening to the Body Electric. Electrophysiology and the Telephone in the Late 19th Century." Accessed May 1, 2020, at http://vlp.mpiwg-berlin.mpg.de/pdfgen/essays/art76.pdf.

Walker, D. P. 1975. *Spiritual and Demonic Magic: From Ficino to Campanella.* Notre Dame, IN: University of Notre Dame Press.

Walker, D. P. 1978. *Studies in Musical Science in the Late Renaissance.* London: Warburg Institute, University of London.

Walker, D. P. 1985. *Music, Spirit, and Language in the Renaissance.* Edited by Penelope Gouk. London: Variorum Reprints.

Walker-Meikle, Kathleen. 2014. "Toxicology and Treatment: Medical Authorities and Snake-Bite in the Middle Ages." *Korot* 22:85–104.

Wallace, Wes. 2003. "The Vibrating Nerve Impulse in Newton, Willis and Gassendi: First Steps in a Mechanical Theory of Communication." *Brain and Cognition* 51:66–94.

Walmsley, Donald Munro. 1967. *Anton Mesmer.* London: Hale.

Warner Classics. 2014. "Diana Damrau Sings *Lucia di Lammermoor* Mad Scene Live." November 7, 2014. https://www.youtube.com/watch?v=BEM3bvdNS_g.

Wedensky, N. E. 1883. "Ueber die telephonischen Erscheinungen im Muskel bei künstlichem und natürlichem Tetanus." *Archiv für Physiologie*, 313–325.

Wedensky, N. E. 1900. "Die fundamentalen Eigenschaften des Nerven unter Einwirkung einiger Gifte." *Pflügers Archiv für die gesammte Physiologie des Menschen und der Thiere* 82:134–191.

Wellmann, Janina. 2017. *The Form of Becoming: Embryology and the Epistemology of Rhythm, 1760–1830.* New York: Zone Books.

West, M. L. 1992. *Ancient Greek Music.* Oxford: Clarendon Press.

White, S. Young. 1948. "Applications of Ultrasonics to Biology." *Audio Engineering* 32 (6): 30, 42–45.

Williams, Elizabeth Ann. 1994. *The Physical and the Moral: Anthropology, Physiology, and Philosophical Medicine in France, 1750–1850*. Cambridge: Cambridge University Press.

Williams, Elizabeth Ann. 2003. *A Cultural History of Medical Vitalism in Enlightenment Montpellier*. Aldershot: Ashgate.

Willis, Martin. 2006. *Mesmerists, Monsters, and Machines: Science Fiction and the Cultures of Science in the Nineteenth Century*. Kent, OH: Kent State University Press.

Willis, Thomas. 1674. *De anima brutorum quae hominis vitalis ac sensitiva est, exercitationes duae*. Amsterdam: Apud Joannem à Someren.

Willis, Thomas. 1681. *The Remaining Medical Works of That Famous and Renowned Physician Dr. Thomas Willis*. London: T. Dring.

Willis, Thomas. 1683. *Two Discourses Concerning the Soul of Brutes*. Translated by S. Pordage. London: Thomas Dring.

Wiltshire, John. 2019. *Frances Burney and the Doctors*. Cambridge: Cambridge University Press.

Winthrop-Young, Geoffrey. 2002. "Drill and Distraction in the Yellow Submarine: On the Dominance of War in Friedrich Kittler's Media Theory." *Critical Inquiry* 28 (4): 825–854.

Wise, M. Norton. 1997. *The Values of Precision*. Princeton, NJ: Princeton University Press.

Wittern, Reante. 1989. "Die Wechselfieber bei Galen." *History and Philosophy of the Life Sciences* 11: 3–22.

Wolf, Rebecca. 2015. "The Sound of Glass: Transparency and Danger." In *Performing Knowledge, 1750–1850*, edited by Mary Helen Dupree and Sean B. Franzel, 113–136. Berlin: De Gruyter.

Wollaston, William Hyde. 1810. "The Croonian Lecture." *Philosophical Transactions of the Royal Society of London* 100:1–15.

Wollaston, William Hyde. 1812. "Ueber die Wirkungsart der Muskeln." *Annalen der Physik* 40:31–47.

Zeitler, William. 2013. *The Glass Armonica—the Music and the Madness*. San Bernardino, CA: Musica Arcana.

Zhmud, Leonid. 2014. "Sixth-, Fifth-, and Fourth-Century Pythagoreans." In *A History of Pythagoreanism*, edited by Carl A. Huffman, 88–111. Cambridge: Cambridge University Press.

Zühlke, L., L. Myer, and B. M. Mayosi. 2012. "The Promise of Computer-Assisted Auscultation in Screening for Structural Heart Disease and Clinical Teaching." *Cardiovascular Journal of Africa* 23 (7): 405–408.

Source and Illustration Credits

Portions of this book appeared originally in the following journals, which have kindly given permission for the appearance of the material here: *Journal of the History of Medicine and Allied Disciplines* (© 2015, Oxford University Press); *19th Century Music* (© 2019, The Regents of the University of California); and *Nineteenth-Century Music Reviews* (© 2020, Cambridge University Press).

Permission for the use of the figures has been kindly given by the following:

Bibliothèque national de France (fig. 11.6); Family of T. G. Brown and the British Medical Ultrasound Society (fig. 14.4); Families of T. G. Brown and Ian Donald (figs. 14.2, 14.3, 14.5); Ceribell Inc. and Josef Parvizi (fig. 19.5a, 19.5b); Collection of Historical Scientific Instruments, Harvard University (figs. 15.9, 15.11); Eric D. Heller (fig. 18.3); Paul and Carl Hubel (fig. 18.2); Roger Keynes (figs. 17.5, 17.8, 17.9, 17.11, 18.1); Hugh Matthews (figs. 17.8, 17.10, 18.1); Esther Ruth Mbabazi (fig. 19.2); Natus Medical Incorporated, natus.com (fig. 19.3); Mark D. Temple (fig. 19.4); Thinklabs Inc. (fig. 19.1); University of Kentucky Library Special Collections (figs. 8.5, 8.6a, 8.7a, 8.8a); University of Toronto Scientific Instrument Collection (fig. 15.9); Wellcome Collection (figs. 4.1, 6.1, 6.2, 7.4, 9.5, 11.5); Nellie Wild (fig. 14.1).

All other figures are either reproduced under a Creative Commons License (figs. 1.2, 3.1, 3.2, 4.1, 4.2, 4.4), the guidelines of fair use, or are in the public domain.

Acknowledgments

My special thanks to Katie Helke, whose encouragement and support have been so important to me, as well as to her colleagues at the MIT Press, whose collaboration, expertise, and enthusiasm have been essential to this work, especially Judy Feldmann and Gita Manaktala, as well as many others. I think it extraordinary and wonderful that the MIT Press has supported me now through seven books and allowed me to complete this trilogy on music and the making of the sciences.

I thank the John Simon Guggenheim Memorial Foundation for a fellowship that supported the early stages of this project. Janet Browne and the Harvard Department of History of Science kindly invited me to present material from this book, as did Viktoria Tkaczyk and Rebecca Wolf at the Max-Planck-Institut für Wissenschaftsgeschichte, Berlin, and David Trippett at the Centre for Research in the Arts, Social Sciences, and Humanities at the University of Cambridge. The discussions on these occasions greatly helped and encouraged me and I thank all those who shared their thoughts and reactions with me.

Andrei Pesic read the whole work and gave me very insightful comments that pushed me to go further and say more. Thanks, Andrei! My gratitude also to Andrew Abarbanel and Alexander Rehding for their careful reading and enthusiastic comments on the whole work. Bevil Conway, Creig Hoyt, Leslie Kay, Mark Pottinger, Mark Temple, David Trippett, Axel Volmar, and Rebecca Wolf read sections of the work and gave me helpful comments, as did three anonymous readers for the MIT Press. I particularly thank Bevil and Leslie for sharing with me their personal accounts of working with audio monitoring, which helped me understand the lived experience behind it. Carl Hubel and Nellie Wild kindly shared memories of their fathers' musical interests.

For over fifty years, Gerald Holton has been my teacher, mentor, and friend. His interest, enthusiasm, and support have been crucial to my development. My affection for him and Nina runs deep. I dedicate this book to him as a small token of my admiration and thanks. Remembering Nina: "What happens to the song when the singer falls silent?"

As always, my family inspired and encouraged me in ways I cannot begin to count. Thanks, Ssu, Andrei, Alexei, Angèle, Belinda, and Felix!

Index

Note: page numbers in italics indicate illustrations.